*Oklahoma Horizons Series*

*Oklahoma Horizons Series*

Oklahoma Heritage Association, Inc.

Oklahoma Heritage Horizons Committee, 1979

William H. Bell, *Chairman*  
Tulsa

Lowe Runkle, *Vice-Chairman*  
Oklahoma City

Kenny A. Franks, *Secretary and Editor-in-Chief*  
Oklahoma Heritage Association

W. David Baird  
Department of History  
Oklahoma State University

Janet Campbell  
Oklahoma City

Jack T. Conn  
Oklahoma City

Arrell M. Gibson  
Department of History  
University of Oklahoma

Jack Haley  
Western History Collections  
University of Oklahoma Library

Robert A. Hefner, Jr.  
Oklahoma City

Henry C. Hitch, Jr.  
Guymon

Paul F. Lambert  
Oklahoma Heritage Association

Edgar Oppenheim  
Oklahoma City

Jerry Richardson  
Oklahoma City

F. A. Sewell, III  
Clinton

Steve Wilson  
Museum of the Great Plains  
Lawton

# McFarlin Library
# WITHDRAWN

| DATE DUE | | | |
|---|---|---|---|
| ~~AUG 1986~~ | | | |
| ~~JUN 17 1993~~ | | | |
| ~~OCT 05 1999~~ | | | |
| ~~JUN 20 2000~~ | | | |
| ~~JAN 04 2003~~ | | | |
| JAN 02 2003 | | | |

GAYLORD — PRINTED IN U.S.A.

# The Oklahoma Petroleum Industry

*Photograph by Fritz Henle, courtesy of Cities Service Company.*

*Oklahoma Horizons Series*

# THE OKLAHOMA PETROLEUM INDUSTRY

## By Kenny A. Franks

Kenny A. Franks, *Series Editor*

*Published for the Oklahoma Heritage Association
by the
University of Oklahoma Press : Norman*

By Kenny A. Franks

(coauthor) *Mark of Heritage* (Norman: 1976)
(coeditor) *Early Military Forts and Posts in Oklahoma*
*(Oklahoma City: 1978)*
*Stand Watie and the Agony of the Cherokee Nation* (Memphis: 1979)
*The Oklahoma Petroleum Industry* (Norman: 1980)

**Library of Congress Cataloging in Publication Data**

Franks, Kenny Arthur, 1945-
   The Oklahoma petroleum industry.

(Oklahoma horizons series)
Bibliography: p. 269
Includes index.
   1.   Petroleum industry and trade—Oklahoma—History.
I.  Oklahoma Heritage Association.  II.  Title.
III.  Series.
TN872.O5F73        338.2'7282'09766        80-5242

Copyright © 1980
by the University of Oklahoma Press, Norman,
Publishing Division of the University.
Manufactured in the U.S.A.
First edition.

*For*
*Jayne*

# Contents

| | | |
|---|---|---|
| Preface | *page* xiii | |
| Acknowledgments | xv | |
| 1. Early Oil Exploration in Indian Territory | 3 | |
| 2. No. 1 Nellie Johnstone | 16 | |
| 3. Red Fork | 27 | |
| 4. The Rush Begins | 36 | |
| 5. Statehood and Regulation | 47 | |
| 6. The Osage Nation, 1896–1920 | 57 | |
| 7. Cushing | 68 | |
| 8. Healdton | 79 | |
| 9. Oil Boomtowns | 89 | |
| 10. The Roaring Twenties | 101 | |
| 11. The Greater Seminole Field | 115 | |
| 12. Oklahoma City | 127 | |
| 13. Regulation and Conservation, 1910–31 | *page* 139 |
| 14. The International Petroleum Exposition and Congress | 151 |
| 15. Martial Law and Proration | 162 |
| 16. Conservation and Cooperation | 173 |
| 17. The War Years | 183 |
| 18. Science and Technology | 194 |
| 19. Oklahoma Petroleum-Related Industries | 207 |
| 20. Oklahoma's Petroleum Legacy | 218 |
| 21. Outstanding Oklahoma Oilmen | 229 |
| 22. The Future | 239 |
| Notes | 249 |
| Bibliography | 269 |

# Illustrations

| | | | |
|---|---|---|---|
| Drilling at Triumph Hill, Pennsylvania, 1871 | page 6 | Cushing gusher | 76 |
| | | The Empire Refinery | 77 |
| Digging for coal | 7 | Robert A. Hefner | 80 |
| Cable tool drilling in the 1870s | 11 | Healdton's largest producer | 81 |
| Pawhuska "on the go" in 1906 | 12 | Halliburton Services in the early days | 83 |
| An Oklahoma oil seep | 14 | Storage tanks ablaze | 85 |
| Nellie Johnstone No. 1, 1897 | 17 | Whittington Hotel office | 86 |
| Oil well in Bartlesville, Oklahoma | 18 | Healdton, Oklahoma | 87 |
| Oil boom in Bartlesville, Oklahoma | 20 | Boomtown | 90 |
| Frank and Jane Phillips | 23 | Crumpled oil well | 91 |
| Oil well supply store in Bartlesville, Oklahoma | 24 | "Shotgun" houses | 93 |
| Pouring nitroglycerine into the tin tube | 26 | Grisso mansion in Seminole, Oklahoma | 94 |
| The discovery well at Red Fork, 1901 | 30 | Seminole, Oklahoma, jail | 95 |
| Oil tank spilling on the ground | 31 | "Bishop's Alley" in Seminole, Oklahoma | 96 |
| Charles F. Colcord | 32 | Company housing | 98 |
| Drilling in Washita County | 34 | Bowlegs, Oklahoma | 99 |
| Drilling in the streets | 35 | Seminole, Oklahoma | 99 |
| Bartlesville, an oil-country hub | 38 | Mud, the curse of all boomtowns | 100 |
| Early Continental Oil Company "transport" | 39 | E. W. Marland | 102 |
| Frank Phillips | 41 | Skelly and Oklahoma | 103 |
| Glenn Pool | 43 | Lew Wentz Oil Company | 106 |
| Production chart, 1900–25 | 46 | Women in the oil fields | 108 |
| Preston, Oklahoma, in the oil boom | 50 | Champlin Petroleum Company | 110 |
| Battery of crude stills | 51 | Three Sands oil field | 112 |
| Oklahoma Territorial Governor Frank Frantz | 53 | *The Pioneer Woman* | 113 |
| Burning gas well in Cushing oil field | 55 | Seminole, Oklahoma, in 1926 | 116 |
| Empire Refineries, Inc. | 56 | Pipeline crew | 117 |
| Equipment for Osage oil field | 59 | Mud, the nemesis of the Greater Seminole field | 118 |
| The Osage Council | 60 | Transportation hub | 118 |
| The Wildhorse Pool | 62 | Okemah, Oklahoma | 119 |
| Four leading pioneer oilmen | 63 | Walter Helmerich and William T. Payne | 120 |
| The Million Dollar Elm | 65 | Downtown Seminole, Oklahoma | 122 |
| The Alfred Mathews Lease | 66 | The loading racks for oil at Seminole, Oklahoma | 123 |
| The Dropright oil field | 70 | Seminole, Oklahoma, at the height of its oil boom | 125 |
| Main Street in Drumright, Oklahoma | 71 | An enclosed derrick | 132 |
| Burning gas well in the Dropright Pool | 72 | Preparing a drilling site | 133 |
| Lakes of oil in the Drumright field | 73 | Firemen battling a wild gas well near the Oklahoma capitol | 134 |
| Drumright fire | 74 | | |

| | | | |
|---|---|---|---|
| Residential drilling | 135 | Claude V. Barrow and Governor Roy J. Turner | 191 |
| Urban oil boom | 136 | Offshore drilling | 192 |
| The Wild Mary Sudik | 137 | Frank Buttram | 196 |
| Prices falling | 142 | Indoor drilling rig used for training | 197 |
| Oil conservation | 143 | A diamond core drill | 198 |
| Oklahoma City, 1932 | 146 | Displaying innovations | 199 |
| Martial law | 148 | Moving into petrochemicals | 201 |
| Curtailing production | 149 | Acid fracturing near Sayre, Oklahoma | 204 |
| Cities Service Oil Company's Bartlesville Experimental Station | 150 | Offshore well cementing | 205 |
| | | Oil survey crew | 206 |
| William G. Skelly | 152 | Erle P. Halliburton | 208 |
| A parade at one of the early International Petroleum Expositions | 153 | Dowell Well Incorporated | 210 |
| | | Reading & Bates offshore drilling platform | 211 |
| International Petroleum Exposition crowds | 154 | The drill ship *J. W. Bates* | 212 |
| "The oil capital" | 155 | The tender *George M. Reading* | 213 |
| Equipment displays | 157 | *Oil Investors' Journal* | 214 |
| International Petroleum Exposition Building in Tulsa, Oklahoma | 158 | Philip C. Lauinger, Jr. | 215 |
| | | John H. Williams | 216 |
| W. K. Warren | 160 | Samuel Roberts Noble | 219 |
| Governor Murray orders out the National Guard | 163 | Eleanor Blake Kirkpatrick | 221 |
| Enforced proration | 164 | Robert A. Hefner and Frank Buttram | 223 |
| The National Guard in the Oklahoma oil fields | 165 | Oklahoma Heritage Center | 224 |
| National Guard camp | 166 | John E. Mabee | 225 |
| Oklahoma City field | 167 | Alfred E. Aaronson and Thomas Gilcrease | 226 |
| Drilling near the state capitol | 169 | Thomas Gilcrease Institute of American History and Art | 227 |
| The January, 1935, oil parley | 177 | | |
| The men who presented the Interstate Oil Compact to President Franklin D. Roosevelt | 178 | Clarence H. Wright | 230 |
| | | K. S. ("Boots") Adams | 231 |
| 1946 Executive Committee of the Interstate Oil Compact Commission | 179 | Ward S. Merrick | 232 |
| | | Six recipients of the Outstanding Oklahoma Oil Man Award | 234 |
| Governors at the 1946 Interstate Oil Compact Commission meeting | 180 | | |
| | | T. H. McCasland | 235 |
| Dedication of the Interstate Oil Compact Commission headquarters | 181 | Robert A. Hefner, Jr. | 237 |
| | | The oil supply outlook | 240 |
| Interstate Oil Compact Commission headquarters | 182 | Oklahoma's oil future | 244 |
| Catalytic cracker | 184 | Offshore drilling | 245 |
| Gas booster station | 185 | Water flooding | 246 |
| Lloyd Noble | 187 | Research and development | 247 |
| West Edmond field | 190 | | |

# Maps

| | | | |
|---|---|---|---|
| Indian Territory, 1885 | *page* 4 | Oklahoma counties, 1907 | 48 |
| Oklahoma oil and gas fields | 13 | Oklahoma natural-gas pipelines | 54 |
| Railroads in Indian Territory | 19 | The Osage Nation | 58 |
| The Creek Nation | 29 | The Oklahoma City field | 131 |
| Oklahoma pipelines | 37 | Oklahoma's seventy-seven counties | 140 |

# Preface

OIL AND OKLAHOMA are synonymous. The discovery of vast petroleum deposits in the state ushered Oklahoma into the twentieth century and touched off the last great mineral rush to the West. Although oil deposits were known across the region for many years before non-Indian settlement and were used by many homesteaders to acquire lubricants for their machinery and as medicine for both themselves and their livestock, the problem of transporting the crude to potential markets prevented its exploitation on a commercial basis. However, as America entered the automobile age and the use of petroleum was expanded, more and more enterprising young men joined in the search for "black gold." As transportation improved and more markets became available, Oklahoma took its place among the nation's leading petroleum producers.

The maturing of the state's petroleum industry marked a significant transformation of the area's economy. Before such fabulously rich finds as Red Fork and Glenn Pool, Oklahoma's economic base rested on agriculture—farming and ranching dominated the scene. However, the great influx of wealth brought about by the huge deposits of crude laid the foundations for many financial and industrial institutions that were to be important to the state's future. From this time onward Oklahoma was never the same because oil transformed practically every aspect of the state's culture.

Conversely, as the discovery of oil opened a new era in Oklahoma's history, it closed one of the most colorful and romantic periods of America's heritage in that the clamor for "black gold" beneath the state's soil marked the final mineral rush of the Great West. Sparked by the gold rush to California in 1849 and heightened by the quest for silver in Colorado and Nevada in the following decades, the search for underground wealth reached its climax in the Oklahoma oil boom. Not only was it the final chapter in that era of excitement, adventure, and romance, it was also the most valuable, exceeding three times the wealth of all the previous gold and silver finds.

Although it was concentrated in a single state, the Oklahoma oil boom had repercussions throughout the petroleum industry. Technological innovations in the Oklahoma oil fields revolutionized petroleum production worldwide and greatly expanded the industry. In addition, the state was the birthplace of many of the giant companies of the industry. Such men as Frank Phillips, J. Paul Getty, William G. Skelly, H. H. Champlin, Erle P. Halliburton, and E. W. Marland launched Oklahoma oil operations that eventually encircled the globe.

Only recently have Oklahomans undertaken the preservation of this fascinating heritage. The Oklahoma Petroleum Council, under its Executive Manager James O. Kemm, and its Historical Committee, chaired by John Steiger, have embarked on a statewide program to mark Oklahoma's significant achievements in the petroleum industry. Begun in 1964, the effort has thus far emplaced two metal highway markers and twelve granite monuments. The metal markers commemorate the Nellie Johnstone No. 1, the state's first commercial oil well, and Oklahoma's first oil refinery. They are located in Bartlesville and Muskogee respectively. The granite monuments commemorate: the birthplace of the Interstate Oil Compact Commission, in Ponca City; the Oklahoma City Field, on the grounds of the state capitol in Oklahoma City; recognition of Tulsa as the "Oil Capital of the World," in the Tulsa Civic Center; the great deposits of oil in the Osage Nation and the Million Dollar Elm, in Pawhuska; the International Petroleum Exposition, at Expo Square in Tulsa; the first gas-processing plant west of the Mississippi River, south of Tulsa on State Highway 75;

Oklahoma's initial waterflood project, on United States Highway 169 about four miles south of Nowata; the Healdton Oil Field, in Healdton; the world's original School of Petroleum Geology, at Gould Hall on the University of Oklahoma campus in Norman; Oklahoma's oil pioneers, at the capital complex in Oklahoma City; the Greater Seminole Field, in Seminole; and the Bartlesville Energy Technology Center, in Bartlesville.

These efforts have been enhanced to a great degree by the work of several individuals across the state, most notably, Ruby Cranor, curator of the History Room, Bartlesville Public Library, and Margaret Teague, the author of a history of the Bartlesville region and its oil heritage. In addition, many of Oklahoma's petroleum companies are currently working hard to preserve the state's oil heritage, and they make their research and archival centers available for additional studies. Nonetheless, it has been several decades since a comprehensive history of Oklahoma's oil industry was produced, and such an undertaking would require several volumes to compile today. Obviously, this work is but a single volume, and as a result many individuals and incidents are not recorded; it is an attempt, however, to portray Oklahoma's oil legacy and those individuals who made it possible.

# Acknowledgments

THIS WORK WOULD NOT have been possible without the full cooperation of the Oklahoma petroleum industry. Among those individuals and companies making this book possible are: Alfred Aaronson, Carl B. Anderson, Apco Oil Corporation, Atlantic Richfield Foundation, Beard Oil Company, William H. Bell, D. D. Bovaird, Horace K. Calvert, Carl Swan Petroleum Company, H. H. Champlin, Cities Service Company, William B. Cleary, Jack T. Conn, Continental Oil Company, Eason Oil Company, Fidelity National Bank, N.A., Hadson Oil Company, Harper Oil Company, Hefner Oil Company, Robert A. Hefner, Jr., Robert A. Hefner, III, T. K. Hendrick, E. C. Joullian, Richard Kane, Kerr-McGee Corporation, Kirkpatrick Foundation, John E. Kirkpatrick, Julius Livingston, T. H. McCasland, Mabee Petroleum Corporation, W. V. Montin, Mustang Gas Company, Walter Neustadt, Jr., William T. Payne, Phillips Petroleum Company, Reading and Bates Offshore Drilling Company, Roulette Oil Company, Samuel Roberts Noble Foundation, Edward A. Smith, Sohio Petroleum Company, Sun Company, Inc., John A. Taylor, The Petroleum Publishing Company, The Williams Companies, Charles F. Urschel, Jr., S. K. Viersen, Jr., W. B. Osborn Operators, Barth P. Walker, Michael R. Waller, L. O. Ward, Warren Petroleum Company, W. K. Warren, and Woods Petroleum Corporation.

In addition, many individuals aided in the compilation of the historical photographs used to illustrate the finished manuscript, and to these I owe a special debt of thanks. Among them are Manon T. B. Atkins, Oklahoma Historical Society; Mrs. Claude V. Barrow; Charles Bowlin, Interstate Oil Compact Commission; Leslie Brooks, Leslie Brooks and Associates; Sloan K. Childers, Phillips Petroleum Company; Ruby Cranor, Bartlesville Public Library; Daniel B. Droege, Phillips Petroleum Company; R. B. Finney, Phillips Petroleum Company; Lewis J. FitzGerald, Reading and Bates Offshore Drilling Company; Beryl D. Ford; Milton B. Garber, Enid Publishing Company; Rex Hudson, Halliburton Services; Robert B. Jacob, Getty Refining and Marketing Company; Ralph W. Jones, 45th Infantry Division Museum; P. C. Lauinger, The Petroleum Publishing Company; Guy Logsdon, University of Tulsa; Stephen R. Malone, Phillips Petroleum Company; Alma Marsh; Keely Marshall, Continental Oil Company; William A. ("Mac") McGalliard, *Daily Ardmoreite*; Allan Muchmore, *Ponca City News*; H. Milt Phillips, *Seminole Producer*; Richard B. Risk, Getty Refining and Marketing Company; Malcolm Rosser, III, Halliburton Services; Roy P. Stewart; Willa Mae Townes, Seminole County Historical Society; Bob Westmoreland, Ponca City Public Schools; and Pendleton Woods.

Special recognition must be given to the Oklahoma Petroleum Council, its Executive Manager, James O. Kemm, and its Historical Committee, John Steiger (chairman), Frank B. Taylor (Vice-Chairman), Leslie Brooks, Bill Burchardt, Charles E. Cummings, E. G. "Ty" Dahlgren, W. M. Murray, Malcolm E. Rosser, III, the late Luther Williams, and Phil C. Withrow. Without the complete cooperation and valuable suggestions of these men, a project of such magnitude never would have been completed. In addition, much of the research found in this work is the result of the diligent efforts of Mac R. Harris, who spent untold hours poring over archival records. Of course, thanks must be given to Dr. Paul F. Lambert, the executive director, Jack T. Conn, the president, and all the directors and members of the Oklahoma Heritage Association for allowing me to undertake this project. Special acknowledgement is given to my good friend Dr. Odie B. Faulk, who worked diligently to launch this project. Much recognition must be given to David G. Campbell, of Leede Exploration, and his wife, Janet, for the tremendous amount of time they devoted to the evaluation and preparation of this book, without whose expertise the work never would have been finished. Finally, my wife, Jayne, deserves much of the credit for this book, for without her willingness to endure my efforts and assume untold extra burdens during its preparation, it would not have been completed.

Kenny A. Franks

*Union City, Oklahoma*

# The Oklahoma Petroleum Industry

## Chapter 1.
# Early Oil Exploration in Indian Territory

LONG BEFORE THE GREAT OIL BOOMS in Oklahoma during the first quarter of the twentieth century, Native Americans had sought out and utilized the state's petroleum deposits. For centuries before the arrival of the Five Civilized Tribes (the Cherokee, Creek, Choctaw, Chickasaw, and Seminole) the Plains Indians had employed the black liquid that oozed from beneath the rocks and accumulated on the surface of spring water as a cure for rheumatism and other chronic illnesses, as well as a treatment for cuts and sores on their animals. When oil accumulated in substantial quantities, it was simply scooped from the water. But when the substance flowed less freely, the ingenious Indians used feathers to skim off the petroleum that had accumulated in the small, shallow reservoirs made by hoof prints of buffalo in the mud surrounding the springs.[1]

Still, the deposits remained generally unknown to non-Indians until after the arrival of the Five Civilized Tribes in the 1830s. White agents who accompanied the five tribes to the area left several written accounts of early oil sites in the region. The Chickasaw agent, A. M. Upshaw, reported that "medicine springs" attracted the Indians in the 1840s, and A. J. Smith, his successor, called attention to several oil seeps used by the Indians in the western part of the Chickasaw Nation near the Wichita Mountains. By the 1850s these surface pools were well known to the citizens of Texas and Arkansas. As popularity of the oil springs continued to increase, Smith reported to his superiors in Washington, D.C. that the "oil springs in this nation [Chickasaw] are attracting considerable attention, as they are said to be a remedy for all chronic diseases." Smith, touting their value as "medical properties," related that "Rheumatism stands no chance at all, and the worst cases of dropsy yield to its effects." "The fact is," he declared "that it cures anything . . . [on which it] has been tried."[2]

Having been closely associated with the white man's economy in the South before their removal, the Five Civilized Tribes were quick to recognize the potential value of the oil springs and soon located others. New Spring Place in the Cherokee Nation, in present-day Delaware County, and Boyd Springs, to the south in the Chickasaw Nation near present-day Ardmore, were two of the best-known sites frequented by the Indians. Situated on Oil Creek, Boyd Springs was a well-known camping area where the Indians often gathered in large numbers. Illumination was provided throughout the night by forcing a tube or gun barrel into the ground and setting fire to the escaping gas. Eventually the Indians began to utilize the underground wealth for a number of practical purposes. Oil discovered by an Indian who was digging a water well near Ada in Pontotoc County was applied as a lubricant for the machinery used to cut the rich prairie grass in the vicinity. Another natural oil spring was found on the Peter Maytubby place, approximately six miles northeast of Caddo, in present-day Bryan County. A health resort that was eventually opened on the site of "Maytubby Springs" catered to many people in the area and surrounding states.[3]

In the summer of 1859, Lewis Ross, brother of the Principal Chief John Ross, was mining salt at the Grand Saline, a large salt spring near the Neosho Crossing on the Grand River in the Cherokee Nation. In attempting to increase the production of salt by sinking a deeper well, Ross was surprised when he struck oil. The natural gas accompanying the petroleum provided enough pressure to cause the well to flow at about ten barrels a day for approximately one year. Coinciding with the drilling of the country's first commercial oil well by Edwin L. Drake near Titusville, Pennsylvania, Ross's discovery undoubt-

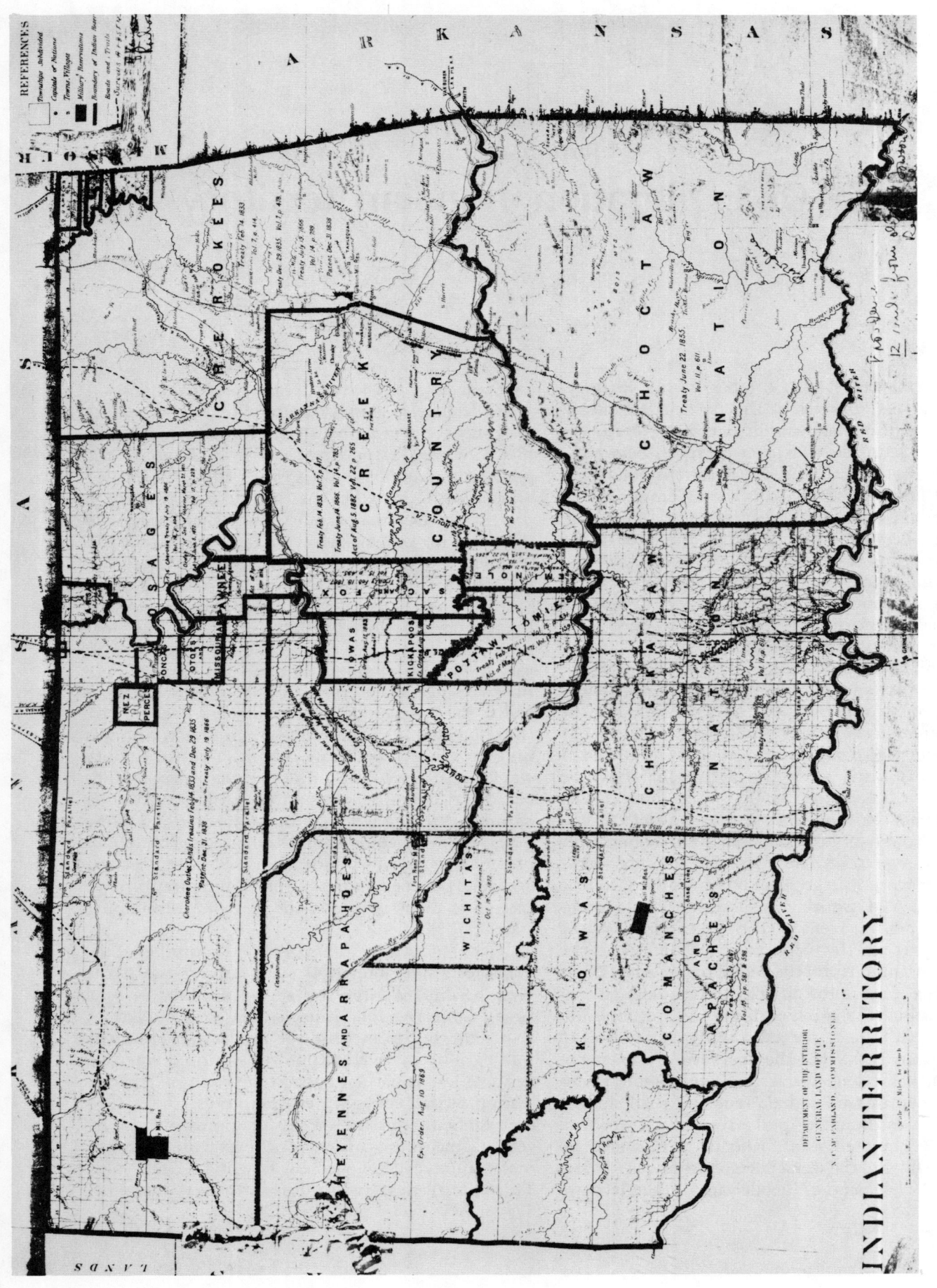

Oklahoma Indian Territory, 1885. During the nineteenth century, while Oklahoma was still the home of many Indian tribes, natural oil seeps were found in several areas, such as New Spring Place, Boyd Springs, and Maytubby Springs. *Courtesy of Oklahoma Heritage Association.*

edly would have caused a great deal of excitement had it not been for the outbreak of the Civil War. The conflict, dividing the Indians much along the same lines as the rest of the nation, effectively ended any further exploration at that time. It was not until the hostilities had ceased that interest was revived in the numerous oil sites in Indian Territory.[4]

Tempers cooled slowly after the Civil War but, as time passed, the death of many of the antagonists allowed the country to return gradually to normal. Concurrently, attention was again directed to prospective oil-producing areas throughout the nation. As word of the celebrated oil springs in Indian Territory spread to other portions of the country, a number of whites entered the Indian lands in an attempt to retrieve the crude. Many of the methods utilized to release the petroleum from the ground, however, were primitive and inefficient. For many years, the inhabitants of the northeastern portion of the Cherokee Nation had been aware of the presence of oil in Cochran Springs, located approximately five and one-half miles southwest of Chelsea. Shortly after the Civil War ended, an entrepreneur from Pennsylvania arrived on the scene intent on drilling an oil well. He sank a hole nearly 300 feet into the spring and set up a large steam engine some distance away to provide pumping power. The engine was connected to the pump with a series of belts. The oil and water were pumped into a large wooden hogshead placed under the pump's spout. With a distinctive gurgle, the oil and saltwater flowed into the container at the rate of about a pint of petroleum every four or five minutes. When the hogshead was nearly filled, a large "bunghole" previously drilled into the bottom of the hogshead was opened to allow the water to escape, while the lighter oil floated on top. Once the saltwater was removed, the petroleum was dipped from the hogshead and placed in nearby containers. Understandably this laborious process did not yield a large amount of oil.[5]

Land in Indian Territory continued to be held in common by tribal members following the Civil War. Individuals were allowed to develop mineral resources, however, and in February, 1872, Robert M. Darden of Missouri persuaded several members of prominent Chickasaw and Choctaw families to meet at the home of Winchester Colbert, one-time Governor of the Chickasaw Nation, to form the Chickasaw Oil Company. Organized under the laws of the state of Missouri, the company planned to acquire leases in several oil-producing areas in Indian Territory. After selecting Darden as president, the group prepared a contract with him giving the Indians "the exclusive right and privileges in the proceeds in the one-half of all oil produced from said entered land," while reserving for the Chickasaw Oil Company "one-half of all oil produced on every quarter-section contracted for, the said oil company being . . . [responsible for] all expenses in boring and barreling the same and the oil to be divided at the wells, the said company taking one-half and the holders of certificates the other half." In exchange for allowing Darden rights to the oil on which they held leases, the Indians were to furnish barrels at the wells and receive one-half of the production on every quarter-section of land, not a bad agreement from the standpoint of the Indians. In addition, it was agreed that after a certain "term of years," the lessors were also to receive the sum of $48,000.[6]

Believing that an "oil ridge" was located near Colbert's home in Pontotoc County, members of the company designated Colbert and Buckner Burns to act on behalf of the Chickasaw Oil Company and secure leases in the immediate vicinity of the formation. Colbert and Burns, having no knowledge of petroleum geology, were paid $5.00 per day for their labors. Stock certificates were also printed and issued during the summer of 1872, and the first oil company formed within the boundaries of the state seemed assured of success.[7]

Their enthusiasm was soon dampened, however, by the United States government. The Commissioner of Indian Affairs, F. A. Walker, and Secretary of the Interior Columbus Delano firmly opposed opening the Indian lands to the whites for any purpose, commercial or otherwise. The attitude of the federal officials, perhaps the first example of government interest in the petroleum industry within the state, effectively curtailed any further development by the Chickasaw Oil Company. In addition, the Missouri, Kansas and Texas Railroad was already building south from Kansas into Indian Territory, and by 1872 attention was diverted from the emerging oil business. Inasmuch as the engines were powered by coal, which was abundant in the Indian lands, there was a sudden switch of interest from oil to coal. A profit was assured in coal mining, while drilling for oil was still a risky and unknown enterprise. This, combined with the position taken by federal officials, ended the efforts of the Chickasaw Oil Company.[8]

Petroleum production in Indian Territory generally reverted to its pre-Civil War status after the demise of the Chickasaw venture. There were efforts by various individuals to extract the mineral from the

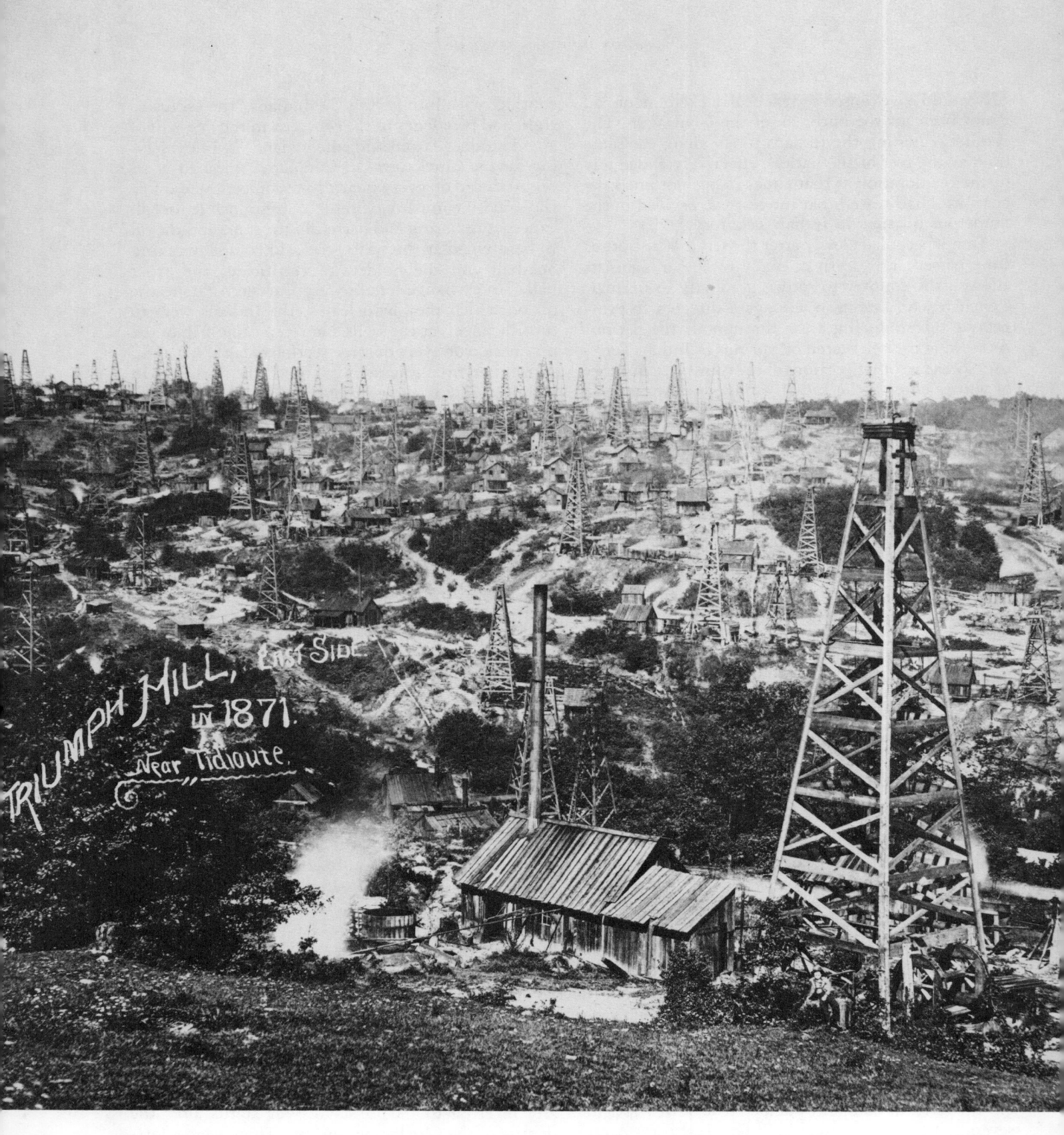

Drilling at Triumph Hill, Pennsylvania, 1871. It was the rapid expansion of such drilling operations as Triumph Hill in Pennsylvania in the 1870s that paved the way for Oklahoma's oil boom. *Courtesy of Interstate Oil Compact Commission.*

Digging for coal. When the railroads began to penetrate Indian Territory, much of the interest in underground wealth switched to coal, which was in great demand as fuel for the engines, and most of the financial support necessary for carrying out drilling operations was shifted to coal production. *Courtesy of Continental Oil Company.*

earth, and a number of areas continued to enjoy popularity as health resorts; however, it was not until the following decade that an interest in oil-producing areas in Indian Territory was rekindled. The new oil business was by that time making a place for itself in the American economy. This was readily apparent by the increase in American exports of crude oil from a total of 17,766,400 gallons, valued at $2,107,800 in 1874 to 67,186,300 gallons, worth $5,303,000 ten years later. Domestic demand was also increasing during this period, and the annual sale of refined petroleum products within the United States increased from about 7,650,000 barrels in 1873–1875, to almost 18,500,000 barrels in 1883–1885. The discovery of oil and gas deposits near Paola, in eastern Kansas, combined with the spiraling market, once again directed oilmen toward petroleum deposits in Indian Territory, despite the fact that previous efforts had been hampered by governmental interference.[9]

Aware of the oil deposits in Indian Territory, Dr. H. W. Faucett of New York, with the financial backing of R. Lenox Belknap, treasurer of the Northern Pacific Railroad, opened negotiations with the Choctaw, Chickasaw and Cherokee nations for oil concessions. On July 9, 1883, Faucett proposed to Allen Wright, the former chief of the Choctaws, that several members of the tribe form a company, secure leases on prospective petroleum producing areas, and sublet the leases to Faucett. In return, Faucett agreed to pay fifteen percent of the total production of each well to the company, which in turn would pay the Choctaw Council five percent and retain the remaining ten percent as dividends for the investors. As negotiations continued, it became apparent that Faucett was interested not only in acquiring drilling rights in Indian Territory but also in securing "exclusive . . . pipeline privilege for transporting oil." This was an absolute necessity if markets outside the area were to be made available to the Indians' oil. Acceptance of the proposed agreement by the Choctaw National Council remained a major stumbling block in the negotiations, as well as the question of whether leases were also to be acquired in the closely allied Chickasaw Nation.[10]

Faucett informed Wright that "It would be impossible to interest capital in the work unless there was some agreement as to the extent of territory that could be had and the specified royalty." He also pointed out that it "would not do to wait until petroleum was found," to settle the questions. Urging Wright to carry the enterprise to its completion at the earliest possible time, Faucett asked him to "please write me fully and explicitly as to what can be done in both the Choctaw and Chickasaw nations." Wright and other interested Indians apparently wasted little time, for on October 23, 1884, Edmund McCurtain, Principal Chief of the Choctaw Nation, approved "Bill No. 12" of the Choctaw National Council creating the Choctaw Oil and Refining Company "for the purpose of finding Petroleum or Rock Oil and increasing the revenue of the Choctaw Nation."[11]

The Choctaw act also recognized that the expense of sinking "prospective wells to a depth sufficient to test the question of its existence in paying quantities" was a great handicap in inducing people to begin exploration. As a result, it stipulated the necessity "to offer inducements to secure the developments of this product." Incorporated by J. F. McCurtain, E. N. Wright, A. R. Durant, and Allen Wright, the Choctaw Oil and Refining Company was to be governed by the rules adopted by its "Directors," and "if found to be necessary for the better development of this product and thus for yielding a larger revenue to the Choctaw Nation, associate with itself other than citizens of the Choctaw Nation." On successful oil ventures, the act also guaranteed the company protection "in securing the products of its labor by the exclusive right of producing oil therein [the Choctaw Nation], for a period of twenty years from the passage of this act." Faucett also obtained the exclusive right "of laying pipelines for its [oil] transportation to the Rail Road or any suitable point for refining."[12]

In return, the Choctaw Nation was to receive five percent of "the product at the place of production or the market value thereof." An accounting of the production was to be presented every month to the national treasurer of the Choctaws, and all money received was to be used for the school fund "or be subject to the actions of the National Council." Work was to begin within one year of the passage of the act, and the company was prohibited from trespassing on the improvements of any citizen "in any manner whatever without their free and written consent." In addition, the law specified that "nothing in this act shall be construed as depriving the people of the free use of the Natural tar or oil Springs," and should the company fail to locate oil in "paying quantities it shall have no claims against the Nation, neither as a Company nor as individuals for the loss of money." In case of failure, however, the Choctaw Oil and Refining Company was allowed to remove all equipment brought into the country but prohibited from taking out any lumber or locally acquired material.[13]

Following the precedent of the Choctaws, the Cherokee National Council passed a similar act on December 13, 1884, approved by Principal Chief Dennis W. Bushyhead, creating the Cherokee Oil Company. The two acts were essentially the same: allowing individuals to organize a company and enter into contracts with others; protection of the property of citizens; use of any money by the school fund or the national council; five percent royalty paid to the Cherokee Nation; denial of any claims against the tribe by the company; and a stipulation that Cherokee officials be allowed to monitor the petroleum production to insure that the tribe was receiving its share. The Cherokee act also provided that the "company shall not prospect or mine any other mineral whatever and if successful in finding oil, shall furnish Cherokee citizens for actual use in quantities not less than one barrel at the lowest market value." Like the Choctaws, the Cherokees granted the company the "exclusive right of producing oil therein, of transporting the same to any suitable point for shipment and of refining," but the Cherokees required the company to post bond "to the satisfaction of the principal chief for the faithful payment of the revenue of the treasurer."[14]

As a result of the actions of the Choctaw and Cherokee councils, Faucett stood to acquire exclusive drilling, transportation, and refining rights to nearly "20,000 square miles or about 13,000,000 acres," undoubtedly the largest oil lease ever made to one party in the history of Oklahoma. The extensive area included that portion of the Choctaw Nation south of the Arkansas and Canadian rivers, and in the Cherokee Nation, "east of the 96th meridian." Faucett needed only to contract with the two Indian companies and begin work.[15]

The Indians wasted little time in taking advantage of the new legislation and, soon after passage of the Choctaw act, met to organize the Choctaw Oil and Refining Company. Charter members of the organization included Samson Holson, McKee James, Charles Winston, Robert Benton, James King, G. W. McCurtain, C. W. Frazier, Thomas Byington and G. C. Dukes, as well as those individuals who had been in contact with Faucett from the beginning: J. F. McCurtain, E. N. Wright, Alexander R. Durant, and Allen Wright. E. N. Wright was elected president of the corporation, with McKee James selected as treasurer and Allen Wright as secretary. Because there was no provision under Choctaw law to grant the power of attorney to another individual to enter into a contract with a third party, they adjourned, with plans to convene again in Sherman, Texas. On November 25, 1885, the incorporators met with Faucett in Sherman to complete the business. After studying the contract carefully, the board of directors, Faucett, and the corporation's members individually signed the contract in the presence of Judge E. P. Gregg of Grayson County, Texas. The members then adjourned, with plans to reconvene at Tuskahoma, in the Choctaw Nation, during the first week of the Choctaw Council in October of 1885.[16]

Although it seemed the legalities of the operation had finally been completed, the federal government was once again interjected into the question when, during its 1884 session, the United States Senate adopted a resolution of inquiry designed to determine the legality of the Indians' actions. Only twelve years previously, the Department of the Interior and the commissioner of Indian affairs had opposed any agreements among members of the Five Civilized Tribes and non-Indians to begin oil exploration and production in the Indian Territory. This time, however, it was a different story, and on January 3, 1885, Secretary of the Interior Henry Teller informed the Senate that as far as his office was concerned, the lease agreements were completely within the Indians' rights. To substantiate his conclusion, Teller pointed to several past treaties with the Choctaws and Cherokees. Article Thirty-three of the Choctaw and Chickasaw Treaty of 1866 provided that any land not selected for occupancy by tribal members "shall be the common property of the Choctaw and Chickasaw Nations, in their corporate capacities, subject to the joint control of their legislative authorities." In addition, Article Forty-three, although it called for the exclusion of non-Indians within the boundaries of the nations, also stipulated that "this article is not to be construed . . . to prevent the legislative authorities of the respective nations from authorizing such works of internal improvement as they may deem essential to the welfare and prosperity of the community." Likewise, Article Five in the Cherokee Treaty of 1835 provided that the United States government guaranteed to "secure to the Cherokee Nation the right by their national councils to make and carry into effect all such laws as they may deem necessary . . . ."[17]

Teller pointed out that "The Cherokees have a fee simple title to their lands and they do not recognize the right of the [Interior] Department to interfere in the management of their affairs with reference thereto . . . ." He firmly supported their contention and declared that they "are quite capable of determining, without the aid of the Interior Department or Congress, what is to their advantage or disadvantage and the government cannot interfere with their rightful use and occupation of these lands, which are as rightfully theirs as the public domain of the United States." In his opinion, Teller informed the senators, "The rights reserved to the United States are clearly expressed in the several treaties, and the right of the United States to control the Cherokee property, and prevent the nation from having the full and absolute control of the products of these lands is not even suggested." He apparently convinced the members of the senate, for plans went ahead to begin, and drilling operations in Indian Territory proceeded without further federal involvement.[18]

It soon became evident, however, that the Indians were becoming more and more apprehensive over Faucett's delay in actually beginning work on the wells. Both the Choctaw and Cherokee companies complained that he was wasting too much time. In answering the allegations, Faucett told Allen Wright in a letter dated August 8, 1885, that "I shall go to Fort Smith [Arkansas] the beginning of the coming week if nothing prevents, and then down into the nation to make a selection of locality and arrange for lumber, sills, etc., for the rig so as to commence drilling." Realizing that the Choctaw legislation required drilling to begin within one year of the passage of the act, Faucett informed Wright, "Don't be alarmed about my allowing the agreement to expire by limitation," for as he assured the Choctaw, "I have spent too much money and time to allow it to go by default."[19]

Soon afterward, Faucett set up a drilling rig on the Clear Boggy River in the Choctaw Nation approximately fourteen miles west of Atoka and about seven miles south of Lehigh. Alum Bluff, located on the Illinois River in Going Snake District, was the site selected by Faucett in the Cherokee Nation. But in December, 1885, Faucett received a severe setback when the Cherokee National Council suddenly repealed the previous legislation which allowed for the formation of an oil company in the Cherokee Nation. The bill passed in 1884 had clearly stipulated that "such company shall begin the work within one year of the approval of this act, otherwise all the rights herein granted shall be null and void." By neglecting to inform the Cherokees of his intentions on Alum Bluff, Faucett had allowed the deadline to expire.[20]

Stunned by the turn of events, Faucett was also advised that the financial backers of the project in New York had withdrawn their support when they learned of the actions taken by the Cherokee Nation. After several attempts were made to correct the situation, Robert L. Owen, a prominent Cherokee lawyer, persuaded the Cherokee Council in 1886 to reconsider and once again provide for the formation of an oil company within the Cherokee Nation. But Faucett was now faced with a new problem. Financiers in New York, previously willing to underwrite the drilling operation, were discouraged by the numerous delays, and the needed financing was not available. Turning to a group of investors in St. Louis, Missouri, Faucett reorganized the company under Illinois law. Dubbed the National Oil Company, the corporation had a capital stock of $10,000,000 with William A. Adams serving as president, Joseph W. Buel as vice-president, George A. Hynes as secretary-treasurer, and Faucett and two others, F. G. Flanagan and E. H. Smith, as directors.[21]

By now, a number of individuals involved in the Choctaw venture were also becoming unhappy over the speed at which the operation was progressing; however, several prominent Choctaw investors, aware that not all of the delays were Faucett's fault, continued their support. Acquisition of the necessary materials was a major problem. After being shipped to Atoka, Indian Territory, via the Missouri, Kansas and Texas Railroad, the heavy machinery was then transported by wagon to the drilling site. Nevertheless, at the annual meeting of the corporation in 1886, members voted to drop Faucett from the firm if he did not begin drilling immediately. E. N. Wright, president of the corporation, advanced funds from his personal account to maintain the workers and keep the drilling operations going.[22]

Finally able to begin actual drilling operations at the Clear Boggy location, Faucett was allowed to continue working with the Choctaw Oil and Refining Company. Although the initial work was slow and a shortage of funds continued to plague the endeavor, Wright always provided money when needed in order that the operation might continue and, by 1888, drilling was well under way. Slate and shale were struck at the 168-foot level and, within a short time, the first indications of oil were encountered at 917 feet; a "show" of oil was found at 1,302 feet; and oil was struck at 1,347 feet, followed by gas at 1,391 feet. A total depth of 1,414 feet was reached by July 23, 1888.[23]

Meanwhile, Faucett had contracted typhoid fever during the first half of the year and died a short time later at his home in Neosho, Missouri. Without his presence at the Boggy River site, work soon stopped and the well was eventually abandoned; however, the "log" or record of this well was often referred to in the later development of oil in Oklahoma.

Faucett's efforts added to the growing excitement over potential oil-producing areas within the boundaries of Indian Territory and, as word of the fortunes being made in the oil fields in the East spread throughout the region, other individuals quickly took up the quest. While searching for stray livestock in 1882, Edward Byrd, an intermarried white citizen of the Cherokee Nation, had stumbled upon "something glittering" on the water of Spencer Creek, located about four miles from Chelsea in the Cherokee Nation. In 1886, Byrd secured a lease from the Cherokee Council to search for oil on a half section of land but, because of the size of the area, could find no one to back the venture. Applying for a larger concession, Byrd was given ninety-four thousand acres, bounded by a line starting "on the east side of Chelsea and running northwest via Pat Henry's farm to Coody's Bluff; thence southwest to Yellow Leaf Crossing on the Verdigris River and down this river to the Clem V. Rogers Crossing; and thence east, via Sageeyah Switch, on the Frisco Railroad, to the point of beginning." This extensive area became known as the "Cherokee Lease."[24]

Shortly thereafter, Byrd, together with William Linn, Martin Hellar, Finley Ross, William Woodman, and Oak Daeson, formed the United States Oil and Gas Company, with a capital of one million dollars valued at one dollar a share. Subsequently, Byrd selected a site near where he had first discovered the oil seep in 1882 and began operations with a

Cable tool drilling in the 1870s. Many of the early-day drilling operations in Oklahoma were rather primitive affairs that employed cable tools. *Courtesy of Continental Oil Company.*

Pawhuska "on the go" in 1906. After Henry Foster began his efforts to acquire drilling rights to the Osage Reservation in the late 1890s, it was not long before such sleepy towns as Pawhuska experienced a mineral rush rivaling the 1849 gold stampede to California. *Courtesy of Oklahoma Publishing Company*.

horse-powered rig. In August, 1889, he struck oil at a depth of thirty-six feet. Producing approximately one-half barrel of "light green oil" a day, the well was sold, after installation of a small pump and tank, as a "dip" to area cattlemen, who drove their livestock through the crude to rid them of ticks. Byrd drilled several other wells in the immediate vicinity, including one which produced fifteen barrels a day. By 1891 the United States Oil Company had completed eleven wells on the "Cherokee Lease." However, because of the limited production of its wells and the lack of a nearby market, the company fell upon hard times and was sold to John Phillips of Butler, Pennsylvania in 1895. Reorganizing the corporation as the Cherokee Oil and Gas Company, Phillips drilled fourteen more wells, but ultimately was no more successful than his predecessor.[25]

In 1894, Michael Cudahy, a meat packer from Omaha, Nebraska, secured a blanket lease for oil and gas production on about two hundred thousand acres of land in the Creek Nation and drilled several wells in the area. One well, located on a town lot in Muskogee, struck oil-bearing sand at a depth of 1,120 feet but was non-productive. It was later abandoned after reaching a depth of 1,800 feet. A short distance to the southwest, another well found "a high-grade oil" at 645 feet and, after being "shot," became an apparent "producer." Primarily drilled for prospecting purposes, however, it was continued to the depth of 1,300 feet before being abandoned.

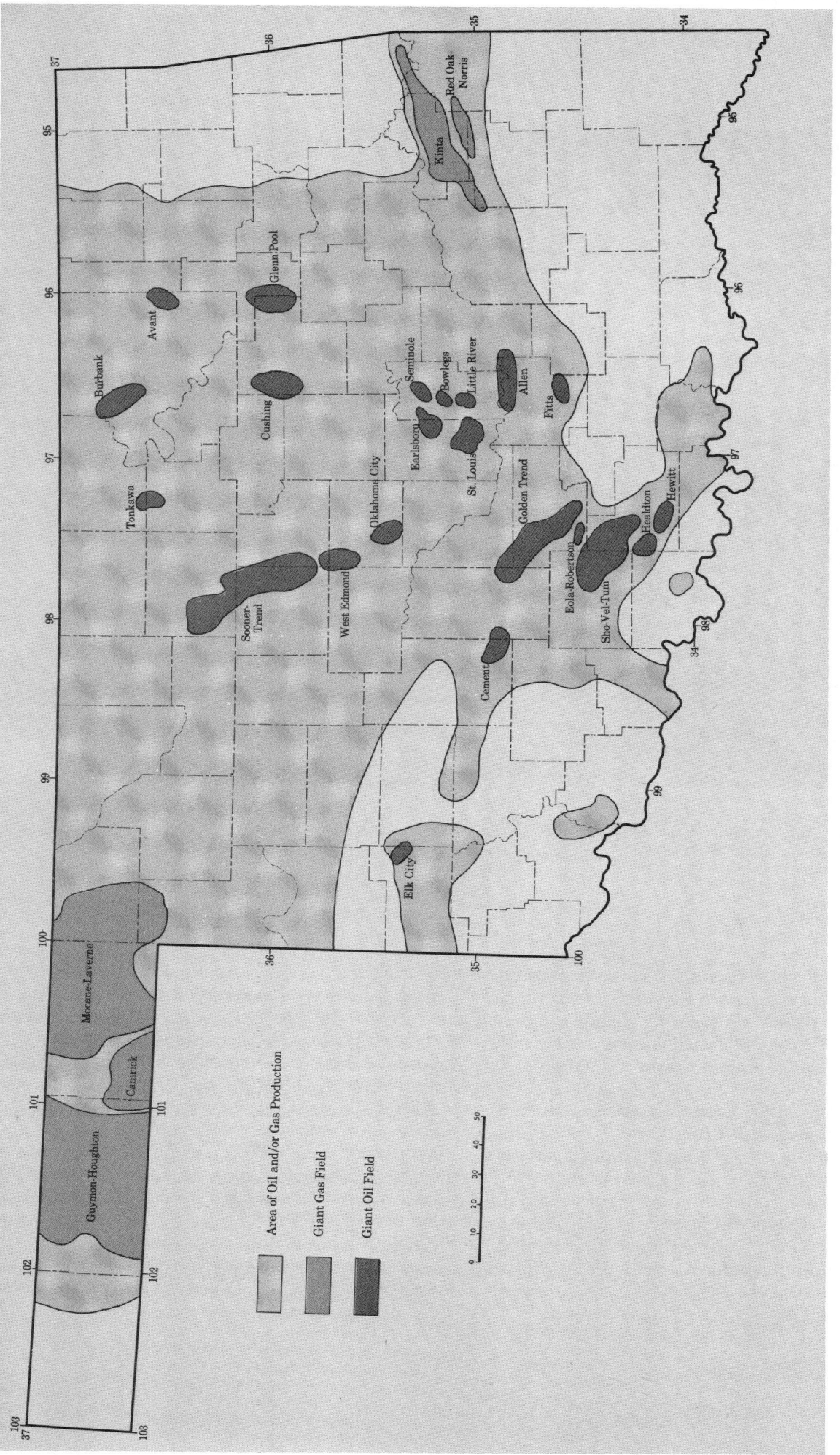

Oklahoma oil and gas fields. In the beginning of the twentieth century the vast underground pool of oil and natural gas beneath Oklahoma was ready to be tapped by pioneer oilmen searching for the liquid gold. *From Morris et al.*, Historical Atlas of Oklahoma.

Bob Stuchel pointing to one of the oil seeps, long known to the Indians, throughout Oklahoma. *Courtesy of William A. ("Mac") McGalliard Collection.*

Cudahy then moved the operations of the Cudahy Oil Company to the area between Red Fork and Sapulpa, near Tulsa, and reopened a well that already had been drilled to a depth of 1,300 feet. He deepened the hole to approximately 1,750 feet before abandoning the site.[26]

On March 16, 1896, Henry Foster signed a blanket lease agreement with Chief James Bigheart and the Osage Council subject to the approval of the secretary of the interior. Under the arrangement, Foster received the right to "mine" petroleum and gas on the Osage Reservation, comprising 1,500,000 acres of land, for a period of ten years. In exchange, the Osage Nation was granted a ten percent royalty on all oil produced and a payment of fifty dollars per year on each gas well. Foster died while negotiations were still being conducted with the Department of the Interior, and the Osage Council granted Edwin B. Foster, Henry's brother, the concession that was approved by Acting Secretary of the Interior John M. Reynolds, on April 8, 1896.[27]

Almost immediately the Phoenix Oil Company was incorporated under the laws of the state of West Virginia, and Edwin Foster assigned his holdings to the new corporation in exchange for 30,000 shares of stock, valued at one dollar a share. Foster was joined by eight others in the venture, and it was agreed that fifty-one percent of the stock was to be held in the company's treasury. As a result, Foster held 5,000 shares, 10 shares were retained as organization stock, and the remaining eight individuals owned a total of 24,990 shares.[28]

By June 20, 1896, the Phoenix Oil Company had begun operations in the Osage Nation with a well located in the NE¼ of Section 13-T29N-R10E. Drilled by contractors A. P. McBride and C. L. Bloom, the well was located just south of the Kansas border, approximately nine and one-half miles from the northwest corner of the Cherokee Nation. Completed the following month, it was abandoned after reaching the 1,100 foot level and producing only a "show" of oil. Not yet discouraged, Foster's company completed a second well in September, 1896, located approximately six miles to the west, in the NW¼ of Section 18-T29N-R10E, but no oil was found. On the third attempt, the Phoenix Oil Company was successful. Working in Lot 32 of the SW¼ of Section 34-T27N-R12E, the drillers struck petroleum at a depth of 1,349 feet in what was later to become famous as "Bartlesville Sand."[29]

Although it struck oil, the Phoenix Oil Company's third well was a "small producer," and the owners of the company were quickly becoming discouraged. The investors had "not been able to place the treasury stock as readily as we hoped," and did not possess a strong financial base. They had hoped to "get into the field at once so we raised a little money among us and did so." With two dry holes and only a "small producer" to show for their investment, they found money growing short. As had been the experience of previous corporations, they suffered from the lack of railroads or pipelines in the vicinity. Supplies and equipment had to be hauled into the Osage Nation by wagon, a slow and costly process; moreover, there appeared to be some objection to the lease agreement in Congress. Because of these difficulties, as well as disappointment over the outcome of the wells previously drilled, the Phoenix Oil Company ceased operations in the winter of 1896.[30]

After Foster's initial attempt to find petroleum in paying quantities in the Osage Nation, his former drilling contractors, McBride and Bloom, took a contract with the Cudahy Oil Company, previously active in the Creek Nation, and moved their equipment to the south bank of the Caney River at Bartlesville in the Cherokee Nation. Although previous efforts to find oil in commercial quantities in Indian Territory had met with only limited success, the fortuitous efforts of McBride and Bloom on behalf of the Cudahy Oil Company resulted in Oklahoma's first commercial oil well: the No. 1 Nellie Johnstone.[31]

## Chapter 2.
# No. 1 Nellie Johnstone

THE PRESENCE OF OIL in the Bartlesville area of the Cherokee Nation had been known for some time. In 1874, Jacob Bartles, an adopted Delaware Indian, opened a trading post on the north side of the Big Caney River, and by 1882, two of his former employees, William Johnstone and George Keeler, had established a store on the west bank of the river. As early as 1875, Keeler had noticed traces of petroleum in the water of Sand Creek when, on numerous occasions, his horse had refused to drink from the stream. He paid little attention to the discovery at the time but, as the oil fever spread into Indian Territory during the drilling activity in Kansas in the 1880s, the importance of the find came quickly to mind, and Keeler and Johnstone became interested in the possibility of drilling an exploratory well in the vicinity of Sand Creek.[1]

After a rather extensive search for oil in the region of Chanute and Neodesha, Kansas, two Kansas oilmen, James M. Guffey and J. H. Galey, moved south in hopes of finding oil in the Cherokee Nation. In 1893, Guffey began negotiations with several businessmen in the Bartlesville area in an effort to execute an agreement with Cherokee officials to drill for oil. Under Cherokee law, citizens of the nation were allowed the right to lease one square mile of land for mineral purposes, and Guffey offered to sublease the area from those individuals for forty-five cents per acre. Keeler, Johnstone, Frank M. Overlees, Robert B. Ross, Major Lipe, William V. Carey, and Jesse Cochrane were among a number of prominent men who accepted his offer and, with the exception of Cochrane, also secured leases in the names of their wives.[2]

By 1895 the Cherokee law was altered to allow each citizen of the tribe the right to lease five square miles of land, subject to the approval of the Cherokee National Council and the Department of the Interior. Eventually, Guffey and Galey were successful in obtaining leases in an area 18 miles square, involving nearly 208,000 acres. The agreement of 1895 also stipulated that the Kansas drillers post a hundred thousand dollar cash bond with the Cherokee Nation, and drill a well on each of the five-mile tracts that stretched south from the Kansas border and west to the ninety-sixth Meridian. Three and one-half percent royalty was to be paid to the tribe on all future petroleum production. When it was discovered that the secretary of the Cherokee Nation, Little Star, had failed to sign the contract, Keeler dispatched a messenger to locate the Indian official, in Texas at the time, and obtain his signature. But this took time, and when the messenger failed to return within thirty days, Guffey and Galey not only withdrew their contract but eventually sold their interest in the project.[3]

Nevertheless, Keeler, Johnstone and the others had no intention of giving up the idea. Michael Cudahy, whose oil company was active in Indian Territory, expressed an interest in acquiring the leases, and after a conference with Cudahy's representative, the leaseholders decided to transfer the agreement to the Cudahy Oil Company.[4]

Wasting little time in arranging for drilling operations, Cudahy ordered McBride and Bloom to move their equipment to the NE¼ of Section 12-T26N-R12E, the site selected for the first well on the banks of the Caney River. Although only approximately seventy miles from the location of their last well in the Osage Nation to the new drill site, the work was being done in mid-winter, and weather conditions made the task extremely hazardous. As the temperature dropped, heavy rains turned to snow and sleet, and Keeler sent fourteen teams of oxen to Red Fork to help the drillers pull the equipment. After the group left for Bartlesville with twenty wagons loaded with

Nellie Johnstone No. 1, 1897. When it blew in on April 15, 1897, this well opened a new era in Oklahoma's history. *Courtesy of the Bartlesville Public Library.*

Several early-day Bartlesville residents pose on a drilling rig on the Caney River bottom in front of George B. Keeler's home. *Courtesy of the Bartlesville Public Library.*

tools, pipe, and other equipment, several of the wagons broke down and the material was simply abandoned by the side of the road. The Arkansas River was frozen over with a crust of ice too thin to support the wagons, but too thick to be easily broken. As a result, several days were spent along the river waiting for the ice to melt. When the ice was thin enough to attempt a crossing, a pathway was cut through and with six animals attached to each wagon, they were able to cross to the other side. After nearly three weeks, the party arrived at Bartlesville, but the trip had cost nearly four hundred dollars.[5]

Drilling operations began on the Caney River in late January, 1897, just northwest of early-day Bartlesville. The crews were delayed by having to wait for the arrival of essential equipment, but by late February operations were progressing smoothly.

McBride and Bloom apparently had little contact with the area's inhabitants. They were described as "a gentlemanly set of fellows, but . . . very closemouthed . . . [and they] have no information to give out." Local interest was high, however, in that most residents realized that a significant oil discovery would greatly boost the settlement's economy and "the next move will be for the railroad to begin building."[6]

As the drilling progressed, the Oswego Lime was located at a depth of from 880 to 942 feet; the Layton Sand between 975 and 987 feet; a gas sand at 1,252 feet; and oil at a depth of 1,303 feet. Drilling continued to 1,320 feet, "where it showed for an oil well." The "shooting" of the well with nitroglycerin was a special occasion. G. M. Perry, an expert "shooter," arrived to oversee the placing of the explosives and, after all the arrangements were made, a large crowd gathered on April 15, 1897, to witness the event. Miss Jennie Cass, Keeler's stepdaughter, was given the privilege of dropping the "go devil," which set off the explosive charge deep within the well. By 3:00 P.M. everything was ready, and Jennie dropped the detonating device down the well bore. The crowd heard a muffled explosion and felt a slight tremor. Very shortly, rocks began striking the crown block at the top of the drilling rig, and a large column of oil, water, and debris blew over the derrick. Cudahy had a producer.[7]

Named for the daughter of William Johnstone, the well produced between fifty and seventy-five barrels of petroleum a day. McBride and Bloom had neglected to prepare storage tanks, however, and the crude formed a "good-sized stream along the ground," which ran toward the nearby river. Cudahy was now confronted with the same situation that had faced drilling operators in the Osage Nation: what to do with the petroleum. Without a railroad or pipeline nearby, there was no economically sound method to ship the oil to market. Although the crude oil was commonly used for illumination, for medicinal purposes, as a dip for livestock, as well as to grease and prevent rust on tools and equipment, the discovery was too much for the local market. As a result, Cudahy capped the well in hopes of a more profitable market in the future.[8]

But the Nellie Johnstone was not properly sealed, and a series of leaks soon developed around the casing and surface fittings. The trickle eventually filled the sump around the well head and overflowed in a small rivulet that made its way toward the Caney River. Residents of the area helped themselves to the pe-

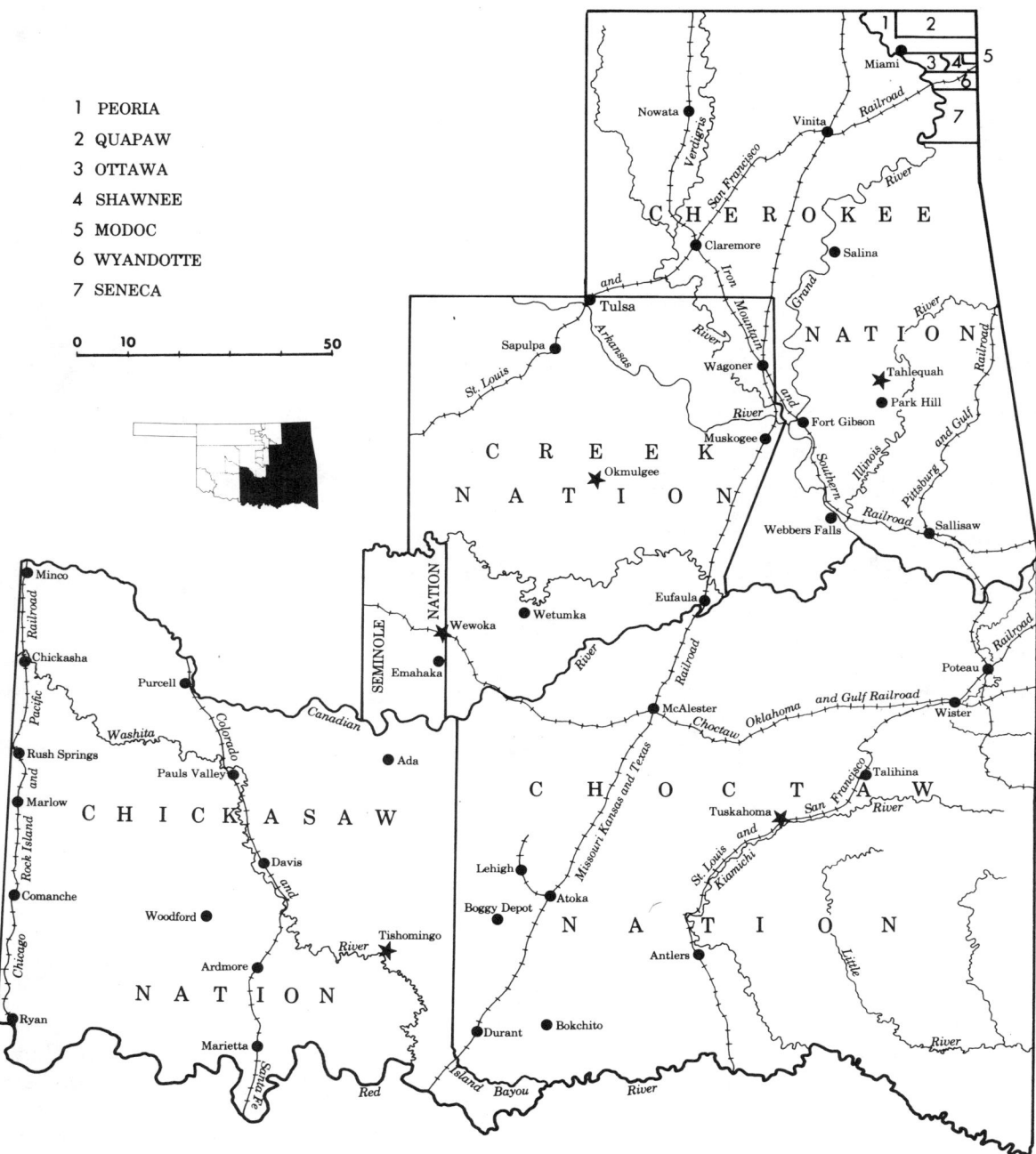

Railroads reach Indian Territory. By the 1890s, Indian Territory had shrunk to approximately one-half its original size, but the introduction of railroads into the region permitted the crude to be transported to refineries in nearby states. *From Morris et al.*, Historical Atlas of Oklahoma.

Within a decade following the drilling of the Nellie Johnstone No. 1, the region surrounding Bartlesville was a forest of oil derricks. *Courtesy of the Bartlesville Public Library.*

troleum that seeped from the well, carrying off the oil in buckets for their own use. Over a period of time, the oil seepage increased, and, during one winter, several of the local residents built a fire on the bank of the river while iceskating. Unfortunately, the fire was located too close to the flowing oil from the Nellie Johnstone, and with a flash, the flames followed the petroleum to its source and quickly engulfed the drilling rig. Thus ended Oklahoma's first commercial oil well, a technical success but a financial failure because of the lack of a market.[9]

Cudahy and the other early oil pioneers had discovered a portion of what would later be called the Mid-Continent Oil and Gas Field. This vast area, comprising a myriad of separate oil and gas reservoirs at varying depths, stretched from the central portion of northeastern Oklahoma into southeastern Kansas, approximately 20,000 square miles, covering 20 counties in present-day Oklahoma. Although early oilmen were forced to plug their wells because there was no readily accessible market, annual production of oil had jumped from approximately 1,000 barrels in 1897 to 43,524,000 barrels in 1907, the year Oklahoma was admitted to the Union. Furthermore, twenty years after the drilling of the No. 1 Nellie Johnstone, the annual crude oil production in the state was a staggering 107,508,000 barrels.[10]

It was now firmly established that there was oil in paying quantities in Oklahoma. The success of the No. 1 Nellie Johnstone generally intensified the search for oil and specifically revived the interest of the Phoenix Oil Company in the area's potential. Still holding the oil concession on the Osage Reservation, the corporation reemployed McBride and Bloom in

October, 1897, and announced a new location west of the 96th Meridian.[11]

Attempting to drill in the same general area as the Cudahy well, Foster ordered McBride and Bloom to locate on Butler Creek, inside the boundary of the Osage Nation, on Lot 32, in the SW¼ of Section 34-T27N-R12E, only approximately two miles northwest of the No. 1 Nellie Johnstone. Oil was found in the Bartlesville Sand at a depth of between 1,323 and 1,345 feet, and it initially produced ten barrels of oil per day. However, in an effort to increase the daily output, the well was "shot," resulting in the influx of a huge volume of water which could not be stopped. The following month, the Phoenix Oil Company drilled another test, the Wilkey No. 2, which proved to be "a much better well." Nevertheless, the company found capital running short, and, in an effort to finance its drilling operations, Foster and the others assigned a large portion of their Osage leases to Samuel C. Sheffield. In turn, Sheffield granted the leases to the Osage Oil Company, composed largely of the members of the Phoenix Oil Company. Even this financial "pump-priming" could not prevent the temporary demise of oil prospecting in the Osage Nation. For, although the Osage Oil Company drilled several additional wells on its Osage leases, it experienced a series of dry holes, and drilling in the region came to a standstill.[12]

The economic factors involved in oil prospecting within Oklahoma Territory and Indian Territory had finally forced a halt to further efforts. With no railroads or pipelines available, potential markets were simply still too far removed to make commercial development feasible. For the crude to be marketed in the East, it would have to be transported overland to railroad terminals in Kansas or sent to Standard Oil's small facility at Neodesha, Kansas, the only refinery in the vicinity. The local market was oversupplied by petroleum from the Kansas fields. Not only was it expensive to ship the oil out, it also was costly to bring supplies in. Most of the equipment was hauled by wagon from Independence, Kansas, ten miles north of the boundary between the Oklahoma Territory and the Indian Territory; however, some of the material had to be brought into the area from Texas, the border of which was more than 150 miles to the south. It was difficult, therefore, to attract capital in sufficient quantity to continue drilling operations in the area.[13]

In an attempt to attract foreign capital, Michael Cudahy and his brother, John, who had joined him in his activities, dispatched W. H. Isom, general manager of Cudahy Oil Company, to London, England, in the hope of finding European investors. Isom met with some success and offered a group of British and Dutch financiers one-half interest in the firm for five million dollars. After listening to Isom explain the potential profit that could be reaped by the development of the large petroleum field in the "Twin Territories," as Oklahoma Territory and Indian Territory were sometimes called, they agreed to send a delegation to the area to investigate Cudahy's offer. The committee arrived in Muskogee during the winter of 1897–1898, and were favorably impressed by the potential offered by Cudahy's proposals. If a refinery were located in the Muskogee area, they reasoned, the petroleum could be sent by pipeline to the Arkansas River, then shipped by water to New Orleans and other markets. It appeared that the British and Dutch financiers were willing to underwrite Cudahy's future drilling operations and that a means might be devised to solve the nagging question of how to get the oil to market.[14]

Unfortunately, any potential development of the region's oil deposits through the introduction of new capital was abruptly halted by the federal government in the spring of 1898. The Choctaws and Chickasaws had, on April 23, 1897, set a precedent in agreeing to the Department of the Interior's claim that it should have the final say in the approval of mineral leases negotiated with the Indians. Section Three of the Atoka Agreement pertained to the coal and asphalt reservations within the Choctaw and Chickasaw nations, and provided that "all coal and asphalt in or under the lands allotted and reserved from allotment shall be reserved for the sole use of the members of the Choctaw and Chickasaw tribes." In addition, it stated "that where any coal or asphalt is hereafter opened on land allotted, sold or reserved, the value of the use of the necessary surface for prospecting or mining, and the damage done to the other land and improvements shall be ascertained under the direction of the secretary of the interior." Therefore, any future leases within the two nations' boundaries would be reviewed by the Department of the Interior and subject to its regulations.[15]

Following the example of the Atoka Agreement, the United States Congress passed the Curtis Act on June 28, 1898, which effectively ended the leasing policy of the Five Civilized Tribes. Section Thirteen of the bill was devoted exclusively to the question, and authorized the secretary of the interior "to provide rules and regulations in regard to the leasing of oil, coal, asphalt, and other minerals in . . . [Indian] Territory." In addition, it specifically stated that "such leases shall be made by the secretary of the interior, and any leases for any such minerals otherwise made shall be absolutely void." Several restrictions were placed on the agreements: "No lease shall be made or renewed for a longer period than fifteen years, nor cover the mineral in more than six hundred and forty acres of land"; "Lessees shall pay on each oil, coal, asphalt, or other mineral claims at the rate of one hundred dollars per annum, in advance for the first and second years"; a payment of two hundred dollars was required in advance for the third and fourth years, and "five hundred dollars, in advance, for each succeeding year thereafter, as advanced royalty on the mine or claim on which they are made"; the money was required "on each claim, whether developed or undeveloped"; and "should any lessee neglect or refuse to pay such advanced royalty for the period of sixty days after the same becomes due and payable on any lease, the lease on which default is made shall become null and void, and the royalties paid in advance shall then become and be the money and property of the tribe." The law also empowered the secretary of the interior to ascertain "the value of the use of the necessary surface for prospecting or mining, and the damage done to the other land and improvements," and required compensation be paid to the owner.[16]

Congress did provide some relief to those companies already engaged in Indian Territory, by declaring that under the provisions of the Curtis Act, "nothing . . . shall impair the rights of any holder, or owner of a lease hold interest in any oil, coal-right, asphalt, or mineral which have been assented to by act of Congress, but all such interest shall continue unimpaired hereby, and shall be assured to such holders or owners by leases from the secretary of the interior for the term not exceeding fifteen years." Even so, they were "subject to payment of advance royalties as herein provided . . . and preference shall be given to such parties in renewals of such leases." The bill also specified that "When, under the customs and laws heretofore existing and prevailing in the Indian Territory, leases have been made of different groups or parcels of oil, coal, asphalt, or other minerals deposits and possession has been taken thereunder and improvements made for . . . [their] development . . . which have resulted in the production of [the minerals] in commercial quantities," the individuals holding the leases "or their assigns" were to be given "preference in the making of new leases."

"Due consideration shall be made for the improvements of such lessees," and the amount of royalty paid by "all lessees shall be fixed by the secretary of the interior."[17]

If Section Thirteen of the Curtis Act was not specific enough to bring about the demise of tribal leases, Section Sixteen of the legislation provided that "it shall be unlawful for any person, after the passage of this Act, except as hereinafter provided, to claim, demand, or receive, for his own use or for the use of any one else, any royalty on oil, coal, asphalt, or other mineral . . . whatsoever." Under its provisions, "all such royalties and rents hereafter payable to the tribe shall be paid under such rules and regulations as may be prescribed by the secretary of the interior, into the Treasury of the United States to the credit of the tribe to which they belong." Exception was made "to any member of a tribe to dispose of any timber contained on his, her, or their allotment." Thus, all past oil concessions were subject to approval of the secretary of the interior before renewal, and future agreements were to be negotiated under the rules and regulations of the Department of the Interior. Under such an arrangement the oil lands of the Five Civilized Tribes were set aside to be leased for the benefit of the entire tribe.[18]

The limitations imposed by the federal government had a stunning effect on the oil industry in Indian Territory, and for the next four years, there was very little prospecting attempted. In April of 1899, a land office was opened at Muskogee to provide a convenient method for the oilmen to apply for leases, but most seemed hesitant to renew their drilling operations. Some of the early petroleum pioneers suffered great losses under the new government policy. For example, the Cudahy Oil Company had thirty-six square miles of leases revoked by the secretary of the interior; however, the firm, in consideration of its efforts prior to the passage of the Curtis Act, was allowed to retain rights on Section 12-T20N-R12E. Such actions did little to encourage a resumption of petroleum exploration in the region.[19]

But while the federal government was studying the future development of petroleum deposits in Indian Territory, several important events were occurring, not only in the general region but throughout America, that would trigger a tremendous oil boom in present-day Oklahoma in the early years of the twentieth century. One of the greatest problems that had faced early oilmen in the area was the lack of transportation facilities to move their crude oil to refineries and markets and, in turn, make drilling operations financially feasible. Railroads had been slowly expanding through Indian Territory during the late 1870s. The Missouri, Kansas, and Texas (MK&T) was the first railroad to penetrate the area, entering the Cherokee Nation in the early 1870s and eventually stretching completely across Indian Territory from Chetopa, Kansas, to the Red River near Colbert's Ferry, in the Choctaw Nation. At about the same time, the Atlantic and Pacific (A&P) constructed a line to join with the MK&T at Vinita, in the Cherokee Nation. During the 1880s, the A&P extended its lines to the Tulsa and Red Fork area on the Arkansas River and, in 1886, to Sapulpa. In 1886 and 1887, the St. Louis & San Francisco Railroad (SL&SF) built a route from Fort Smith, Arkansas, across the Choctaw Nation to Paris, Texas. During the same period, the Atchison, Topeka and Santa Fe (AT&SF) was developing an extensive rail network through present-day Oklahoma. This rail boom grew even larger in the 1890s and during the first decade of the twentieth century, and the tremendous expansion of the railway system provided a convenient means of transporting Oklahoma crude to markets outside the region.[20]

Frank and Jane Phillips. Seven years after the Nellie Johnstone No. 1 was brought in, Frank and Jane Phillips arrived in Bartlesville and launched one of the most famous successes of the early Oklahoma oil boom. *Courtesy of the Oklahoma Heritage Association.*

Although railroads offered one method of getting oil from the wellhead to the consumer, the most efficient means was by pipeline. Shortly after the penetration of the region by railroads, efforts were made to provide them access to producing wells via pipelines. In the summer of 1899, the AT&SF offered service to the Bartlesville area, opening the way to ship petroleum to refineries in nearby states. The following year, a two-inch pipeline was constructed from the Phoenix Oil Company's wells in Lot 32 to the Bartlesville railway depot, where a loading rack was built. Eventually, the pipeline system was expanded to include some other wells owned by the Almeda Oil Company on a sublease in the Osage Nation. Tank cars to transport the crude were furnished by the Neodesha refinery, and in May of 1900, the first oil was shipped out of Indian Territory by rail.[21]

In 1889 a portion of Indian Territory known as the "Unassigned Lands" was opened by presidential proclamation to homesteaders, and the following year, on May 2, 1890, the United States Congress passed the "Organic Act" providing for the creation of the Territory of Oklahoma in the recently opened region. Over the next seventeen years, Oklahoma Territory was expanded into what became the western half of the present-day state. As Indian title to this land was abolished by a series of land runs, lotteries, and tribal allotments, it became available to those oilmen who wished to negotiate drilling leases with the new owners.[22]

Perhaps one of the most important factors leading to the impending oil boom was the tremendous growth of technology and mass-manufacturing methods between 1890 and 1910. For the first several decades of the petroleum industry in America, oil was utilized primarily to supply an inexpensive illuminant: kerosene. By the late 1890s, technology had developed some 200 byproducts of petroleum such as naphthas, petroleum wax, lubricants, and others, which accounted for nearly one-half of the industry's annual sales. Then came the greatest technological revolution of all, the practical application of the internal combustion engine, which provided a cheap and dependable means of power. From the 1870s to the 1890s, internal combustion engines patterned after those developed by Nikolaus A. Otto in Germany were utilized in the United States to provide a source of stationary power. But with the application of the principles of Rudolph Diesel in 1895, the

Oil-well supply store in Bartlesville. Because of its accessibility to railroads, Bartlesville quickly became a major supply center for drilling activities in the northern part of Indian Territory. *Courtesy of the Bartlesville Public Library.*

development of powerful yet comparatively inexpensive prime movers quickly spread throughout the transportation and manufacturing industries in America. Despite these technological breakthroughs, it was the evolution of the gasoline-powered automobile that provided an almost insatiable demand for petroleum products.[23]

Beginning in Europe, automobile development quickly spread to the United States, and, by one estimate, more than 300 automobile and parts manufacturing companies were organized prior to 1900. Moreover, Henry Ford's application of the theories of mass-production and mass-consumption provided automobiles by the hundreds of thousands at a price that was affordable to the average individual. In 1899 the total output of vehicles in the United States numbered a paltry 3,723, powered by an assortment of engines; a decade later, in 1909, American manufacturers poured out a total of 126,593 vehicles, of which 120,392 were powered by gasoline engines; and by 1919, that total had increased to 1,683,916 vehicles, with 1,680,327 of them dependent upon gasoline.[24]

With the emergence of the gasoline-powered automobile, there was a corresponding growth of petroleum markets. Before 1899 few would have predicted such an increase, and some oilmen were fearful of a decrease in petroleum markets because of the competition of coal as a source of domestic energy. With the advent of the automobile, however, this all drastically changed. Gasoline was easily portable, provided a large amount of energy in relation to its bulk, did not evaporate rapidly, yet was a perfect fuel for internal combustion engines at normal air temperatures. The effect on the petroleum market was phenomenal: of the approximately 6,200,000 barrels of gasoline sold in the United States before the automobile, practically all of it was used in cleaning establishments as a solvent, although a small amount was utilized as a fuel for household stoves and portable space heaters; by 1909, the annual production of gasoline amounted to 11,300,000 barrels, with 25 percent used as fuel for automobiles; and a decade later, in 1919, there was a total domestic distribution of gasoline of 87,500,000 barrels, with 85 percent marketed as fuel for internal combustion engines.[25]

Thus, at the beginning of the twentieth century, Oklahoma oilmen, although controlled by federal rulings in Indian Territory operations, had at their disposal a relatively reliable means of transporting their crude to market, as well as the potential of an expanding market to tempt financiers to provide capital for their projects. All that seemed to be lacking was the motivation to renew their prospecting in the region, and that was provided in January of 1901, with the discovery of oil at Spindletop in Texas. In slightly less than one year after the initial gusher, the Spindletop region produced approximately 6,100,000 barrels of oil, and the oil rush to the American West was on.[26]

That there was oil in commercial quantities within the boundaries of present-day Oklahoma had been well established. Not only had petroleum been located in several sites by the early oil pioneers, but the 1898 *Report* of the Governor of Oklahoma Territory, Cassius M. Barnes, acknowledged the fact. Commenting that "There has been much speculation in the past year or two as to the existence of either oil or gas in paying quantities in the territory," Barnes alluded to the fact that "A well put down at Pawhuska a year ago struck some oil, but was immediately sealed up, the drillers declaring that the yield was too small to make it pay commercially." The governor noted, however, "that the company that put down this well afterwards leased large tracts of land in the Osage Reservation and in Pawnee and Payne Counties." Barnes also related that "At Bartlesville, just over the line in the Indian Territory, oil of an excellent quality has recently been struck in paying quantities, and at Muskogee and Eufaula, in the Creek Nation, several paying wells have lately been drilled." In light of these discoveries, he suggested that "The finding of gas and oil northeast of here in Kansas and along the eastern line of the territory, coupled with the fact that oil and gas fields always trend in a belt running from northeast to southwest, leads us to believe that both gas and oil can be found in paying quantities at least in central and eastern Oklahoma."[27]

In his message to the secretary of the interior the following year, 1899, Territorial Governor Barnes was even more optimistic. "There are unmistakable evidences of oil and gas at many points in the territory," he declared. "In Payne and Pawnee counties are several springs where the water is polluted with oil, and in the Chickasaw Nation and the Kiowa and Comanche reservations are oil springs and large deposits of asphaltum, which are but the residue of great fields of oil" Barnes continued. He recounted that "The well put down to a depth of 1,100 feet in the Osage Reservation, near Pawhuska, two years ago was recently opened and made a run of 300 barrels,"

Pouring nitroglycerine into the tin tube. Many of the early wells in Oklahoma were "shot" with nitroglycerine to increase their output; however, it was often a dangerous undertaking. Here the man, or "shooter," has two and a half gallons of nitroglycerine in his hands and five additional gallons on the floor. One-half gallon of the explosive could blow a hole in the ground large enough to bury a railroad freight car. *Courtesy of the Oklahoma Heritage Association.*

and that "Paying wells are being put down in all parts of the Creek country . . . [and] A local company is putting down a well in Guthrie." Barnes pointed out that "These known deposits of oil, together with the fact that part of Oklahoma lies in a direct line from the oil fields of Kansas to those of Texas, all indicate that at some place in the Territory gas and oil will be found in paying quantities."[28]

Oil was known to be present in the "Twin Territories," its transportation to markets was now possible, and a steadily expanding demand for gasoline insured the potential profit necessary to attract capital. All that remained was a discovery that would touch off a rush for Oklahoma's "black gold," a stampede that would rival the mineral rushes to the gold fields of California and the silver mines of Nevada in the nineteenth century. Such a find was not long in coming, for on June 25, 1901, a strike was made on the Sue A. Bland allotment at Red Fork, just to the southwest of present-day Tulsa, that was reported by the Kansas City *Times* as a "GUSHER FIFTEEN FEET HIGH." This was the beginning of the odyssey which forever altered Oklahoma's history.[29]

## Chapter 3.
# Red Fork

THE EVENTS WHICH LED to the drilling of the first well at Red Fork, four miles southwest of Tulsa, remain enmeshed in controversy. There are two major theses offered in describing the events and the participants involved in the historic venture. One faction contends that two local physicians, John C. W. Bland and Fred S. Clinton, sank the well that started the oil boom in Oklahoma. Yet, another group adamantly insists that the first well at Red Fork was the work of Jesse A. Heydrick and J. S. Wick, both experienced oilmen. Indeed, time has shrouded the facts, and the dispute may never be settled; nevertheless, this incident launched the era of black gold in the Sooner State, and it is necessary, therefore, to examine both sides of the question.[1]

Bland and Clinton, two of the community's leading citizens, "were always on the alert for some industry to aid in the development of this section and the coming state." At the turn of the century, Red Fork rivaled Sapulpa and Tulsa as the economic center of the region. In fact, because of the presence of the Frisco Railroad's terminal in the town, it was in a position to outgrow its competitors. In reminiscing about his participation in the event, Clinton recalled that in May of 1901, Dr. Bland asked him to stop by one morning to discuss a 500,000-acre oil and gas lease rumored to have been recently approved by the Creek National Council. When Clinton arrived, Bland also informed him of the presence of a drilling rig on a flatcar at the railroad siding near the town. He told Clinton that the owner of the rig, P. L. Crossman, was short of money and unable to pay the freight charge to the railway. In addition, the local freight agent would not allow Crossman and his crew to unload the equipment.[2]

Bland's wife, Sue, was a citizen of the Creek Nation, and Bland proposed that the two doctors raise the money to release Crossman's drilling rig, drill a well on Sue's allotment, and enter the oil business together. Neither of them was knowledgeable about the petroleum industry, nor did they have the necessary funds; however, Bland argued that the discovery of oil in the immediate vicinity of Red Fork would create such an economic boom that the future of the community would be assured.[3]

Clinton apparently was persuaded by Bland's arguments, and plans began for their proposed well to be located in the NW¼ of the SW¼ of Section 22-T19N-R12E. After obtaining his wife's consent to allow the well to be drilled on her allotment, Bland persuaded Crossman to do the drilling, provided that enough money could be raised to secure the release of his equipment from the railroad. While Bland was working out the arrangements with his wife and Crossman, Clinton attempted to raise the necessary funds. He prevailed on H. H. Adams, the Frisco Railroad agent at Red Fork, to loan them $300, "which was enough to free the equipment and get the drilling started." In their excitement, however, they overlooked one important fact: Sue A. Bland had not yet been granted her allotment by the Creek Nation and did not have title to the property when drilling began on May 10, 1901.[4]

Drilling progressed slowly, and Crossman, who was overseeing the work, was faced with several pressing problems. Food for the drilling crew proved difficult to provide and, when available, it was very expensive. The only public eating places in the area were temporary tents erected to serve the oil field crews. Drinking water was also in short supply and just as expensive. Colonel Robinson's Hotel was the only sleeping establishment available in Red Fork, but fortunately, a number of workers found food and lodging in private homes. Bland paid a large portion of the salaries of the drillers and sometimes assigned them lots adjoining the drill site in order that work might continue. Influential investors were offered accommodations in the homes of Clinton and Bland, but because they could not guarantee title to the land, they were unable to sell any adjoining property to finance the project. Thus, Bland contributed a considerable amount of his personal fortune towards the completion of the well.[5]

During the following month, anticipation mounted as the well neared completion. Bland was stricken with appendicitis but steadfastly refused to leave the activity. Suddenly, during the night of June 24–25, 1901, the drilling crew encountered gas, and excitement reached a fever pitch. Bland's health, however, forced him to ask Clinton "to take full charge" of the operation shortly after 1:00 A.M. Later that morning, on June 25, oil came into the well bore from a depth of approximately six hundred feet and spewed over the top of the derrick; it was Oklahoma's first real gusher.[6]

Accompanied by an attorney and a notary, Clinton hurried to tell the ailing Bland of the strike, taking along a quart of the oil flowing from the Sue A. Bland No. 1. After Sue Bland signed over the power of attorney to her husband, Clinton raced by buggy to Tulsa in an attempt to catch a train to Muskogee, headquarters of the Dawes Commission, and file for an allotment on the site. He arrived at the station "just in time to drop the lines and jump onto the train," and asked a "friend either to take my buggy home or to notify my wife." Traveling to Vinita on the Frisco, Clinton changed to the Katy line for the remainder of the journey and, in the same oil-soaked clothes he had worn when he left the site of the Sue A. Bland No. 1, arrived in Muskogee late that night. Despite the late hour, Clinton hurried to the residence of Dr. F. B. Fite and informed him of the discovery. Skeptical, the Muskogee physician suggested that they test the substance in his backyard. Once outside, Clinton poured the oil on a pile of shavings and, when his friend struck a match, it burned. Clinton suggested a second test: a new lantern was filled with the liquid, the wick was saturated and a match ignited. It burned. That it was oil, and of a high grade, Fite had no doubt.[7]

Fite and Clinton then hurried by buggy to the residence of Allison Aylesworth, a member of the Dawes Commission, empowered by the government to divide the lands of the Five Civilized Tribes into individual allotments. After being told of the series of events, Aylesworth instructed the two doctors to be at his office when the commission opened at 8:00 A.M. the following day. On the morning of June 26, 1901, Clinton was at Aylesworth's office at the appointed time and presented Bland's power of attorney. Aylesworth had made all the necessary arrangements, and Clinton soon had title to the allotment of Sue A. Bland.[8]

Before leaving on the return trip to Red Fork, Clinton made a farewell visit to Dr. Fite's office and met Dr. J. L. Blakemore, who expressed a desire to inspect the well. Clinton consented, and preparations were made for the return journey. Blakemore rented a hack and invited an acquaintance, A. Z. English, to accompany the party. Traveling at night, they missed a turn in the road and ran into a barbed wire fence on the property of the I. X. Ranch. One of the horses was cut very badly, but the owner of the ranch, Bluford Miller, provided a new horse, and the party was able to continue the journey. Early on the morning of June 27, less than forty-eight hours after the well blew in, Clinton and the others arrived in Red Fork—along with title to the land on which the Sue A. Bland No. 1 was drilled.[9]

The other version of the events which preceded the gusher at Red Fork centers around two promoters, Jesse A. Heydrick and John S. Wick. Heydrick, from Butler, Pennsylvania, had allegedly been drawn into exploration for oil in Indian Territory by Wick, who was forming the John S. Wick Company. Because of stringent federal regulations in the latter part of the nineteenth century, petroleum promoters encountered great difficulty in securing leases on which to drill exploratory wells. Nonetheless, several prominent citizens of the Creek Nation: Thomas J. Adams, Wash Adams, Lewis Adams, James Sapulpa, William A. Sapulpa, Samuel C. Davis, John I. Yargee, L. C. Perryman, D. L. Berryhill, and Lilah D. Lindsey jointly acquired a lease to 410,000 acres of land in the Creek Nation for the purpose of oil exploration. The boundary of the lease stretched south from the Arkansas River to the Frisco Railroad on the east and south, then west to the border of Oklahoma Territory. Signed on November 9, 1899, the lease was to run a period of ten years, or for as long as oil, gas or minerals were extracted in profitable quantities. Because the Dawes Commission was in the process of allotting the Creek Nation to individual tribal members, however, there was some question of the legality of the agreement.[10]

The Creek citizens were anxious to begin drilling operations immediately. Heydrick instructed Wick and J. M. Dunn, another experienced oilman, to examine the lease area carefully and select the best potential site for petroleum deposits. In addition, a consultant, John Q. A. Kennedy, accompanied Heydrick from Pennsylvania to examine the lease and assist in selecting a drilling site. Kennedy was also to approve the project for several Pennsylvania investors who wanted to put money into the operation. This took time and, according to the terms of the original lease, they had only six months to begin drilling operations. Nevertheless, Heydrick wanted to be as sure as possible before attempting to lure prospective investors into the project. As a result the deadline was missed. Faced with the task of negotiating a new agreement, Wick secured another lease on July 16, 1900, which carried the names of Sue A. Bland, Dr. John C. W. Bland, Timmie Fife, Eli E. Hardridge, J. H. Land, E. L. Simley, Albert P. Owen, and several other citizens of the Creek Nation. Fifty shares of stock, valued at one hundred dollars each, were sold to provide the necessary capital, and plans were made to begin drilling operations on the site selected in the NW¼ of the SW¼ of Section 22-T19N-R12E.[11]

Late in the spring of 1901, Heydrick ordered Crossman to move his drilling equipment from Sapulpa to Red Fork; however, the Red Fork railroad agent refused to accept Heydrick's New York draft to pay for the freight charge and would not allow the drilling rig to be unloaded. It was at this point that Bland and Clinton agreed to cash the New York draft, pay the freight agent, and allow the drilling operation to proceed. It has been claimed that, out of gratitude for the doctor's intervention, Heydrick gave Bland and Clinton each a share in the John S. Wick Company. In any event, after payment of the railroad's freight charge, the drilling equipment was released and Heydrick started work at the well on May 10.[12]

Although drilling operations had finally begun, some of the eastern stockholders of the John S. Wick Company raised the question of the validity of the lease arrangement with the Creek citizens. On May 15, 1901, when the drilling crew had reached a depth of 150 feet, Heydrick returned to Pennsylvania to meet with the investors. He hoped to reassure them that all was going according to plan, in order that they would continue their financial support in the venture. Taking advantage of Heydrick's absence, Crossman returned to Joplin, Missouri, for a weekend visit and left his son, Luther, in charge of the operation. Drilling continued in their absence and, shortly before midnight on June 24, 1901, the crew reached a "stray gritty lime" at a depth of 534 feet. When the bit penetrated the formation, it

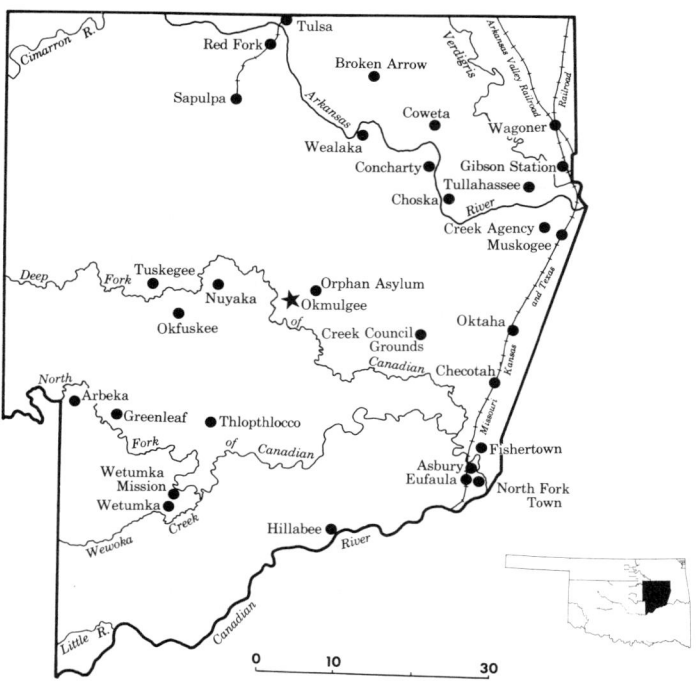

Map of the Creek Nation. Red Fork, just to the southwest of Tulsa in the Creek Nation, was the site of a discovery well in 1901 that brought many oil seekers to Oklahoma. *From Morris et al.*, Historical Atlas of Oklahoma.

struck a gas pocket, and oil shot 30 feet into the air.[13]

In the absence of Heydrick and the elder Crossman, Wick was asleep near the rig when the well came in. Completely taken by surprise, Wick was elated over the strike. Although previously instructed by Heydrick to cease operations immediately should oil be found during his absence and attempt to keep the find quiet, Wick telegraphed the elder Crossman to "Send packer, oil is spouting over the top of the derrick." The message was heard by telegraph operators all the way to Joplin, and the secret was out. Relayed by word of mouth, as well as by wire, the word spread rapidly, and the boom was on.[14]

At this point the two versions of the events surrounding the drilling of the first well at Red Fork become compatible. Of all those involved in the operation, it appears that Clinton was the only one who kept his head over the discovery. In the absence of Bland, Heydrick, and Crossman, Wick did not show good judgment in handling the situation. On the other hand, Clinton had hurriedly obtained the

The discovery well at Red Fork, 1901; Sue A. Bland No. 1. *Courtesy of the Oklahoma Publishing Company.*

ailing Bland's power of attorney and, with the help of a prominent friend, quickly secured title to Sue A. Bland's allotment from the Dawes Commission in Muskogee.[15]

Regardless of where the credit for the discovery lies, Wick's imprudent telegram produced pandemonium. Newspapers across the state carried accounts of the discovery at Red Fork in large headlines. In bold type, the *Tulsa Democrat* proclaimed "A GEYSER OF OIL SPOUTS AT RED FORK," and the *Kansas City Times* headlined "OIL WELL GUSHER FIFTEEN FEET HIGH." In describing the discovery, the Tulsa newspaper announced that the oil "is a dark green, light oil, free from lumps or dirt." Calling the discovery "A Gusher," the *Muskogee Weekly Phoenix* reported that "Oil men who have seen every phenomena in the oil world, say that this gusher is close in a well defined oil territory and that the strata of oil-bearing sand extends under thousands of acres, of which Tulsa is the center."[16]

Newspaper coverage of the operations at Red Fork set off a stampede to the area reminiscent of the gold rushes to California and Colorado. Within a short time, every route leading to the Red Fork community was crowded with horses and wagons carrying the curious and the ambitious, as well as an assortment of shady characters, all wanting to procure a share of the black wealth. "Hundreds of visitors were there to see the well, and the woods are full of people trying to get leases," reported one journalist. "A never-ending stream of wildly excited men are flocking into Red Fork, the nearest point to the scene of the great strike," wrote another. Trains brought legions of new recruits who quickly flooded the existing facilities of the town, creating a sprawling, overcrowded tent city. "The surrounding country is practically depopulated, every known means of transportation being used to reach the scene of the strike. . . . Farms have been abandoned in the mad rush, by men who have been stampeded by the greatly exaggerated stories sent out from here." Lawyers swarmed to the new oil field to offer their services in securing drilling leases. As described by one reporter, "A suggestive coincidence is the presence of several scores of attorneys who have scented future fees in the legal difficulties sure to arise." However, "The majority of those here are without money to invest, and have come through mere curiosity or looking for positions in the companies. . . . Conservative investors are going carefully over the legal status of property rights in the Creek Nation before risking their money."[17]

As word of the strike spread to Oklahoma City, a special train made the trip to Red Fork carrying such individuals as Charles F. Colcord, Robert Galbreath, U. T. B. Wilson, Lee Van Winkle, and William Pettee. Undoubtedly, these prominent citizens were vitally interested in the impact the discovery would have on the growth of the region. When Crossman returned to the drilling site, he found a "carnival town," teeming with a mass of humanity, and described Red Fork as "one of the vilest spots in the territory." Crossman reported that one train brought in a special car loaded with "gamesters," and by early July, more than 1,000 oilmen from other drilling companies were in the area. Petroleum speculation was rampant, dealers in fraudulent stock were in abundance, land promoters offered property at an inflated price, and, in general, chaos ruled.[18]

The excitement resulting from the Red Fork find was of great concern to the Indian officials in Washington D. C. Clifton R. Breckenridge, a member of the Dawes Commission, hastened to inform the secretary of the interior, Ethan A. Hitch-

Tank spilling oil on the ground. With the discovery at Red Fork such scenes as this overflowing oil storage tank became commonplace across the state. *Courtesy of Cities Service Oil Company.*

cock, of the petroleum discovery within the bounds of the Creek Nation. Hitchcock in turn telegraphed the United States Indian Inspector for the Indian Territory, J. George Wright, who was empowered to "perform any duties required of the secretary of the interior," and ordered him to "investigate the report promptly regarding Red Fork, Creek Nation." Wright called upon Inspector Guy P. Cobb, in the vicinity of Red Fork at the time, to report back on his impression of the oil discovery. Arriving at the site on the afternoon of June 27, 1901, Cobb related that "there was a considerable flow of gas issuing from the top of the casing pipe, accompanied by a babbling, rumbling sound, and at intervals of perhaps from 15 to 30 minutes each, a small quantity of oil was forced up by the gas to a distance of perhaps two feet above the top of the pipe." He also called attention to the use of a "bailer or cleaner" that was used to produce a gush of oil from the well to impress prospective buyers. The drillers fastened a "canvas bag large enough to completely fill the drill hole . . . [to] the top of the bailer and lowered [it] slowly to the bottom of the well and permitted [it] to remain there for a sufficient length of time to confine the gas and to permit the accumulation of oil; then the engine was started up at full speed, and the bailer and canvas sack drawn up as rapidly as possible, causing the sudden discharge of gas and oil to be thrown up as rapidly as possible." The spurt of oil and gas was thrown some three to ten feet above the top of the derrick and according to Cobb the ingenious scenario was used to impress spectators and newspaper reporters.[19]

The chicanery did not impress Cobb, however, and Wright, in his report to the secretary of the interior, acknowledged the "considerable agitation near the town of Red Fork . . . concerning the discovery of what was believed to be a valuable flowing oil well." Upon investigation of the circumstances, he concluded that "the reports concerning [the well] . . . were much exaggerated, and that it would require a considerable further expenditure of money to ascertain the character and quantity of such oil." Nonetheless, the members of the Dawes Commission reported, "Unusual activity in the matter of selecting allotments was displayed during the month of June, [1901] when the discovery of petroleum was made

Charles F. Colcord, early-day oilman. When word of the strike at Red Fork reached Oklahoma City, Charles F. Colcord and several other citizens boarded a special train to the site so that they could observe the discovery firsthand. *Courtesy of the Oklahoma Heritage Association.*

near the town of Red Fork." The Commission pointed out that "For the most part these applications for allotments in this vicinity were stimulated by the actions of speculators, who desired to secure leases from the citizens who might secure the lands in that vicinity in allotment." The commissioners also declared that they "exercised all possible care to see that the best interests of the Creek citizens were subserved." The lease question was of great concern to all involved, for according to Wright, "any leases for the purpose of extracting such oil would not be recognized by the department."[20]

Federal officials were not the only ones concerned over the oil strike at Red Fork. Pleasant Porter, Chief of the Creek Nation, declared, "I think it unfortunate that oil, in apparent commercial quantities, has been found there, as it may, to some extent, embarrass the work of appraising the allotments." It was the opinion of the Creek Chief that "those who live in the supposed oil field and who have taken their allotments and have received certificates therefor, that the treaty confirms their allotments . . . the property will belong to them." He was concerned with the "Consequences of the Untimely Location of the Valuable Product, Before the Titles to the Allotments are Clear." As with the federal officials, Porter wanted to insure that the Creek claims to the land and the oil were not overlooked.[21]

Wright contacted Chief Porter on June 19, and called his attention to Sections Seven and Thirty-seven of a Creek Treaty signed on March 1, 1901. Section Seven discussed "Restrictions Upon Alienation of Allotted Lands," and stated that "Lands allotted to citizens hereunder shall not in any manner whatsoever, or at any time, be encumbered, taken, or sold to secure or satisfy any debt or obligation contracted or incurred prior to the date of the deed to the allottee therefor and such lands shall not be alienable by the allottee or his heirs at any time before the expiration of five years from the ratification of this agreement." Section Thirty-seven dealt with leases, and stated that the "Creek citizens may rent their allotments, when selected for a term not exceeding one year, and after receiving title thereto without restriction." The secretary of the interior was empowered by the terms of the agreement to grant exceptions to Section Seven if he so desired.[22]

Porter replied that "It appears to me that these sections are plain enough for a person of ordinary intelligence to understand, and will require no elucidation." In commenting on Cobb's report of the "various methods [used] by . . . persons with a view to securing the control of the future disposition of selected lands of the Creek Nation," Porter declared, "I am of the opinion that each and every one of them are void, for the reason that they are in violation of the terms of the agreement [Creek Treaty of 1901]." In addition, Porter pointed out that the leases were "at variance with the policy of the United States and the Creek Nation," seeking to distribute the Indian lands to the citizens of the tribes "in order that the allottees may be taught the value of their lands." This was the reason that the Creeks were prohibited from selling their allotments for a period of five years. Such a policy was wholeheartedly supported by the Creek Chief. "It is scarcely necessary for me to say that it is the policy of the government to distribute the lands of the Creek Nation to the citizens thereof, and to teach them the use of it, and to clothe them with United States citizenship, and all the responsibilities and protection that inures to any citizen of the United States, and not to provide [a means] by which the Indians may be divested of his landed estate," he declared. In conclusion, Porter contended that "I think this policy is commendable, and should be adhered to with consistency."[23]

Despite the apparent dissatisfaction of both federal and tribal officials regarding the legality of the leases, Heydrick continued to express optimism. Upon his return to the Red Fork region, Heydrick wrote his son, James A., that the output of the well in its "present state . . . is not more than 10 bbls. per day, all things considered." Nevertheless, he stated that "it looks like an oil field," and reported an offer of twenty thousand dollars for the well and forty surrounding acres. But in 1901, the nearest refinery was located at Neodesha, Kansas, and Heydrick was faced with severe economic problems. The oil was valued at one dollar a barrel at the wellhead, but the cost of transporting it to the refinery was approximately ninety cents, leaving a margin of only ten cents per barrel. Furthermore, the cost of production had to be paid before any profit could be shown to investors. Such stark economic factors cooled the ardor of potential stockholders; nonetheless, he apparently made plans to drill another well in the vicinity.[24]

His eastern backers were, however, becoming alarmed over the question of the validity of the agreement allowing oil exploration within the Creek Nation. The objections of Creek Chief Porter, together with inquiries by Cobb, Wright and other government officials, had caused concern over the security of their financial contributions, and Heydrick was eventually forced to suspend his operations because of the question of ownership regarding the lease.[25]

Indeed, their fears appeared to be justified when Indian Inspector Wright made public the opinions of many officials of the Department of the Interior on the question of the Creek leases. Pointing to the provisions of the 1901 Creek agreement, Wright declared that "the lands of the tribe . . . shall be allotted to the citizens of such tribe so as to give each an equal share of the whole in value." All allotments made to Creek citizens by the Dawes Commission prior to the ratification of the agreement were considered binding; however, "lands allotted . . . shall not in any manner whatsoever . . . be encumbered, taken or sold to secure or satisfy any debt or obligation contracted or incurred . . . except with the approval of the secretary of the interior." Once they had received allotments, Wright contended, the agreement allowed the Indians to lease their property only "for a term not exceeding one year." In reviewing the provisions of the treaty, he stated that, after careful consideration of the "right of Creek citizens to dispose of oil, coal or other mineral in or under their prospective allotments, and also the leasing of such allotment," the secretary of the interior had ruled that "a Creek citizen who attempts to make a lease by the provisions of which he disposes of . . . [the] oil or other mineral, in or under his allotment, illegally attempts to dispose of a part of the fee."[26]

In explaining the reasoning behind the decision of the secretary of the interior, Wright stated that the framers of the agreement clearly indicated that "they desired to prevent the Creeks from disposing of the mineral in or under their lands unless they receive a reasonable compensation." Furthermore, it was Secretary Hitchcock's intention to "determine whether or not such compensation is reasonable and just," as provided under the provisions of the treaty. As a result, it was the policy of the Department of the Interior that "in case of a valuable oil well . . . deposit being discovered in or under any of the Creek lands prior to the final appraisement of such lands, the value of such mineral deposit will be taken cognizance of by the Commission to the Five Civilized Tribes in appraising the lands, and should the land upon which such deposit exists be valued at more than the standard value of an allotment, the nation would have a lien upon the rents and profits . . . until the excess had been paid." In concluding, Wright gave public notice that "the Department will not, at this time, permit any citizen of the Creek Nation, or other person, to take or extract from the lands of the Creek Nation, in any manner whatsoever, coal, oil or other mineral substance."[27]

Wright followed his public accouncement explaining the position of the Department of the Interior on the question of oil leases with another pronouncement on August 13, 1901, in which, in reaction to the complaints of many Creek citizens, he declared that under federal law the Indian agent was obligated to "put each [Indian] citizen in unrestricted possession of his land, and remove therefrom all persons objectionable to him." At the direction of the secretary of the interior, Wright gave notice that,

33

Drilling on the Wilcox farm in Washita County. Not all the searches for petroleum at the turn of the twentieth century were centered in Indian Territory. This early drilling operation was just to the north of Canute in Oklahoma Territory. *Courtesy of the Western Oklahoma Historical Society.*

in compliance with the law, "all persons in possession of lands who have not leases with the citizens who have filed on or selected such land, must remove therefrom not later than October 1, 1901." Anyone wishing to remain on Creek property should "make satisfactory leases with the citizen to remain in possession, otherwise," the Indian inspector declared "it will be necessary for the Indian agent to take prompt action."[28]

All leases agreed upon between oil promoters and Creek citizens were, of course, subject to the approval of the secretary of the interior, and Hitchcock considered the leases made following the discovery at Red Fork void. Heydrick made every effort to have his lease recognized, but once again the secretary of the interior withheld his approval of the agreement, and the lease was nullified. Such action by federal officials had a devastating effect on the plans of oilmen in the region and, undoubtedly, curtailed drilling activity in Indian Territory. But some drilling continued and, during the same year, John Davidson, working with C. H. Donohue, drilled a well in the general area.[29]

Although the federal government had impeded oil exploration in the eastern portion of present-day Oklahoma, the western half of the state, Oklahoma

Territory, was undergoing a drilling boom. North of Granite Mountain in southwestern Oklahoma, about three miles north of the town of Granite, oil was found on Watt Armstrong's farm by the Oklahoma Natural Gas, Light and Heat Company. Although predictions were made that the discovery "will produce giant gushers that will rival the famous Spindletop gushers in Texas," the well produced only about eight to ten barrels of petroleum per day. The drillers did, however, employ an unusual method of bringing the crude to the surface: a rope was attached to a sheet-iron bucket and passed through a pulley, the bucket was lowered into the well, and when filled with petroleum, it was pulled to the surface by a horse attached to the opposite end of the rope. Hopes were high for the new discovery. A. J. Greiner, secretary-treasurer of the oil company, exclaimed that "the character of the land and strata in which this oil is found is exactly similar to that of Beaumont, Texas," and he believed that "Oklahoma is bound to be one of the greatest oil fields in the United States." A small refinery was constructed at Gotebo to accommodate the anticipated flow of crude, but the output was so small that the crude was simply sold to area farmers for "greasing livestock, mowing machines, and other farm machinery."[30]

Some exploration was also attempted near Fort Sill on the east side of Cache Creek, but the local homesteaders were opposed to the drilling operations and some "machinery . . . [was] burned by them and much litigation had." Another location known to contain deposits of oil was the Asphaltus Spring, some four miles southeast of Fort Sill; however, the location was within the boundaries of the Kiowa, Comanche, and Apache Reservation, therefore "inaccessible to the prospector." A number of wells in the Fort Sill region had been drilled to a depth of 600 feet, with oil "usually found at about 125 feet, and frequent layers of oil sand . . . struck in going down." The oil strata in the area "are heavily charged with gas, which has been occasionally utilized as fuel for engines in sinking other wells." However, Territorial Governor Thompson B. Ferguson reported that "All operations in this section have been retarded by litigation with homesteaders."[31]

In 1901, the Red Fork Land and Investment Company, based at Lexington in Pottawatomie County, was incorporated under the laws of the Territory of Oklahoma with a capital stock of thirty thousand dollars reserved for "developing mineral and oil claims at Tulsa and Red Fork, Indian Territory." Other parts of eastern and northern Oklahoma Ter-

Drilling in the streets. The area around Lawton in southwestern Oklahoma Territory was the scene of several exploratory wells, such as this one being sunk on one of Lawton's streets. *Courtesy of the Oklahoma Historical Society.*

ritory also were witnessing efforts to find oil. In 1904, Territorial Governor Ferguson reported that several natural gas wells near Blackwell and Newkirk had found gas at various depths ranging from 600 to 850 feet, and one well, approximately nine miles southeast of Newkirk, had "been producing small quantities of gas and a high-grade crude oil for two years." Other wells were sunk near Oklahoma City, Ponca City, McLoud, Chandler, Shawnee, Cushing, and Guthrie; however, no major discoveries were made at this time.[32]

By 1902, the Department of the Interior had withdrawn some of its objections to oil leases in Indian Territory and allowed petroleum exploration to begin anew. In 1904, a significant strike was made near Cleveland in Pawnee County, Oklahoma Territory. Then, in November, 1905, the famous Glenn Pool discovery well, Ida Glenn No. 1, roared in about ten miles south of Tulsa, in the Creek Nation, and once again a steady stream of men rushed to the scene in an attempt to strike it rich. There was, however, a difference between earlier oil strikes and the Cleveland and Glenn Pool discoveries: pipelines. This cheap and efficient method of getting the crude from the Glenn Pool and Cleveland strikes to refineries proved a significant boon to the petroleum business in Oklahoma and encouraged the development of the first major oil fields in Oklahoma.[33]

## Chapter 4.
# The Rush Begins

EXPLORATION AND DRILLING in the "Twin Territories" continued into the early years of the twentieth century. In December, 1901, the Indian Territory Illuminating Oil Company was incorporated. The remnant of Henry Foster's Phoenix Oil Company and its subsidiary, the Osage Oil Company, it had a capital stock of three million shares, valued at one dollar a share. Little was accomplished during the first few years of its existence, however, because of the continued stringent policy of the Department of the Interior in dealing with Indian leases, as well as a power struggle within the company. The internal strife was eventually resolved, and the ITIO, still holding a blanket lease on the entire Osage Nation, 2,298 square miles, began to sublease portions of its vast holdings. ITIO signed subleases with numerous petroleum companies, including the Patton Company, the Asphalt Oil, Mining and Manufacturing Company, and the Sandfork Gas and Petroleum Company. In addition to retaining a royalty interest, most of the subleases contained bonus provisions for the sale of oil produced and reserved the sole right to take over any gas-producing wells at actual cost. Even so, by New Year's Day in 1903, only thirty-one wells had been drilled on the Osage Reservation, of which seventeen were producing oil, two were producing gas and eleven were dry holes. Nonetheless, two important events were soon to give impetus to the development of petroleum exploration in what is now Oklahoma.[1]

The Dawes Commission was nearing the completion of its task of allotting the lands of the Five Civilized Tribes to members of the tribes by the summer of 1904. As a result, the Department of the Interior began to be more susceptible to oil and gas lease applications from oilmen, and drilling activity was no longer curtailed to the same degree it was in earlier years. In addition, transportation of the crude to refineries at a cost low enough to justify its production had become available in the form of pipelines.[2]

The first pipeline constructed in Oklahoma was a two-inch line built by the Phoenix Oil Company in 1900, connecting the company's wells to the Bartlesville railroad terminal. Although the system was later expanded to nearby producing wells, it remained necessary to haul the crude by railroad to refineries in Kansas, a costly process. By June, 1903, the Prairie Oil and Gas Company had erected a 35,000-barrel storage tank near the railroad station in Bartlesville and connected it with producing wells in the Osage Nation. Still, it did not take long to overtax the limited storage facilities, and the company was forced to install an unloading station at Caney, Kansas, which was joined by pipeline to the Neodesha refinery. Although this was a major improvement to the existing facilities, the crude still had to be shipped by train from Bartlesville to Caney.[3]

It was not until 1904 that the Prairie Oil and Gas Company extended a six-inch pipeline from Independence, Kansas, to Bartlesville and Ramona, in the Cherokee Nation. The delay of such an obvious necessity resulted from the need to secure an act of the United States Congress to allow construction of the pipeline through Indian Territory. Opening an entirely new market for crude oil in Indian Territory, the line stretched from Independence through Kansas City, Missouri, to Whiting, Indiana, and from there to outlets on the Atlantic Seaboard. The following year an even larger pipeline was planned, and a pumping station was later established approximately eighteen miles southwest of Bartlesville. Called "Bible," the station tied the Osage wells into a series of gathering lines which funneled the crude into the Bartlesville connection.[4]

The new oil industry continued to lure a number of

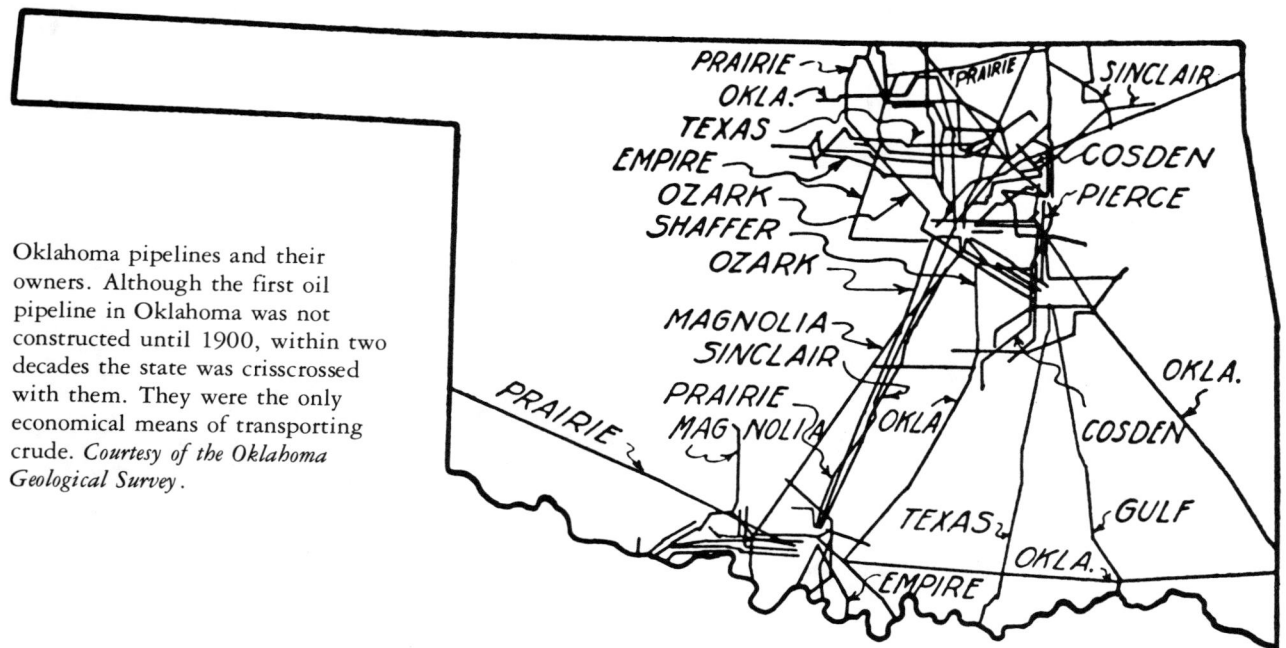

Oklahoma pipelines and their owners. Although the first oil pipeline in Oklahoma was not constructed until 1900, within two decades the state was crisscrossed with them. They were the only economical means of transporting crude. *Courtesy of the Oklahoma Geological Survey.*

adventurous and ambitious young men with business acumen to Indian Territory, and two brothers from Creston, Iowa, were destined to have a tremendous impact on Oklahoma's oil industry: Frank and L. E. Phillips. Frank Phillips originally came to Bartlesville in 1903 as a "young bond salesman with an eye for opportunity." Captivated by the economic prosperity of the region because of the discovery of oil on the nearby Osage Reservation to the west, Frank was also convinced that a bank "would probably prosper with the new town." In 1905, he built the Citizens Bank and Trust Company and became its president. As their banking interests and oil leases flourished, the two brothers established the Phillips Petroleum Company on June 13, 1917, incorporated under the laws of Delaware with assets listed at three million dollars. There were twenty-seven employees on the payroll. According to Dr. O. S. Sommerville, Frank's personal physician during the organizational period of the company, L. E. spent much of the time scouting for potential well sites in the area, while Frank worked out the details of drilling operations at the office. According to Sommerville, "L. E. brought the business in and Frank took care of it." From this inauspicious beginning, the company would become one of the major petroleum corporations in the world.[5]

Earnest E. Traywick, one of the company's longtime employees, recalled driving the two Phillips brothers to lease sites in a Model-T Ford. After Frank and L. E. had examined the location and agreed upon the best location for a well, Traywick, outfitted with a hatchet and a bundle of stakes, marked the site. There was no well-trained geological staff, only Frank and L. E. Phillips, and their unerring hunches. Production steadily increased during these early years, and large reserves of raw materials were accumulated and stored, including natural gas liquids, which would prove an important factor in the future development of petrochemicals. In October, 1917, the firm's initial "Plant to extract liquids from natural gas" was built north of Bartlesville. Natural gas was first viewed as a nuisance by oilmen. Millions of cubic feet had been allowed to escape into the atmosphere daily; however, the engineers and chemists of the Phillips Petroleum Company soon developed a means of "transforming these waste gases into marketable products." Although the early facility produced only approximately 350 gallons of natural gas liquids daily, it enabled the company's "newly hired engineers and scientists . . . [to learn] much about improving methods of manufacturing gasoline, reducing waste and bettering product quality." As a result, by 1924, Phillips Petroleum Company was "the leading producer of natural gas liquids."[6]

Bartlesville: an oil-country hub. Because of its railway connections, Bartlesville was the terminus of many of the early pipelines. *Courtesy of the Bartlesville Public Library.*

In 1925, a research department was formed, and the company's operations were expanded into the Texas Panhandle. The scientists proved their worth when, a short time later, the engineers of the company developed fractionation, a method by which "the mysterious natural gas liquids could be 'tamed' " by removing the highly volatile elements. During the same year, in order to determine the best use of "these removed by-products," the firm also established its first "organized laboratory and pilot plant" near Shidler, Oklahoma. In 1926, a fractionating tower was installed at the Shidler facility which produced Philgas, the Phillips trademark for liquid petroleum gas. R. C. Alden, a supervisor in the research and development department at the time, recalled that great emphasis was placed upon the refinement and improvement of Philgas soon after its introduction into commercial marketing outlets. Between 1927 and 1930, practically all of the research facilities of the firm were devoted to the project.[7]

As a result of these pioneering endeavors, the assets

Early Continental Oil Company transport. By the time of Glenn Pool several of the country's major oil producers had expanded their operations into the western portion of the United States, as is indicated by this early Continental Oil Company wagon. Although Continental had its beginnings in Colorado and Utah, it was to play a major role in the development of several Oklahoma oil fields. *Courtesy of Continental Oil Company.*

of Phillips Petroleum Company totaled $121,000,000 by the end of the first decade, and it embarked on a rigorous program of diversification: refinery operations in Borger, Texas, which eventually became "a vast manufacturing complex"; a multi-product pipeline from Borger to East St. Louis, Illinois, which made the firm "a leader in promoting the first long-distance gas transmission pipeline"; and, in 1935, a thermal polymerization plant at Borger manufactured Phillips 66 Polygasoline, a product made by a new process which converted lighter hydrocarbons into high quality motor fuel. By the time the country emerged from World War II, Phillips Petroleum Company was the "Fastest Growing Major Oil Company" in the nation.[8]

By 1904, the more relaxed policy of the Department of the Interior regarding oil and gas leases in Indian lands had, together with the extension of pipelines into the region, a profound effect on the resumption of drilling operations in present-day Oklahoma. Once again, pioneer oilmen were lured to the area in the quest for black gold. So great was the interest in finding petroleum in paying quantities that the Indian Territory Illuminating Oil Company and its subleases reported a total of 361 wells at the end of 1904, of which 243 were producing oil, 21 were producing gas, and 97 were dry holes.[9]

The first fruits of the new surge in exploration were reaped near Cleveland, just across the Arkansas River from the Osage Reservation, in Pawnee County, Oklahoma Territory. On May 27, 1904, the Minnetonka Oil and Gas Company "spudded in" on the William "Uncle Bill" Lowery farm just south of town. Drilled by W. J. Fellows and John Schell, the well created quite a stir among the townspeople and, by the time drilling operations were ready to begin, more than one hundred citizens were gathered at the scene to watch the operation. Apparently not all of the residents of Cleveland greeted the prospect of finding oil with the same enthusiasm as supporters of the venture. On June 10, the Board of Trustees for the community enacted an ordinance "Prohibiting the Boring for Gas and Oil, and the Erection and Construction of Tanks, Vats and Other Receptacles for the Keeping or Storing of Crude Oil in the Limits of the town."[10]

Work continued on the well, however, with two-men crews working two twelve-hour shifts around the clock. Well samples revealed that the formation was "identical with that of developed fields" and convinced the drillers that "something would be shown before the well was abandoned." Sand, filled with saltwater and a trace of oil, was encountered at a depth of approximately 600 feet. This confirmed the prospects of finding crude and work continued at an accelerated pace. On June 28, a sand found between 900 and 1,000 feet "carried a light flow of gas." Expectations mounted and on the following day a third layer of sand was struck at approximately 1,100 feet and "with it came gas in abundance." Christened the Uncle Bill No. 1, the well spewed gas strong enough to be seen "ten or fifteen feet above the mouth of the hole." By the time the drill reached a depth of 1,250 feet, the flow of gas was "variously estimated between three-quarters of a million to three million cubic feet" per day. Excitement soared, and representatives from several oil companies were on hand to witness the well's progress.[11]

Their expectations were fulfilled. Another zone of gas, more prolific than the previous finds, was encountered on July 2, 1904. Because of the estimated flow of from ten to twenty million feet of gas per day, many believed that the Uncle Bill No. 1 was destined to be a "gasser." Oil was noticed on several feet of the rope when drilling resumed, however, and when Mrs. E. G. Todd placed her hand over the well opening to feel the flow of gas, "oil was sprayed over her shirt waist." The drilling crew quickly withdrew the tools, and the Uncle Bill No. 1 began producing oil at a rate of about ten barrels per day.[12]

Drilling continued, however, and by July 21, a depth of 1,625 feet had been reached, with oil sand present throughout the final 55 feet. Schell decided to "shoot" the well in order to determine the potential size of the reservoir and to improve the well's production. J. C. Moore with the Kansas Torpedo Company was employed by the company, and on the evening of July 23, the explosive was detonated. "The shot tore the casing up pretty badly and blew part of the packer in the bottom clean to the top," related a local reporter. Immediately after the shot, the Uncle Bill No. 1 produced crude at a rate of approximately 250 barrels daily, but shortly thereafter, this amount dropped off to between 100 and 150 barrels and eventually stabilized at about 50 barrels per day.[13]

Excitement generated by the discovery well at Cleveland swept throughout the petroleum industry. Oil companies rushed to the site, and the Cleveland Oil, Gas and Manufacturing Company immediately announced a location on an adjoining lease. The field quickly grew toward the south and west, as numerous wells were drilled shortly thereafter. Territorial Governor Ferguson reported a total of 255 wells drilled in the area by July, 1905, of which only 28 were listed as "dry and non producers," and 7 were listed as gas wells. The remaining 220 wells had a daily production of "about 11,000 barrels." The crude was stored in tanks excavated from the earth for a time, but this method proved unsatisfactory because of the large amounts of oil lost to seepage and evaporation. Within the next several years, steel and wooden tanks capable of storing 1,500,000 barrels of oil were constructed in the Cleveland vicinity.[14]

The find proved a tremendous boon to the town of Cleveland. The population swelled from seven hundred to more than seven thousand, and a total investment of $1,500,000 was spent in the development of the oil field. Moreover, it had an even greater impact on the history of petroleum in the state, in that the Uncle Bill No. 1 touched off a revival of exploration for oil in both of the "Twin Territories." So great was the excitement in Oklahoma Territory that Territorial Governor Ferguson reported that the "effect and influence of this great oil development in the Cleveland oil field has been felt all over the Territory, and has encouraged other localities to prospect for oil and gas." Although several additional exploratory wells were drilled in Oklahoma Territory, none produced "good results," and the only promising indications were found in the "northern and southwestern part of the Territory." However, the fever quickly spread into Indian Territory and, in 1905, Robert Galbreath and Frank Chesley drilled a well on the Ida Glenn farm, about ten miles south of Tulsa in the Creek Nation, a well that was to open the greatest of all early-day Oklahoma oil fields.[15]

Previously involved in the activity at Red Fork, Galbreath wanted to drill a wildcat approximately ten miles to the southeast on the Glenn farm. When his original backers withdrew their financial support for the speculative venture, the former resident of Ohio enlisted the aid of Frank Chesley, who, like Galbreath, was a citizen of Tulsa. Galbreath and Chesley selected a site near the center of Section 10-T17N-R12E, where a sandstone outcrop had convinced Galbreath of the potential of the location.[16]

After obtaining a lease for the Glenn property, Galbreath and Chesley secured the services of Roy Dodd and Shorty Miller, two of the best-known drillers in the area, and work began in the fall of 1905. Although several strikes had previously been found in Indian Territory, the spectacle was still a novelty to the citizens of the vicinity, and the activity attracted onlookers in great numbers. "Farmers . . . and the inhabitants for miles around, including the cowboys from the ranches, would visit the operations daily and watch with wonder the drillers at work." Limited in finances, Galbreath, Chesley, Dodd, and Miller lived at the drilling site, each taking turns firing the boiler in order that drilling might continue around the clock in an effort to see the well completed before funds expired. In an attempt to economize, they drilled the well without casing.[17]

Although no great difficulties were encountered in the drilling, apprehension rose as the bit tore deeper into the earth without finding oil. Not only Galbreath and Chesley, but Dodd and Miller as well, stood to suffer heavy financial loss should the well prove to be a "duster." They had hoped to encounter the Red Fork sand at a depth of between 1,300 and 1,400 feet and, as the bit neared that level in late November, 1905, they devised a ruse in order to maintain control of the situation should petroleum be found. They announced that the drillers were suspending work and operations were ceasing, and told the spectators that there was nothing for them to see if they remained at the site. It worked. The crowd, present almost continuously during the drilling activity, then dispersed.[18]

Frank Phillips, one of Oklahoma's leading early-day oil pioneers. *Courtesy of the Oklahoma Heritage Association.*

Much to their disappointment, the bit reached a depth of 1,448 feet on November 21 without locating any gas or oil, and the specter of financial ruin loomed greater with every passing hour. Their perseverance was rewarded when, shortly before 5:00 A.M. on November 22, 1905, after penetrating only ten additional feet, they saw oil spurt over the derrick. The Ida Glenn No. 1 produced approximately seventy-five barrels of crude daily and ushered in the first major oil field in Oklahoma—the fabulous Glenn Pool.[19]

Having witnessed the chaos at Red Fork, Galbreath was keenly aware of the confusion and legal tangles that resulted from conflicting lease claims. Hoping to avoid such problems at the discovery well at Glenn Pool, Galbreath, Chesley and the others posted armed guards to screen the well from the prying eyes of potential competitors. Nonetheless, by December 22, 1905, the *Tulsa Democrat* was reporting that the "Galbreath & Co.'s 'Mystery' well" was a producer and flowing at a rate of "twenty barrels a day." Another problem facing Galbreath and his companions was the lack of casing in the Ida Glenn No. 1. In an effort to provide additional operating capital, Galbreath and Chesley accepted Charles F. Colcord of Oklahoma City and J. O. Mitchell of Tulsa as partners in forming the Creek Oil Company. Colcord had previously underwritten several of Galbreath's wells in the Red Fork area; however, when initially approached by Galbreath with the proposal for the wildcat that eventually became the Ida Glenn No. 1, Colcord had declined because he viewed the proposal as too risky. With adequate financial backing available, the casing was quickly obtained and soon in place, and by the end of March, 1906, the well was completed, producing at a rate of about eighty-five barrels of crude per day, rated at 32-38° gravity, with a paraffin base.[20]

Several nearby towns, including Mounds to the southwest, Tulsa to the northeast, and Sapulpa to the northwest, were eager to profit from an influx of oilmen by naming the new discovery after their community. However, in a meeting at the old Reeder Building in Tulsa, Galbreath and the others named the discovery in honor of Mrs. Ida E. Glenn, on whose property the well had been drilled. This decision gave birth to the name "Glenn Pool" and, by July, 1906, the find was being referred to as the "Glenn wells."[21]

Galbreath and the others quickly realized the potential of their discovery and strongly suspected that they had discovered a large petroleum field. Their armed barricade prevented anyone from getting within a mile of the well or land they hoped to lease. But the Creek Oil Company was not the only firm active in the area. J. C. Eldred and M. H. Mosier had completed an oil well near the Glenn lease at approximately the same time as Galbreath's discovery. That well flowed at the rate of about seventy-five barrels a day. Nonetheless, the prime land was controlled by Galbreath and his associates, who had leases on approximately 600 additional acres "of the choicest selections in the Glenn Pool." This included not only holdings on the Glenn farm, but property owned by Nevada Berryhill and E. E. Brock as well, which was destined to become the "most prolific oil-producing properties of the continent."[22]

Soon after completion of the Ida Glenn No. 1, Galbreath and his partners drilled the Glenn No. 2 approximately three hundred feet from the discovery well. The second well blew in on March 15, 1906, with an initial flow of 1,700 barrels of oil during the first twenty-four hours. It was under such force that the crude flowed over the derrick for the first hour; however it was eventually capped and the flow dropped to approximately 800 barrels per day. In January, 1906, Galbreath and Chesley began work on the Brock farm, but the initial well was delayed when the tools became caught in the hole. Nonetheless, arrangements were made to drill an additional sixteen wells on the lease. The Brock No. 2 was brought in with a strong flow of gas in April, 1906. Unfortunately, the gas was detonated by a light being used by the drillers, seriously burning Frank McGuinn, one of the crew. Spreading to the oil-soaked wooden platform of the derrick, the flames ignited the structure and destroyed the rig. Undaunted, Galbreath constructed another rig and reopened the well.[23]

The tremendous success of the Creek Oil Company continued. The Gailey No. 2, a 500-barrel-per-day producer, was brought in on April 19, 1906, and the following month, the Gailey No. 3 was completed with a flow of 1,000 barrels daily. The huge volume of petroleum pouring from the company's wells created a bottleneck in the transport of crude to market; it simply overtaxed the existing facilities. Galbreath, in an effort to ease the problem, hastily constructed a number of earthen reservoirs to hold the oil; however, the supply soon exceeded the capacity of the makeshift storage tanks and overflowed their banks. Likewise, all the wooden and steel storage tanks were quickly filled. By February 16, 1906, work was completed by the Prairie Oil and Gas Company on a two-inch pipeline connecting the Ida Glenn No. 1 with the larger pipeline at Red Fork, and through it to the refinery at Whiting, Indiana. As drilling continued, additional pipelines were laid to connect the wells to available storage facilities, but the amount of crude overwhelmed the effort. It required two pipes just to carry the petroleum produced by the Gailey No. 3 to the storage tanks.[24]

It was obvious that a major oil field had been discovered. Glenn Pool was an oilman's dream in many respects: the crude was found at a fairly shallow depth, thereby reducing the cost of drilling; the natural gas found with the oil facilitated the extraction of the petroleum from the earth; and the quality of the crude was very good, with a low content of asphalt. It was, however, in Indian Territory, and some of the old bureaucratic traits of the Department

Glenn Pool, one of the greatest early-day Oklahoma oil fields. *Courtesy of the Beryl D. Ford Collection.*

of the Interior were still in evidence: all leases were subject to approval of the secretary of the interior, and all royalties, ten percent on all petroleum sold, were paid to the federal Indian officials in Muskogee. One interesting aspect involved the leases obtained from freedmen in the Creek Nation, who held deeds to the land and had the right to sell it if they so desired. Nonetheless, by the summer of 1906, the productivity of the discovery was no longer in doubt, and new leases were rapidly becoming scarce.[25]

Property still available varied greatly in value. In the case of minors, lease arrangements were subject to the jurisdiction of the local courts, and the price fluctuated between $1.55 per acre and $27.00 per acre. But as word of the strike spread, the price of acreage rose accordingly and, in January, 1907, W. E. Campbell, a lease and real estate broker from Tulsa, reported an offer of $2,500,000 for a choice tract was rejected. From the beginning, Galbreath and his associates continued to expand their holdings in the Glenn Pool area, at one time delivering $30,000 in gold coins in "an old 'gunny' sack" for a group of leases purchased from several Indian minors. In addition to the lease payments, bonuses were paid to many of the Indian children who held allotments in the area. In June, 1906, a total of $691,860 in bonuses was paid by petroleum firms for leases covering 3,200 acres, and two minors, Elma Glenn and Earl Berryhill, both aged three, were paid $43,000 for 20 acres and $25,000 for 160 acres respectively.[26]

Oilmen flocked to Glenn Pool from throughout the United States, and drilling continued on an unprecedented scale. Supposedly wells were to be drilled near section lines approximately 150 feet from the road with a maximum of five wells to the half-mile. However, such was not always the case. In their haste to draw the maximum amount of crude from the earth in the shortest amount of time, many drillers sank wells as fast as possible. More than a thousand wells were quickly drilled in the Glenn Pool area, and only a very small percentage proved to be "dusters." By January 1, 1908, only fifteen wells had been dry holes. The extensive drilling soon established the general limits of the field as approximately two miles in length, north and south, and less than one mile wide, east to west. The greater part of the petroleum was located in the eastern parts of Townships 17 and 18 North, Range 12 East.[27]

Another of Oklahoma's major petroleum companies owed a large portion of its eventual success to the Glenn Pool, the Sinclair Oil and Gas Company. Born in Wheeling, West Virginia, Harry Ford Sinclair had been trained by his father to be "a small town druggist," but instead founded one of the nation's largest oil companies. Sinclair began his career in the petroleum industry by selling lumber for oil derricks and dealing in leases on the side. However, he soon attracted the attention of several pioneer oilmen in Oklahoma, such as J. M. Cudahy, for whom Sinclair organized small companies around single leases for a share of stock in each enterprise. By 1904, Sinclair had netted a hundred thousand dollars from a drilling syndicate near Kiowa, Oklahoma, providing cash which he promptly invested in the area of Glenn Pool, his first big strike.[28]

Sinclair's success was phenomenal: by 1913, he headed sixty-two oil companies; owned eight drilling rigs; was a partner with his brother, Earl W., in a Tulsa bank; and was producing from every Mid-Continent field. By 1916, Harry Sinclair was the largest independent oilman in the Mid-Continent region, engaged in production, pipelines, transportation, refining, and marketing of petroleum products. On May 1, 1916, he established the Sinclair Oil and Refining Corporation and "joined the big powers of the oil industry." A remarkable man, Sinclair built his holdings by assembling "bits and pieces of depressed property, five small but profitable refineries, and many [areas of] untested production." Five months after establishing the refining company, he announced plans to build a new eight-inch pipeline from the Cushing field to East Chicago, with refineries located at the Chicago ship channel and at Kansas City, Missouri, thus providing his company with a marketing outlet that allowed him to challenge the "industry giants." At the same time the firm's sales were expanded in the United States, until by the end of 1916, the Sinclair marketing area stretched from Oklahoma north to Iowa and from Denver, Colorado, to Albany, New York. Under his direction, the Sinclair interest continued to grow and expand until the company was recognized as one of the nation's leading petroleum concerns.[29]

During its greatness, Glenn Pool crude was never quoted higher than forty cents a barrel; countless thousands of barrels were sold as low as twenty-five cents each, and the field was developed at an average price of thirty cents a barrel. The depressed prices could be blamed on the lack of transportation facilities during the early development of the state, as well as lack of control over well spacing, which rapidly reduced the gas pressure and increased the cost of withdrawing the oil from the ground. But regardless of the reason for the price of the crude, the average cost of a well drilled at Glenn Pool was five thousand dollars, and it took a vast number of barrels priced at thirty cents to return the oilmen's initial investment.[30]

The need for pipelines to carry the tremendous volume of crude flowing from Glenn Pool was recognized early in the field's history, and the Prairie Oil and Gas Company wasted little time extending its line from Red Fork to the area. However, the early attempts did not maintain pace with the rapidly growing production figures, and several efforts were made to increase the facilities for shipment of the petroleum by railroad. In November, 1906, the Gulf Pipe Line Company began work on an extensive system of gathering lines throughout the oil field, which merged at the Frisco Railroad depot at Keifer. Here the company constructed a loading rack some one thousand feet in length and capable of handling twenty-five tank cars. Provisions were also made to increase the loading facilities should the need arise. By early 1907, the Gulf Pipe Line Company announced that it was ready to ship between five thousand and ten thousand barrels daily from the Kiefer facility to its refinery at Port Arthur, Texas, where they anticipated receiving seven million barrels annually from Glenn Pool wells.[31]

The Prairie Oil and Gas Company also increased the size of its pipeline running north from Tulsa to eight inches, and replaced the two-inch line from Red Fork to Glenn Pool with an eight-inch line. In

44

## Production of the Glenn Pool, 1907–14 (*in barrels*)

| Month | 1907 | 1908 | 1909 | 1910 | 1911 | 1912 | 1913 | 1914 |
|---|---|---|---|---|---|---|---|---|
| January | 385,939 | 1,796,461 | 1,362,602 | 1,745,206 | 1,099,192 | 882,385 | 792,336 | 829,483 |
| February | 572,414 | 1,897,054 | 1,410,878 | 1,543,660 | 967,924 | 867,566 | 718,580 | 769,809 |
| March | 1,084,636 | 2,098,411 | 1,543,463 | 1,974,514 | 2,584,464 | 924,144 | 807,022 | 871,334 |
| April | 1,716,079 | 1,968,761 | 1,467,179 | 1,674,709 | 1,570,947 | 898,527 | 823,645 | 849,316 |
| May | 1,923,262 | 1,630,111 | 1,590,720 | 1,676,366 | 1,069,863 | 927,182 | 850,607 | 987,397 |
| June | 1,971,387 | 1,051,045 | 1,809,898 | 1,573,578 | 958,519 | 816,028 | 816,789 | 852,901 |
| July | 1,922,387 | 1,914,134 | 1,856,524 | 1,557,869 | 965,122 | 880,906 | 787,274 | 828,350 |
| August | 2,003,607 | 1,770,819 | 1,699,486 | 1,609,702 | 891,936 | 927,675 | 734,476 | 535,027 |
| September | 2,309,205 | 1,639,252 | 1,570,167 | 1,593,986 | 937,886 | 794,958 | 773,847 | 431,051 |
| October | 2,441,662 | 1,832,033 | 1,602,988 | 1,521,794 | 969,247 | 921,736 | 817,628 | 584,178 |
| November | 1,971,595 | 1,404,234 | 1,039,342 | 1,400,118 | 864,519 | 768,254 | 753,115 | 504,397 |
| December | 1,625,127 | 1,491,998 | 1,393,392 | 1,365,412 | 910,489 | 886,157 | 794,551 | 514,346 |
| Total | 19,927,300 | 20,494,313 | 18,946,740 | 19,236,914 | 13,880,118 | 10,495,518 | 9,469,870 | 8,677,589 |

SOURCE: Oklahoma Geological Survey.

addition, it began work on an approximate million-barrel storage facility of steel tanks at Jenks, where it would have an outlet to the Midland Valley Railroad. It was also rumored that The Texas Company was planning to construct a pipeline from its refinery at Humble, Texas, to Glenn Pool, thus providing a link with The Texas Company refineries at Port Arthur and Port Neches, Texas. In December, 1906, Galbreath announced preparations to form a company financed with twenty million dollars of British and Eastern capital to undertake the construction of a pipeline from Glenn Pool to the Gulf of Mexico. However Galbreath's enterprise, the Mid-Continent Port Arthur Pipe Line and Refining Company, did not develop as planned, and the line was never laid. By February, 1907, construction on an eight-inch cast-iron pipeline was approaching Oklahoma from the Gulf Production Company and The Texas Company facilities in Texas, and a sixteen-inch gas line was in the process of being laid from Tulsa southwestward toward Oklahoma City. In October of that year, The Texas Company's line was ready for operation and relieved the field of some excess oil.[32]

Drilling and production continued to increase, doubtless stimulated by the availability of pipeline and storage facilities for storing and shipping the crude. Several firms, the Prairie Oil and Gas Company, the Gulf Pipe Line Company, Associated Producers Company, the Oklahoma State Oil Company, and others hurriedly built storage tanks for the Glenn Pool oil and by August, 1907, facilities capable of holding a total of 1,475,000 barrels were available. Even so, the estimated production of the field for the month of August was 2,635,000 barrels and it seemed as though the supply of oil at Glenn Pool was inexhaustible as production continued during the summer of 1907.[33]

Problems were beginning to appear, however. The field suffered a series of disastrous fires, underwent a decline in the older portion of the pool, and a drought created such a shortage of water that it forced the curtailment of drilling operations. Moreover, during the latter months of 1907, Glenn Pool felt the effects of a drastic reduction in the price of crude, which necessitated laying off 250 men working on storage tank construction, decreasing rail shipments, and reducing drilling operations. By January 1, 1908, the worst had apparently passed and work pushed forward.[34]

At the end of 1907, there were in excess of ninety-five companies operating in the Glenn Pool field. These firms had invested $4,548,000 in wells, $5,000,000 for lease agreements, paid out $3,000,000 in salaries, and purchased $15,323,125 worth of timber and steel for construction. This was a total in excess of $18,000,000 spent by the petroleum industry in Glenn Pool alone, by far the largest amount invested in Oklahoma oil production to that time. In exchange, they had taken an estimated 27,377,337 barrels of oil from the field by the end of January, 1908, including 19,926,995 barrels taken during 1907. The field still contained a vast amount of crude, however, and the peak year of production came in 1908, with a total of 20,494,313 barrels of oil produced.[35]

In January of that year, it became apparent that the huge amount of crude flowing from the wells could not be adequately handled, and the only method of alleviating the situation was a cutback in productivity. A vast amount of petroleum was being wasted because of the inability of the pipeline, storage, and railway shipping facilities to maintain pace with production. Black smoke was frequently seen against the sky over Glenn Pool as the overflow from the storage containers burned. Polecat Creek, located near the field, was often on fire for miles, and it was estimated that five thousand barrels of petroleum were lost daily in the field. Being astute businessmen, the oil producers quickly realized that such waste was not good business practice and set about correcting the situation. A voluntary reduction of drilling operations was attempted in 1908, but because many of the new wells were good producers, little was accomplished in reducing the overall supply of crude. That same year, the Taneha Pool was discovered in the northwest portion of the field.[36]

Glenn Pool eventually grew from an eighty-acre tract to a field of almost eight thousand acres, establishing Oklahoma as one of the leading petroleum-producing regions in the United States. This had, of course, a strong impact on the state, but Glenn Pool also had a great impact on the oil industry throughout America that resulted in two important developments: it was discovered at a time when the production of the Gulf Coast field, along the coast of Texas, had begun to decline. This prompted the Gulf Coast operators to extend their pipeline trunks northward across Texas and into Oklahoma, thereby opening up a new producing area to the petroleum industry; and, in addition, the oil from Glenn Pool was rich in gasoline content, easy to refine and very good for fuel oil, which made it easier for big refining and producing companies (The Texas Company, Gulf Oil, Standard Oil, and others) to shift from the use of

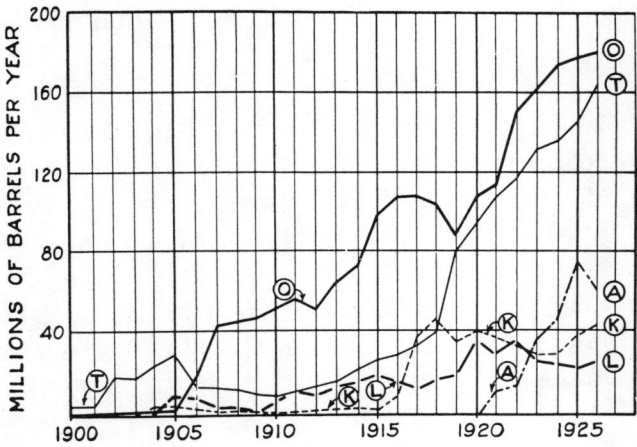

Oil production of the principal oil-producing states, 1900–1925. By 1910, Oklahoma had taken the lead. The letters in the diagram refer to the states of Oklahoma, Texas, Arkansas, Kansas, and Louisiana. *Courtesy of the Oklahoma Geological Survey.*

oil for illumination to the primary utilization of petroleum for energy.[37]

Glenn Pool continued to expand and by 1910 the size of the field had more than tripled from that of 1907. It had grown from less than a mile in length to more than four, and its width was approximately two miles. It continued to be a major producer of Oklahoma crude for many years and in 1910 eighty-four wells were completed in the area, with an average daily output of 62.2 barrels of oil per well. Annual production declined slowly from a peak of 20,494,313 barrels in 1908, to 18,946,740 in 1909 and 19,236,904 in 1910. However, the decline was more rapid after 1910, and by 1915 had fallen to 5,993,628 barrels. Nonetheless, the impact of Glenn Pool on the petroleum industry of Oklahoma lived far beyond the heyday of the field. It was Glenn Pool that brought the pipelines to Oklahoma and opened new marketing outlets; it was Glenn Pool that caused the great influx of capital into Oklahoma that would make continued development of the petroleum industry possible; and it was Glenn Pool that focused the eyes of the nation on the petroleum industry in Oklahoma.[38]

## Chapter 5.
# Statehood and Regulation

FOLLOWING GLENN POOL, it was several years before Oklahoma witnessed the discovery of another major oil field. Several small pools were found during this time, however, and the drilling fever did not abate. Of greater significance, it was during this period that Oklahoma Territory and Indian Territory were joined to form the State of Oklahoma. Subsequently, the rules governing the activities of oilmen in the Territory of Oklahoma, roughly the western half of the newly formed state, and those of the federal government in Indian Territory, the eastern half, were replaced by the regulations of a single government. This had a tremendous impact on the petroleum business within the boundaries of Oklahoma.[1]

Although overshadowed by the phenomenal strike at Glenn Pool, there was a series of discoveries of smaller pools throughout the Okmulgee area in the Creek Nation some twenty-five miles south of Glenn Pool. The first strike in the vicinity of Okmulgee was made in 1906 on the Booch farm in Section 20-T13N-R14E, but it produced only ten barrels of crude daily and was not commercially developed. However, the Tulsa Fuel and Manufacturing Company drilled a well in the same section early in 1907 and, at a depth of 1,486 feet, struck oil that flowed at the rate of five thousand barrels per day. This set off a flurry of drilling activity in the Morris Pool, as the field was named.[2]

That same year, the Lucky Pool was also discovered near Okmulgee, in T13N-Rs12&13E. Again, crude was found at a fairly shallow depth, ranging from one thousand to two thousand feet, and the production was marginal, amounting to only twenty-five barrels per day for two wells and between fifteen and three hundred barrels daily for a third. Gas was also found in several wells with flow rates ranging from one million to nine million cubic feet per day. Nonetheless, these discoveries caused barely a ripple in the petroleum industry.[3]

In 1907, the Coalton Pool was opened with the completion of the "picnic well" in Section 22-T12N-R13E. Although its initial production was sixty barrels per day from a depth of eighteen hundred feet, this was not great enough to trigger a boom in the area. Even so, it was the area's first discovery of oil in s significant quantities, and several of the wells eventually produced as much as six hundred barrels daily. Of greater significance was the Schulter or "shoe string" pool, opened the same year in T12N-R13E, where production was as high as four thousand barrels per day in some wells. However, the Bald Hill Pool, opened in 1907–1908, was by far the most enticing to oilmen in the area. In the spring of 1908, Joe Burns and Lou Caton completed a well near the center of the NE¼ of Section 6-T14N-R14E that flowed four hundred barrels of oil per day. This well precipitated active development of the region, and was soon followed by a well drilled by Robert Galbreath, of Glenn Pool fame, in Section 22-T15N-R14E, which found sand at a depth of approximately seventeen hundred feet. Various wells in the Bald Hill Pool produced a dark green high gravity crude with flow rates ranging from ten to fifteen hundred barrels per day. Another promising discovery was made at Turkey Pen Hollow, where the Tiger Oil and Gas Company completed a well which flowed at the rate of two hundred seventy barrels per day in Section 10-T12N-R12E. Drilled to an approximate depth of twenty-three hundred feet, the well produced oil until all available storage tanks in the area were full, and it was then capped. Unfortunately, when this well was reopened two years later, it proved to be dry.[4]

Although relatively insignificant when compared

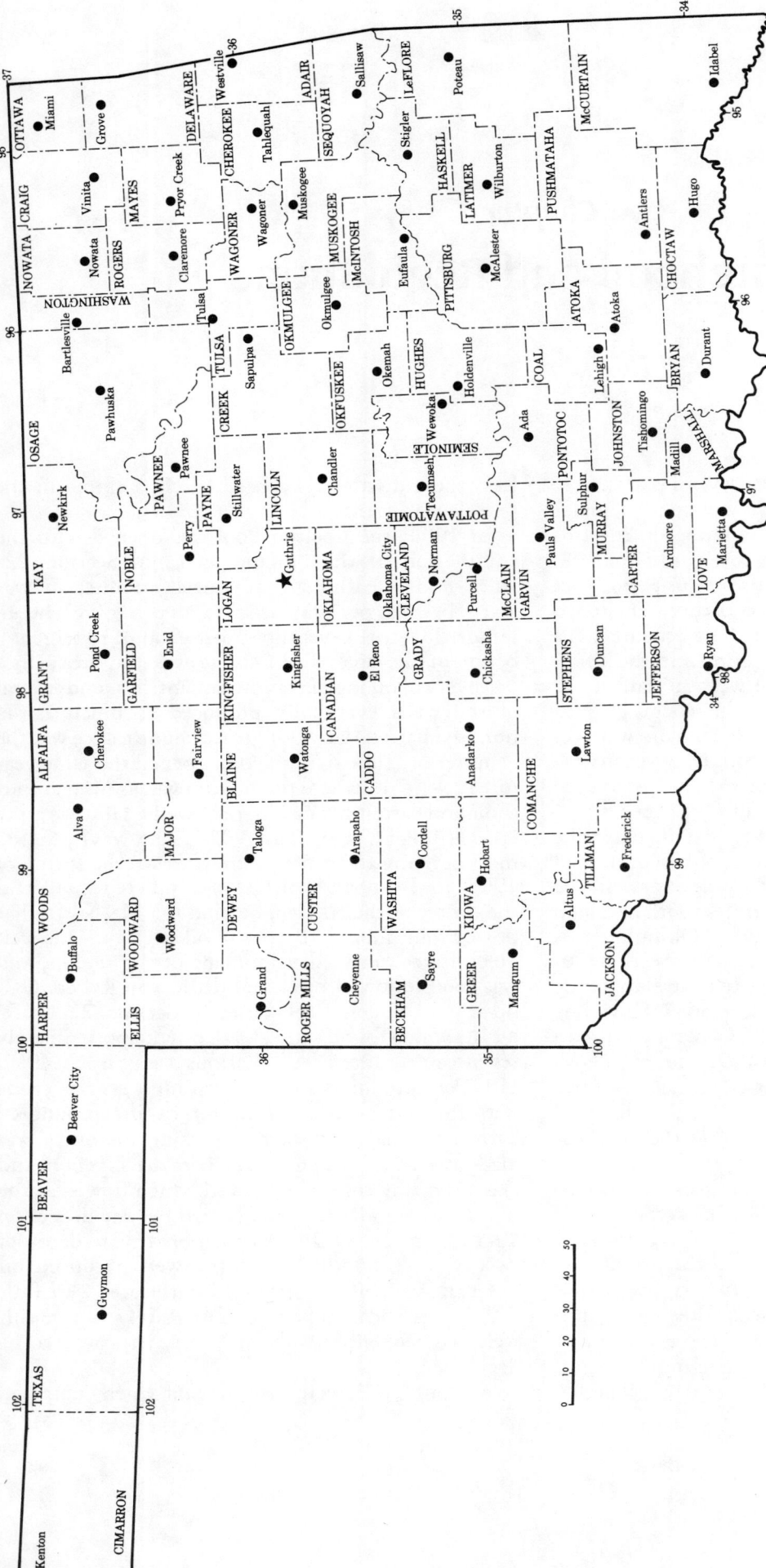

Oklahoma counties in 1907. At the time of statehood Oklahoma and Indian territories were joined and organized into seventy-five counties—two more would later be formed—and the search for oil was being carried on throughout the new state. *From Morris et al., Historical Atlas of Oklahoma.*

to the more productive drilling to the north, activity continued in the Okmulgee area for several years. The year 1908 saw the opening of the Independent Pool in T16N-R12E, which showed promise at first, but proved a disappointment when subsequent wells flowed only between forty and one hundred barrels daily. In August, 1909, Alex Preston completed a well in Section 11-T14N-R12E which tapped the Hamilton Switch Pool. The flow was moderate, ranging from ten to two hundred barrels per day, but a subsequent well drilled by Preston in close proximity to the earlier well, flowed up to a thousand barrels of oil a day. The Okmulgee area continued to be the site of extensive exploration after the discovery of the Preston Pool and by 1910 three major strikes were reported: at Beggs, where approximately forty-four wells were drilled during the years 1910 and 1911, a number of which produced as much as a thousand barrels a day; at Henryetta, T11N-R12E, where in 1910 the discovery well flowed at a rate of six hundred barrels daily, a subsequent well produced eighty million cubic feet of gas per day, and development continued intermittently for the next several years; and Salt Creek, opened in October, 1910, on the Tobler allotment in Section 25-T12N-R11E, when the Prairie Oil and Gas Company, having paid twenty thousand dollars for the lease, completed a well that was drilled to a depth of 2,367 feet and produced two hundred barrels of crude daily.[5]

Thus far, nothing discovered had been of sufficient magnitude to generate a boom in the area like that at Glenn Pool; however, just enough oil had been found to encourage continued exploration. In 1912, M. C. French drilled a combination oil and gas well in Section 8-T11N-R12E and two years later the Natura Pool was tapped by the Kingwood Oil Company in Section 22-T15N-R13E. First known as a gas field, producing as much as sixty million cubic feet of gas daily, the pool also produced crude at a rate of sixty barrels per day. The year 1915 saw the discovery of the Brinton Pool in Sections 9 and 16, T13N-R12E and the Youngstown Pool in Ts13&14N-Rs11&12E, which flowed at a maximum rate of a thousand barrels of oil daily.[6]

Although the exploration in the Okmulgee area produced both oil and gas in marketable quantities, most of the "small" finds were treated as temporary diversions for oil operators, who were still in awe of the much larger production capability of Glenn Pool.[7]

Of greater importance was the transition of the separate governments in the "Twin Territories," a combination of federal and local laws, to the status of statehood, comprising a legal system of regulatory laws and agencies. This was, in fact, a crucial era in the petroleum industry of Oklahoma, as the new state undertook the formation of its own policy toward the oil business. The regulatory statutes of the federal government, the Territory of Oklahoma, and the Five Civilized Tribes gave way to a more centralized control.[8]

From the beginning of the oil industry until 1907, there had been three distinct governmental bodies creating policies regulating petroleum production in present-day Oklahoma. Of these two, the Indian tribes and the federal government were more concerned with the monetary advantages available to the Indians, than with the regulation of the industry. For the most part, the Indians accommodated themselves to the wishes of the pioneer oilmen in order to insure the receipt of royalties to the tribe. The earliest legislative actions of the Indians occurred shortly after the arrival of the Five Civilized Tribes in Indian Territory, and basically iterated the fact that precious minerals were the property of the tribe. Although Cherokee law allowed individual members of the tribe to utilize the minerals for personal use, all other individuals were prohibited from extracting minerals from the land. Following the Civil War, the expansion of railroads into the region generally allowed commercial development of minerals upon payment of royalties to the tribe as a whole. This was especially true of the Choctaw and Chickasaw nations.[9]

By 1872 the first oil company, the Chickasaw Oil Company, had been organized within Indian Territory; however, the federal government intervened when the secretary of the interior refused to approve the company's charter. Thus, the question of petroleum production within present-day Oklahoma was then subject to the legislation of the Indian tribes contingent upon approval of the federal government. The result was an overlap of responsibility which led to confusion in obtaining leases. At the same time, there was tribal pressure placed upon the federal government to allow the exploitation of minerals within the Indian lands for the purpose of increasing the revenue of the tribes. This led to the alteration of tribal law which allowed non-Indian development of oil deposits inside Indian Territory.[10]

In 1882 the Cherokee Nation imposed a tax, ten cents for each forty-gallon barrel, on oil production within its boundaries. Later, the law was further modified to require a five thousand dollar bond by oil companies, and included a fine and forfeiture of the

Preston in the oil boom. This town in Okmulgee County was first known as Hamilton Switch and was fairly typical of turn-of-the-century oil towns. *Courtesy of the Oklahoma Publishing Company.*

lease and equipment for failure to provide regular quarterly reports and royalty payments. As more and more drillers applied for permission to drill wells, however, the Cherokee officials also required a certification of their expertise in the petroleum industry. Nonetheless, by the mid-1890s, most of the tribes in Indian Territory had adopted legislation that encouraged oil exploration within their boundaries.[11]

Throughout this same period, the United States Congress also began to take a stronger role in providing federal supervision of Indian petroleum leases. The culmination of the congressional effort was "An Act for the Protection of the People of the Indian Territory, and for other Purposes," sponsored by United States Representative Charles Curtis of Kansas. Commonly known as the Curtis Act, the provision gave the secretary of the interior control over mineral leasing of tribal lands. Specifically, the law stated that "the secretary of the interior is hereby authorized and directed from time to time to provide rules and regulations in regard to the leasing of oil, coal, asphalt, and other minerals" in Indian Territory. In addition, any leases not approved by the secretary of the interior were declared void, and those agreements approved were limited to no longer than

A battery of crude stills. El Dorado, Kansas, was a major refining point for much of Oklahoma's early oil output. *Courtesy of Getty Refining and Marketing Company.*

fifteen years and were not to exceed six hundred forty acres. Advance payments were required on a prorated basis, ranging from one hundred dollars the initial year to five hundred dollars for every fifth year and additional years thereafter. Leases were declared void if, after a delay of sixty days, royalty payments were not made, and the lessee was bound to pay for surface damages incurred by the lessor. For all practical purposes, the Curtis Act removed the tribal governments from lease negotiations.[12]

At the turn of the century, the federal government implemented the distribution of tribal lands in individual allotments according to the 1887 Dawes Act. Between 1897 and 1902, allotment agreements were finalized with the citizens of the Five Civilized Tribes and by the end of 1902 federal officials were solely responsible for regulating the petroleum industry in Indian Territory. The transition from tribal to federal control resulted in confusion among the oilmen regarding leases, and in at least one instance, a drilling company's equipment was seized. The Cherokees eventually secured an exemption from the section of the Curtis Act relating to government approval of tribal leases, but even so, the secretary of the interior, with the approval of the Attorney General of the United States, continued to insist on the federal government's right to approve leases among the Cherokees. When the Cherokees appealed the case to the Supreme Court, it ruled that it had no jurisdiction over the congressional or executive acts affecting the Indians.[13]

Thus the secretary of the interior retained supervision over all leasing in Indian Territory at a time that coincided with the tremendous expansion of the petroleum industry in the region. New leasing regulations adopted by the federal government in 1902 limited the lease of any one individual to forty-eight hundred acres, and required the lessee to declare his intention to seek oil on the lease. In 1903, regulations concerning Cherokee leases were altered to state "carry on operations in a workmanlike manner to the fullest possible extent . . . [and] to commit no waste." The following year, the Department of the Interior issued definite rules governing the leasing of Cherokee and Creek land. They required an applicant to present his lease request to the Union Agent at Muskogee; stipulated that the lessee must have a minimum amount of five thousand dollars; required that all royalties be paid to the Union Agent, not the allottee; declared leases by intermarried non-Indians subject to the same procedure as Indians until their tribal citizenship was established in court; voided all leases on which work had started prior to approval by the secretary of the interior; established a maximum of fifteen years' duration on all leases; and regulated those leases with minors, in addition to other limitations.[14]

These efforts by the federal government were not always appreciated by the Indians or by oilmen. Opposition to governmental interference grew, and in 1904, an assistant United States Attorney General ruled that the Choctaw and Chickasaw Indians might lease their land without federal approval. In April of that year, Congress removed several allotment restrictions placed on members of the Five Civilized Tribes and all restrictions on intermarried non-Indians and freedmen, unless they were minors. In 1905, the Department of the Interior placed the royalty on oil wells at ten percent of the value of the oil produced and demanded an annual charge of one hundred and fifty dollars for gas wells.[15]

At its aborted convention in 1905, it is interesting to note that the Constitution of the State of Sequoyah, proposed to be formed from Indian Territory, roughly the eastern half of present-day Oklahoma, made little mention of regulation of oil companies within its boundaries. The federal government, however, made no reduction in its regulatory policy and, in April, 1906, Congress passed a very detailed law governing the leasing of Indian lands. But regardless of the numerous federal rules governing oil exploration and production in Indian Territory, practically nothing was done by government officials to combat the extravagant waste of petroleum associated with the Glenn Pool discovery. In fact, some of its regulations actually contributed to the waste. An example was the rule that called for the drilling of one well on each lease annually, which added to the problem of over drilling. However, by 1906, the oilmen themselves had undertaken the task of regulating the obvious overproduction at Glenn Pool and established an organization of petroleum producers in an attempt to control the excess production.[16]

While the eastern half of present-day Oklahoma had been governed by five tribal governments and federal officials, portions of the western half of the state had been organized as the Territory of Oklahoma since 1890. There was little success in the search for petroleum deposits in Oklahoma Territory during the early years of the 1890s; however in 1895, the Territorial Legislature passed an act creating the office of oil inspector. Appointed by the territorial governor, the oil inspector was to hold office for two years and was to inspect all "coal oil, carbon oil, petroleum, kerosene oil, gasoline, or any other product of petroleum used as illuminating or burning fluids by whatever name known, which may be manufactured or offered for sale in this Territory." In addition, the act specified the manners of testing fluids with "standard instruments": oils were to be tested by "*First*, The water cup shall have sufficient water in it to rise two-thirds up the side of the oil cup. *Second*, Fill the oil cup with oil to be tested to within one-eighth of an inch of the top. *Third*, Suspend the thermometer so the bulb is just under the surface of the oil. *Fourth*, Use an alcohol lamp to heat the water bath, and before placing the light under the water bath cup, test the oil in the oil cup by bringing a lighted match in contact with the surface of the oil." If the oil "does not ignite, place the lump under the water cup [and] slowly heat the oil, not slower than one degree of the termometer in a minute, nor faster than two degrees of the thermometer in a minute, moving a lighted match across the surface of the oil, at each degree the thermometer rises, not more than three-eighths of an inch from the surface of the oil." During this process, "If the oil should flash (that is, a little gas burn on the surface) and go out again, the degree indicated by the thermometer is the flash test of oil. The flame moved across the surface of the oil should not exceed that of any ordinary match."[17]

If the oil flashed under the lighted match at a temperature lower than 120°, the oil inspector was to brand it "Rejected," and if it did not flash, it was to be marked "Approved Standard Oil." However, if it had no flash test, it was marked "Specific Gravity— 0." In addition, the inspector was to mark all containers that had no fire test as "Highly Inflammable." Fines were imposed on any dealer, manufacturer, or producer of oil who refused to cooperated with the oil inspector and on persons who sold the oil without the inspector's approval.[18]

Intended as a consumer protection act, the law was praised by the territorial governors as an example of the area's progressive outlook. It was opposed by the oil industry, however, and several efforts were made to repeal the act. Statutes annulling the law were passed by the territorial legislature in 1895 and 1897, but were vetoed by the territorial governors. In 1899, the 1895 act was amended to give it greater strength and to update the criteria established for testing the oil. In 1903, the Oil Inspector Act was again upgraded and expanded, and eventually, twenty-seven deputy oil inspectors were stationed near major points of entry into the territory, in order to test oil products as they crossed into the region.[19]

In 1905, Scott Ferris, a member of the territorial house of representatives, secured passage of legislation regulating the installation of casing in wells and the plugging of abandoned holes. This was in reaction to the apparent waste of crude from wells in the Cleveland and Glenn Pool areas, and provided that prior to drilling, a driller should "incase the well with good and sufficient casing, and in such manner as to exclude all water from above from penetrating the oil or gas bearing rock." Should the well be drilled to an even greater depth, the driller was required to continue the casing into the new formation. In addition, the owner of an abandoned well was required to "fill the well with sand or rock sediment to a depth of ten feet above the top of each oil or gas bearing rock, and drive therein a round, tapered, seasoned wooden plug at least two feet in length, and in diameter equal to the full diameter of the well below the casing." Once this was done, the casing could be withdrawn; however, the owner was then obligated to refill the hole to a depth of five feet, plug the well with a three-foot wooden peg and place another layer of rock or sand sediment twenty feet deep on top of the second plug.[20]

The Ferris legislation also prohibited the use of natural gas for "illuminating purposes other than in what is known as storm burners." Anyone using gas in such a manner in the open area surrounding drilling rigs was prohibited from lighting them between 8:00 A.M. and 5:00 P.M. In addition all gas wells brought in had to be utilized for "light, or fuel, or steam power" within ten days, or the well was capped. This was designed to prevent unnecessary waste of natural gas, which was often allowed to simply escape into the air or was burned.[21]

In general, the regulatory legislation of Oklahoma Territory was designed to prevent the obvious waste of crude and natural gas in Indian Territory and to provide protection for consumers against inferior petroleum products. Some additional regulations were imposed upon the industry through court decisions, but these were directed more toward control of corporations than the petroleum production. During the pre-statehood era, the largest and most productive finds were in the eastern part of the state under federal and tribal control. Therefore, the western portion of present-day Oklahoma was not faced with the same

Oklahoma Territorial Governor Frank Frantz. When he left office at the time of Oklahoma's statehood, many of the duties and responsibilities of the federal government in the supervision of oil leases on Indian lands were transferred to state officials. *Courtesy of the Oklahoma Heritage Association.*

problems as the eastern section. As a result, there was less regulation, not from lack of concern but from lack of necessity.[22]

The Enabling Act, signed into law by President Theodore Roosevelt on June 16, 1906, provided for the joining of Oklahoma Territory and Indian Territory as the State of Oklahoma. With the change, many of the duties and responsibilities of the federal government in connection with the supervision of oil leases in Indian Territory were transferred to state officials. In addition the civil cases then pending in federal courts were assigned to state courts, including many that were of great importance to the oil industry. The legislation also prohibited the sale of any oil lands granted to the new state by the federal government prior to January 1, 1915; however, state officials could lease the lands for periods of up to five years by competitive sealed bidding.[23]

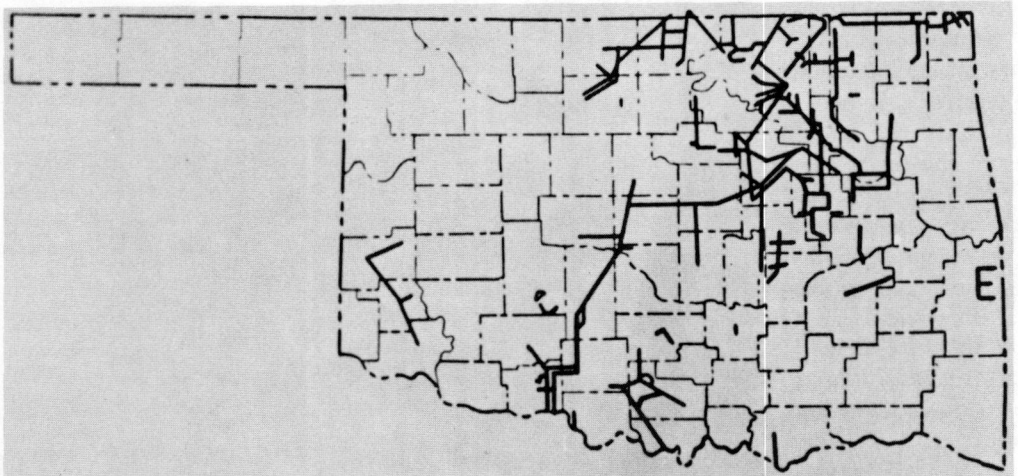

Oklahoma's natural-gas pipelines. One of the first actions of Oklahoma Governor Charles N. Haskell was to order the Oklahoma National Guard to patrol the state's borders to prevent the construction of pipelines into Oklahoma from adjoining states. Although this map dates from the mid-1920s, it can easily be seen that much of Oklahoma's gas production was transported through pipelines entering the state from Kansas and Texas. *Courtesy of the Oklahoma Geological Survey.*

The Constitution of Oklahoma made no specific reference to regulation of the oil industry, but instead placed a great deal of control over all corporations. Included were provisions against price fixing and a complicated system of corporate regulatory duties by various state agencies. Specifically, oil pipelines and other transportation systems were prohibited from exercising the right of "eminent domain . . . until it shall have become a body corporate pursuant to or in accordance with the laws of this State." Oil pipelines and the others were guaranteed the right to "construct and operate its line between any points in this State, and as such to connect at the State line with like lines." A myriad of agencies and commissions were provided to regulate oil companies and other corporations, and the list included the office of the chief inspector of mines, oil, and gas, as well as other executive and elective offices.[24]

Oklahoma was admitted to the Union on November 16, 1907, and within the following month and a half, the first legislature of the state passed a series of "emergency laws," one of which affected the oil industry and a second which dealt specifically with it. Senate Bill Number 11 transferred all cases "from the Courts of the Territory of Oklahoma and the United States Courts in Indian Territory" to Oklahoma jurisdiction. This complied with the enabling act and influenced those oil leases then in litigation. House Bill Number 78 regulated the "Laying, Constructing and Maintaining and Operating of Gas Pipe Lines for the transportation of natural gas within the State of Oklahoma." In addition to defining the term "eminent domain" in regard to pipeline construction, the law provided damage payments to property owners during construction, required that pipelines be constructed under the direction of skilled individuals approved by the chief mining inspector, and limited the pressure used for gas pipelines to three hundred pounds per square inch. Surface lines constructed for the purpose of transmitting gas from wells in process of being drilled were, however, exempt from the regulations. In 1911, the United States Supreme Court handed down a decision that the law conflicted with the interstate commerce provisions of the United States Constitution.[25]

The regulation of pipelines resulted from the fact that several companies had already completed lines to the border of the new state and were awaiting statehood, when when federal restrictions on pipeline construction in the area formerly called Indian Territory would no longer apply. As soon as Oklahoma's constitution was approved, Governor Charles N. Haskell ordered the National Guard to patrol the borders of the state in order to control the situation and prevent the extension of pipelines across the state's boundaries. His orders were based on provisions of the constitution requiring pipeline companies to be chartered within the state, and his action, reinforced by House Bill Number 78, prohibited the shipment of natural gas "to a point outside or beyond this State."[26]

Federal recognition of the authority of state courts to deal with Indian leases came in late 1907, when the secretary of the interior informed officials of the Union Agency in Muskogee that Indians in the process of leasing land should make the necessary affidavits before district or county courts rather than with the agency. But a number of questions arose that were of great concern to oilmen in the state. For example, although statehood had removed the authority of federal courts regarding questions of guardianship, no law had given the right to state courts. This problem was eventually worked out, and jurisdiction of minor allottees was placed in the hands of county judges. Nonetheless, in 1909, the federal government employed district agents to insure fair treatment of minors in the state's courts.[27]

The 1907 Oklahoma Constitution created a corporation commission and gave it the "power and authority . . . of supervising, regulating and controlling all transportation and transmission companies doing business in this State;" however, prior to 1909, the commissioners had exercised little regulation over the oil industry. That year, the state legislature enacted Senate Bill 168, which placed pipelines under the supervision of the corporation commission and required them as a "common purchaser . . . [to] purchase all the petroleum in the vicinity of, or which may be reasonably reached by its pipelines, or gathering branches, without discrimination in favor of one producer or one person as against another." Any company not adhering to the provisions of the law was prohibited from owning "directly or indirectly any oil well or wells, oil leases, or oil holdings or interest in this State."[28]

The 1909 legislature also approved House Bill Number 238, regulating the use and preservation of oil and gas within the state. It specifically required anyone engaged in the production of natural gas "immediately after penetrating the gas-bearing rock . . . [to] shut in and confine the gas . . . until such time as the gas . . . shall be utilized for lights, fuel or power purposes." Four days were given from the completion of a well until compliance with the law was required. The use of natural gas in flambeau lights was prohibited, and gas lights around drilling rigs were to be turned off between 8:00 A.M. and 5:00 P.M. All abandoned wells, including both oil and natural gas, were required to be plugged with alternating sections of crushed rock or sand, plugs of "seasoned pine," cement and sand, then another wooden plug. The state mine inspector was authorized to appoint deputies "of at least five years' practical experience in operating and drilling oil and gas wells," to supervise the "using and operating of natural gas . . . and the proper observance of the law . . . dealing with the drilling and production of oil

One of the early acts of the Oklahoma legislature was to prevent the waste of natural gas through such fires as this one at a well in the Cimarron riverbed near Cushing. *Courtesy of the Petroleum Publishing Company.*

Empire Refineries, Inc., at Okmulgee, Oklahoma. *Courtesy of Cities Service Oil Company.*

and gas or the piping or storage or purchase or use thereof."[29]

House Bill Number 238 also required the inspection of plugged wells, as well as a drilling log of the wells furnished to the inspectors. In addition, the act warned any individual not to "maliciously . . . set fire to any gas or oil escaping from wells, broken or leaking mains, pipes, valves, tanks or other appliances," or to interfere with any equipment of gas or oil companies, unless they were acting under the direction of the company's officers or proper legal authorities. Gas pipeline firms were required to provide the corporation commission with plats, showing all pipelines owned or operated by them within the state, as well as all proposed new pipelines before construction. It also authorized the companies to acquire land for pumping stations along pipeline routes and to utilize public rights-of-way for pipelines if the work was done under adequate supervision.[30]

The 1909 legislative action was the extent of state regulation of the oil industry for several years; however the federal government continued to "muddle through" the question of Indian leases for some time. Nonetheless, it was more concerned with insuring the Indians their rights than with controlling the industry. For the most part, the early efforts at regulation of the oil business within present-day Oklahoma could be described as a maze of often conflicting tribal, federal, territorial, and state actions which, in many cases, overlapped and were almost always a source of confusion to the oilmen. The most successful exercise in restraint of production of the era was that shown by the oilmen themselves, who, when it became apparent that overproduction was a threat, attempted to regulate the flow of crude.[31]

The exploration and production of petroleum did not wait for politicians to find answers to the questions, however. The revival of drilling operations in the Osage lands in 1910 resulted in wells that flowed at the rate of 4,000 barrels a day. The strike at Cushing in 1912 produced approximately 236,000,000 barrels, approximately three percent of the total production in the world, in only seven years. These new finds were to keep Oklahoma in the limelight of the oil industry and maintain its position as a leading oil-producing state.[32]

## Chapter 6.
# The Osage Nation, 1896–1920

THE OSAGE NATION, later Oklahoma's largest county, played a unique role in the development of the state's petroleum industry. For administrative purposes, the Indian reservation was included within the boundaries of Oklahoma Territory, but it was the federal government, not territorial officials, that controlled the leasing arrangements. The 1906 congressional act for "the division of the lands and funds of the Osage Indians in Oklahoma Territory," provided in Section Three that "the oil, gas, coal, or other minerals covered by the lands . . . are hereby reserved to the Osage tribe for a period of twenty-five years from and after the eighth day of April, nineteen hundred and six." All leases made for the extraction of minerals were to "be made by the Osage tribe . . . through its . . . council . . . [and] the royalties to be paid to the . . . tribe . . . ." The "royalty received from oil, gas, coal, and other mineral leases . . . shall be placed in the Treasury of the United States to the credit of the members of the Osage tribe of Indians . . . and . . . distributed to individual members of the said Osage tribe." Provisions were also made to set aside oil and gas royalties to finance schools for Osage Children and for "agency purposes." Upon the termination of a twenty-five-year period dating from January 1, 1907, "the lands, mineral interests, and moneys . . . held in trust by the United States shall be the absolute property of the individual members of the Osage tribe." At the advent of statehood, the Osage Nation was to be incorporated into the new state as one county, and "the present boundaries [were] to remain unchanged until all the lands of the Osage Tribe of Indians shall have been allotted . . . ." Leasing procedures, however, remained under the jurisdiction of federal officials.[1]

Drilling operations on the reservation began in 1896, when Edwin B. Foster secured a blanket lease on the region. Following Foster's death a short time later, his brother, Henry, organized the Phoenix Oil Company and continued exploration. Although their first well was a disappointment and gave only a "show of oil, their third well, located in the SW¼ of Section 34-T27N-R12E, showed promise when completed in June, 1897; however, the returns did not repay the initial investment and operations in the Osage Nation ceased.[2]

The task undertaken by the Phoenix Oil Company was renewed by its subsidiary, the Osage Oil Company, and drilling contractor A. P. McBride. Securing oil rights to the entire Osage Nation in 1899, the Osage Oil Company drilled several wells on the north side of Butler Creek near Bartlesville. They, too, encountered a series of dry holes and small producers, and several backers expressed grave doubts about continuing the operation. Nevertheless, McBride argued that their mistake had been in drilling on the north side of the creek. He insisted that the best prospects for oil were on the south side and persuaded his backers to finance one more well. Consequently, the Osage Oil Company's seventh well, drilled on the south side of Butler Creek, brought in one of the most productive wells in the area up to that time.[3]

Although exploration prior to the twentieth century had produced no startling discoveries of oil in the Osage Nation, enough petroleum had been located to convince most oilmen that deposits in paying quantities were present in the area. Nonetheless, despite an agreement made in 1900 that allowed subleasing of portions of the reservation by other oil firms, there were only five producing wells in the area by 1901. In December of that year, the Osage and Phoenix companies assigned their leases to the newly-formed Indian Territory Illuminating Oil Company, which immediately began an extensive exploration program in the area. More successful than its predecessors, ITIO drilled a well that, after pene-

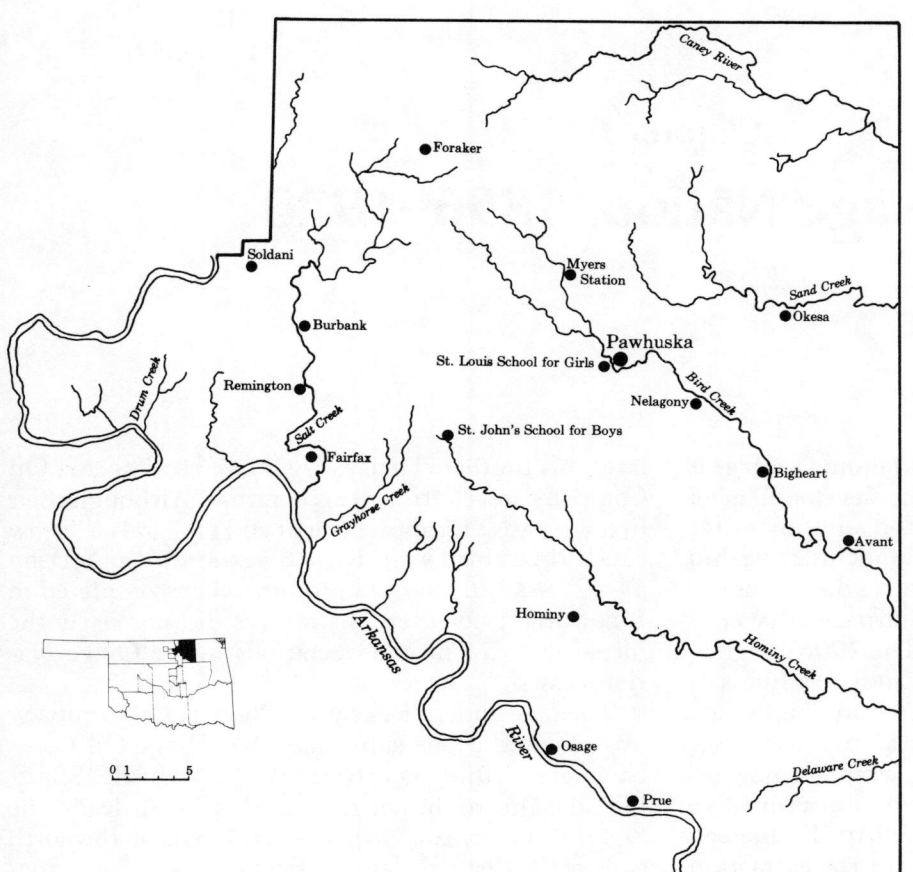

The Osage Nation. This territory in north-central Oklahoma would become one of the richest oil-producing regions of the nation, with huge pools of petroleum in a wide band across the eastern portion of the region from the Caney River to Okesa, Nelagony, Bigheart, Avant, and on south to Hominy and Delaware creeks and running west to Pawhuska, Hominy, and Osage. Other pools were found to the north of Burbank. *From Morris et al.,* Historical Atlas of Oklahoma.

trating only six feet of an oil-bearing formation, produced a flow of approximately fifty barrels daily. Before the well could be drilled deeper, however, the hole caved in and was filled with two hundred feet of debris. Even so, pressure forced crude through the filling, and the well continued to flow at a rate of forty barrels per day. After the well bore was cleaned out, casing was set to prevent future cave-ins, and drilling operations were renewed in an oil-bearing formation that proved to be nearly forty feet thick. The success of the No. 10 well convinced ITIO to continue its operations in the area and make "the fullest and most complete development of the fields."[4]

Nonetheless, drilling operations continued at a rather slow pace. In the summer of 1902, the Almeda Oil Company, under a sublease, completed a producer in the SE¼ of Section 22-T26N-R12E. By the end of 1903, there were thirty wells drilled on the reservation, eleven of which were dry holes. The total annual production amounted to 52,217 barrels of high grade crude, free of sulphur and asphalt, and easy to refine, just what oil men needed for a profitable return on their investment.[5]

In June, 1904, the Workman Oil and Gas Company discovered the Okesa Pool in Section 16-T26N-R12E, with a well that flowed at the rate of twenty-five barrels a day. In November of that year, the Osage City Pool, in Section 20-T21N-R9E, was opened by the Barnsdall Oil Company. Drilled to a depth of 1,075 feet, the discovery well produced a meager five barrels daily; however, the firm drilled another well in November, 1905, that had an initial flow of twenty barrels per day. Still, it was not until 1911 that the pool reached its full potential. The Avant and Wiser pools were also discovered in 1904–1905. The former, located in T23N-R12E, was the southern extension of the Bartlesville Field along the eastern border of the Osage Nation. Wells in the Wiser Pool, occupying the southeast portion of T27N-R12E, flowed at an initial rate of ten to sixty barrels daily. In addition, the Boston Pool on the

As the oil pools of the Osage Nation were developed, Bartlesville became a major staging area for equipment and men. *Courtesy of the Bartlesville Public Library.*

Arkansas River, proved to be a geographically small but very productive extension of the Cleveland Pool.[6]

Much of the activity in the Osage Nation prior to December 31, 1904, resulted after the Indian Territory Illuminating Oil Company had subleased 687,000 acres of land. By retaining only the gas rights and subleasing the oil rights on 537,000 acres, or approximately 80 percent of this block, ITIO encouraged many smaller petroleum companies to explore the Osage area. On March 3, 1905, the United States Congress approved the extension of the lease holdings of ITIO for an additional 10 years which again gave the company and its sublessees access to more than 680,000 acres of land. The renewed activity resulted in a total of 1,868,260 barrels of oil being produced on the reservation in the year 1905.[7]

Excitement generated by the drilling activity in and around the Osage Nation spread throughout the United States, and among the ambitious young men drawn to the area during this period was Herbert R. Straight, the son of a Pennsylvania oilman. Although employed first by T. N. Barnsdall, Straight later joined Henry L. Doherty when Doherty purchased the Barnsdall Oil Company and other producing property interests in the area. These properties, which included Barnsdall's interest in ITIO, were consolidated into the Empire group of companies in 1912, with their operational headquarters in Bartlesville. This consolidation was the origin of the petroleum energy operations of what is now Cities Service Company. As head of production, Straight was mainly concerned with the location of a supply of natural gas for the areas surrounding Wichita, Kansas, and Joplin, Missouri. The situation became serious when the firm's known reserves of gas began to run low. Although geology was still a new science in the petroleum industry, Straight employed two geologists, Charles N. Gould and Everett Carpenter, to examine the area between the Osage Hills and Wichita in an effort to locate new accumulations of gas. Upon their recommendation, the firm leased 12,000 acres of land near Augusta in eastern Kansas, a decision that proved to be extremely profitable.[8]

In 1913 the first well was drilled for gas at Augusta. At a lower depth, oil was encountered the

The Osage Council. *Seated, left to right:* Principal Chief Fred Lookout; D. E. Murphy, superintendent of the Osage Agency; Harry Kohpay, assistant chief; and Charles Brown. *Standing, left to right:* Frank Lessert; Francis Revard; Dick Petsomoie; Franklin Revard, secretary to the Council; Clement Denoya; George Alberty; and Sam Kennedy. The value of leases negotiated by the Osage Council ran into the millions before the oil boom was over. *Courtesy of the Oklahoma Publishing Company.*

following year. Although this did much to eliminate the shortage of gas, Straight was not satisfied and, in February, 1916, he brought in the discovery well at the El Dorado field in Kansas. Eventually Cities Service expanded its operations until it became a major producer in Oklahoma.[9]

By 1906 drilling activity had shifted to the eastern portion of the Osage Nation near the boundary of the Cherokee Nation. Here, in TS20&21N-RS12&13E, the Bird Creek-Flat Rock field was opened with initial production as high as 400 barrels of oil and one million to three million cubic feet of gas daily. Before 1906 the Osage Nation had produced a total of 4,176,164 barrels of petroleum; however, the Bird Creek-Flat Rock strike, combined with production from other fields on the reservation, catapulted the total production for the year 1906 alone to an astonishing 4,514,004 barrels of oil, more than all previous production combined. During the next few years, production in the area continued to climb at a fairly steady rate, added to by new discoveries at Sperry in 1909 and Skiatook in 1911. The latter, drilled by the Barnsdall Oil Company, initially produced 50 barrels of oil per day at a total depth of 1,514 feet.[10]

Generally the Osage Indians had reacted to the

discovery of oil on their lands with little interest. They had no knowledge of petroleum operations and did not understand the wells or the pipelines carrying the gas to markets. In fact, during a blizzard in 1910 they cut holes in the gas pipelines that crossed the area and set fire to the escaping gas in order to provide a source of heat for their livestock roaming the countryside.[11]

The federal legislation of 1906 regulating leases in the Osage Nation had not only reserved mineral rights for the tribe, but also had given the Osage Council power to negotiate mineral leases in the western half of the reservation. When it became known that oil deposits were present in that area, steps were immediately taken to make that portion of the reservation available to oilmen. However, dissension within the tribe regarding methods of assigning leases delayed action on the question until November, 1912. Bidding was stringently controlled: each bid required an accompanying deposit of ten percent of the bonus offered, as well as the rental payment for the first year in advance; if a bidder did not execute within thirty days, the lease was forfeited; no common carrier was permitted to bid on a lease; and no individual or company was allowed to hold more than twenty-five thousand acres. In addition, advance rental payments were set at fifteen cents an acre the first year, thirty cents for the second, fifty cents for the third, and a dollar annually for each additional year.[12]

The secretary of the interior set the royalty rates at one-sixth of the gross production (based on the actual market price of the crude) or not less than sixty cents per barrel. Gas also carried a royalty of one-sixth of the market value. In the hope of attracting industry to the reservation, the Osage Council initially considered restricting the export of gas from the region; but as more and more became available, it was apparent that there was an abundance and the tribe relented. Lessees were prohibited from drilling wells on individual homesteads or plots of land that had been cultivated within thirty days prior to the lease application, as well as drilling within fifty feet of a highway or section line. Sediment containers were required to be a safe distance from oil tanks, and crude was not allowed to remain in earthen tanks for more than fifteen days. Osage schools were to have free use of natural gas. All leases were to be in effect for ten years, dating from its approval by the secretary of the interior; however, none were to be in effect after April 8, 1931.[13]

Oilmen reacted strongly to the one-sixth royalty payment. Previously, the tribes in Indian Territory had received only one-eighth; however, the Osage Indians argued that the amount was the same as that being paid ITIO for its subleases. The Osage National Council also opposed the new regulations because of the fact that not all available Osage land was being offered for lease, because of insufficient protection provided to the surface rights of Osage allottees, and other reasons. Nonetheless, the tribe received $39,436 for the leases, approximately $1.60 per acre, and drilling resumed at an even greater pace.[14]

Production in the region climbed at an astounding rate. By 1911, royalties paid the tribe had almost tripled the $200,000 paid during the preceding four years. In August, 1912, the Wildhorse Pool was discovered in TS21&22N-RS10&11E by the M.B. & K. Oil Company and the Barnsdall Oil Company. Within five years, wells in the field were producing from as many as six different formations, with an initial production ranging from ten to seven hundred barrels daily. In March, 1917, the Sand Springs Home Oil Company began activity in the southern extension of the pool, and numerous wells were drilled in Sections 33 and 34, T22N-R10E. So great was the production that, after 1919, acreage in the region sold for between $170,000 and $430,000 per quarter-section. The continued activity pushed the total Osage royalty payment in 1912 to more than $600,000.[15]

In July, 1913, the Roxana Petroleum Corporation found the Pond Creek Pool, located in TS28&29N-R10E. Their discovery well, drilled to a depth of 1,065 feet, flowed at a rate of one hundred barrels daily. After 1925, the pool was extended northward and linked to the South Elgin Pool. In 1914, the Hickory Creek, Landon, and Quapaw pools were discovered in the Osage Nation. Despite the fact that most of the wells drilled in the Quapaw Pool were under the provisions of the old Foster lease and did not contain a bonus provision, they produced over more than one million dollars in royalties for the Osage tribe in 1913 and 1914.[16]

By 1915 it was clear that existing regulations needed to be changed regarding the blanket lease negotiated by Foster before the turn of the century and later acquired by the Indian Territory Illuminating Oil Company. A public hearing on the question, held by the Department of the Interior in Washington, D.C., was attended by members of the Osage National Council, interested oilmen, and officials of the Indian administration. Oklahoma's Governor Charles N. Haskell spoke on behalf of ITIO,

The first well in the Wildhorse Pool. Opened in 1912, the Wildhorse Pool produced from as many as six different horizons within the first five years of operation. *Courtesy of the Bartlesville Public Library.*

maintaining that ITIO would develop the mineral resources of the Osage Nation in an orderly manner. If the leases were made available on a competitive basis, he argued, the petroleum industry in the area would flood the prospective markets with such a supply of oil that all facilities would be overtaxed. Representatives of several of the petroleum firms holding subleases on the reservation replied that the terms of the original agreement awarded all natural gas discoveries to the holder of the blanket lease; therefore, when markets were not readily available, ITIO had monopolistic control of gas production. The Osage National Council readily agreed and expressed its desire to work directly with the companies operating on the reservation, without ITIO acting as an intermediary.[17]

Those voices raised in defense of the actions of ITIO were drowned out by the smaller companies calling for its demise, and at the end of the hearing in July, 1915, the Department of the Interior declined to renew the blanket lease and relegated ITIO to the status of a company operating within the limits of the Osage Nation, thereby breaking the Foster monopoly. The eastern part of the reservation, previously under the old Foster lease, was placed under new regulations to comply with provisions of the western part of the Osage lands: divided into 160-acre tracts, with a limit of no more than 4,800 acres of land to be leased by an individual or company, except those areas capable of producing twenty-five barrels of oil per day as of July 1, 1915; public lease auctions were to be conducted by the Osage tribe under the supervision of the Department of the Interior, and all oil leases were to be made for a period of five years from the date of approval by the secretary of the interior; royalty was set at one-sixth of the oil produced, with the exception of those leases which produced in excess of one hundred barrels of oil per day, in which case royalty was set at one-fifth; sublessees of ITIO could retain rights to those quarter-section units they were currently developing if the total amount did not exceed 4,800 acres of land.[18]

Gas rights remained separate from oil rights. Nonetheless, the royalty on natural gas was also set at one-sixth, with all contracts "for the sale or use of gas" subject to the approval of the secretary of the interior. Acreage for the current gas leases was to be granted by the Osage Council "covering all or part of their present holdings and for such periods as the Secretary of the Interior may determine."[19]

The revised regulations were intended to open approximately 70,000 acres of producing territory

Four of Oklahoma's leading pioneer oilmen. *Left to right:* H. R. Straight, W. Alton Jones, H. V. Foster, and Burdette Blue. *Courtesy of Cities Service Oil Company.*

and about 165,000 "nonproductive" acres within the Osage Nation. The area classified as "nonproductive" was subject to an annual rental fee of $1.00 per acre "in lieu of development." Those quarter-sections capable of averaging twenty-five barrels of oil a day, as well as approximately 16,000 acres of additional producing territory, were made available for lease "at public auction to the highest bidder, under such rules and regulations as the Secretary of the Interior may prescribe." There remained 430,000 acres of unleased "nonproducing territory," and this area was to be "leased at such times and under such rules and regulations as the Secretary of the Interior may prescribe." On March 16, 1916, the Foster lease, inherited by the Indian Territory Illuminating Oil Company, expired, and a vast area known to have productive capability was thrown open for bid.[20]

In 1915, production in the Osage Nation dipped from more than 11,000,000 barrels of oil the previous year to just over 7,000,000 barrels. Royalty paid to the Osage tribe also declined from nearly $1,000,000 in 1914 to approximately $600,000 the following year. During the same period, only one new field, the South Elgin located in T29N-R9E, was opened in the area. It was an extension of the Elgin Pool in Kansas. However, renewed activity occurred on the reservation in 1916 with the discovery of three pools: the Myers Dome, Barnsdall, and Hominy. A gas well drilled by the American Pipe Line Company on Myers Dome had a flow in excess of three million cubic feet daily. The Minnehome Oil and Gas Company drilled the first oil well in the pool located in Section 1-T26N-R8E, which had an initial production of 50 barrels per day. At one time, leases in the Myers Dome area sold for as much as $140,000 per quarter-section. These three finds helped boost the annual production of the Osage Nation in 1916 to 9,805,477 barrels, with royalty payments totaling $858,008.97—a headright of $384.97 for every Osage Indian.[21]

When the United States entered World War I, the military's increased demand for oil spurred the activity of oilmen in the area and in 1917–1918 five additional oil sites were discovered. In March, 1917, ITIO struck natural gas at a depth of 813 feet in Section 6-T24N-R10E, opening the Pershing Pool. Producing gas at the rate of 5,500,000 cubic feet daily, the initial well was only a sample of what was to come. In July, 1918, the William M. Graham Oil and Gas Company completed the first oil well of the field in the same section, and in January, 1919, the Carter Oil Company brought in a well nearby (within the same square mile) that flowed 5,500 barrels per day. The Domes, Nelagoney, and Wynona pools were also opened in 1919, the Almeda Pool near the Cherokee-Osage border having been discovered the previous year. Production increased to slightly more than one million barrels of oil between 1916 and 1918; however, royalty payments increased ten-fold —rising to $3,672.33 per headright.[22]

The heightened activity begun during World War I resulted in several additional strikes in 1919, including the Pettit, Pearsonia, Pawhuska, X686, and Y686 fields. In June, 1919, the Marland Oil and Refining Company paid $85,000 for the SE¼ of Section 20-T23N-R8E and, at a depth of 725 feet, brought in a gas well with an initial production of an incredible sixteen million cubic feet daily. After the discovery, well prices soared to as much as $150,000 per quarter-section, even though it was not until October, 1923, that oil was found. Although natural gas had been found at the Pearsonia site in 1916, the American Pipe Line Company did not find oil until August, 1919. In the fall of that year, the New England Oil and Pipe Line Company brought in the discovery well of the Pawhuska Pool. Although it produced only 150 barrels per day, as much as $620,000 was paid for a quarter-section lease at the site. The X686 and Y686 finds were mainly gas, and as much as twenty-three million cubic feet of gas per day was produced by the X686 strike.[23]

The early-day oilmen were handicapped on the reservation in much the same manner as their counterparts in other portions of Oklahoma Territory, as well as Indian Territory. There was simply no economical means of transporting the crude to market. As in other instances, the oil was first stored in wooden and steel tanks or, if these facilities were not available, in earthern excavations until it was transported by railroad to the nearest refinery, usually in Neodesha, Kansas. As the fields were developed, the Indian Territory Illuminating Oil Company connected several of its wells in the Osage Nation with storage facilities at Bartlesville, and from there, at a cost of sixteen cents a barrel, it was shipped to refineries. Another line, connected with a storage facility in the Creek Nation, was extended into the Osage area by the firm Guffey & Galey, but this did not prove profitable. By 1905, the Prairie Oil and Gas Company had a forty-mile six-inch pipeline in operation through the Osage and Cherokee nations. In addition, several smaller liines linked the Osage wells to area refineries. By 1907, the Uncle Sam Refinery in Tulsa was in operation and processing some of the Osage crude, and the following year, the Prairie Oil and Gas Company reported handling in excess of four million barrels of Osage oil. In 1910, the Gulf Pipe Line Company also entered the competition, thereby offering another market outlet. Nonetheless, the inadequate transport system continued not only to depress the price of Osage oil, but curtail development of the fields.[24]

The lack of improved roads was also a great handicap in moving the heavy oil field equipment. H. E. Koopman, the first secretary of Phillips Petroleum Company and president of the firm's purchasing department for many years, recalled that one of Phillips' initial investments was a four-wheel drive Jeffry Quad truck which was the only thing capable of hauling loads over the primitive roads. Another early-day Phillips employee, Melvin B. Heine, remembered that the roads were so bad it was often necessary to cut a large limb from a tree and fasten it to the two rear wheels of a buckboard in order to get the conveyance down some of the hills in the Osage region. Only then could equipment be lowered down the inclines without the wagon overrunning the team pulling it.[25]

As a tribe, the Osage Nation profited greatly from the 1906 "Allotment Act," which allowed them to hold their land intact, although not in common. An allotment of 160 acres of land was parceled out to each of the 2,229 members of the tribe. After all members of the tribe had selected an allotment,

The Million Dollar Elm. This historic auction site is on the grounds of the Osage Council House in Pawhuska, Oklahoma. *Courtesy of the Oklahoma Publishing Company.*

approximately 498 acres of "surplus land" was also awarded to each man, woman, and child. Thus, each Osage Indian received approximately 658 acres of land. In addition, the agreement provided that the tribal funds were to be divided on a pro rata basis. The act also called for the election of an eight-member tribal council, an elected principal chief, and an assistant chief, with the authority to execute oil leases, subject to the approval of the secretary of the interior. Such an arrangement proved to be of great financial benefit to the individual tribal members.[26]

After the initial twenty-five year period, the reservation of mineral rights for the common benefit of the tribe was continued by congressional action and eventually extended to April 8, 1983. The amount of money received by members of the Osage tribe was large. Amounting to $384.93 for each Osage headright in 1916, it had increased by 1919 to $3,672.33; and in 1923, reached a total of $12,400 for every man, woman, and child in the tribe. Inherited, the headrights were passed on from generation to generation. Approximately two thousand members shared equally in the royalty payments at first, but the shares were divided as time passed or, in some instances, added to other headrights. As a result, the

The Alfred Mathews Lease in Osage County, Oklahoma, which was purchased for $1,990,000. *Courtesy of Cities Service Oil Company.*

amount of money distributed varied from individual to individual, especially during later years.[27]

In 1906 the headright money was distributed to the full-bloods, holding approximately one-third of the rights, on the first two days of each quarter, after which the mixed-bloods received their money. Sometimes taking as long as a week, the actual task of issuing the funds occupied the time of eight officials. Perhaps it was predictable that the sudden acquisition of such large sums of money would lead to excesses on the part of some tribal members. Money meant little and vast sums of money were spent on luxury items. It was reported that on a single day in 1927, one Osage woman spent $12,000 for a fur coat, $3,000 for a diamond ring, $5,000 for a new automobile, $7,000 for furniture plus $600 to ship it to California, and $12,800 for some land in Florida, a total expenditure of more than $40,000. The Osage, dressed "in their gypsum-rubbed leggings, their moccasins, silken shirts, wide beaded belts, wampum necklaces, silver arm bands," and wearing "otter-tail pieces, the traditional scalplocks on their heads," were often seen being chauffeured along the dirt roads on the reservation in new automobiles. Many others, however, took advantage of their new wealth to insure a proper education for their children.[28]

It has been estimated that more than three hundred million dollars was paid to the Osage tribe for leases, bonuses, and rentals during three-quarters of a century of oil development on the reservation. For many years, the lease sales, conducted by Colonel E. Walters, were held under the "Million Dollar Elm" just north of the Osage Agency in Pawhuska. While prospective bidders lounged on the agency grounds or took seats in bleachers arranged around the tree, the auction often resulted in spirited bidding for the more promising acreage. In 1919, a family of eleven, two parents and nine children, received a total of almost $55,000 dollars in cash bonuses and royalties during a ten-month period. In 1922 for the first but not the last time, a single 160-acre tract carried a bonus of $1,000,000.[29]

Indeed, it appeared there was no end to the "black gold" that flowed from beneath the earth. Production continued to rise on an almost unbelievable scale. In 1923 a total of 41,810,178 barrels of oil flowed from the reservation to refineries, resulting in a headright payment of more than $12,000. On one occasion, the Midland Oil Company paid $1,990,000 for 160 acres located in the NW¼ of Section 14-T27N-R5E in the Burbank Pool. Unfortunately, however, a large number of Osage Indians ultimately lost their valuable holdings. After obtaining certificates of competency, many of the mixed-bloods, as well as the full-bloods, sold their land and, after an initial spending spree, found themselves without land or money. It became obvious that many of the Indians could not manage their financial affairs, and, in an attempt to correct the situation, congressional legislation dated March 2, 1921, greatly restricted the issuance of competency certificates, without which an Osage Indian could not dispose of his holdings.[30]

The first two decades of the twentieth century had not tapped all petroleum reserves in the Osage Nation. Yet to be discovered were the vast deposits of the Burbank Pool, where some leases produced twenty thousand barrels of oil daily per acre, and the

Prue Pool, containing some of the best gas wells on the reservation. The discovery well of the Prue Pool had an initial open flow of up to sixty-five million cubic feet of natural gas per day. The frenzy of money and oil continued on the Osage lands for many years, peaking in the 1920s, only to boom again in more recent years. These discoveries were, however, harbingers of the future for the petroleum industry in Oklahoma.[31]

By no means was the oil activity during the first two decades of the 1900s restricted to the Osage region of Oklahoma. To the south, under the provisions of the Dawes Act, the Creek Indians had also been forced to accept individual allotments. Some members of the tribe resisted the division of their homeland, but to no avail. Those who refused to select their allotments were arbitrarily given plots in the western part of the Creek Nation near the border of Oklahoma Territory. Considered almost worthless, the land was valued as low as fifty cents per acre in some instances. Nevertheless, within five years after statehood, this region would be the site of an oil discovery of unprecedented scale. Known as the Cushing field, the pool was producing a daily average flow by 1915 of more than three hundred thousand barrels, and by the end of 1919, seventeen percent of all the oil marketed within the United States was gushing from Cushing's seemingly unlimited reserves.[32]

## Chapter 7.
# Cushing

BY THE TIME OKLAHOMA entered the Union in 1907, Frank Wheeler had already purchased a farm about twelve miles east of the community of Cushing in Section 31-T18N-R7E. There, he shared a small log house with his wife, eight daughters, and one son. The rough terrain was described as "sixty acres of plow land and one hundred acres of rocks and scrub." It was difficult for Wheeler to wrest a living from the barren earth, and he spent much of the time seeking employment as a mason to supplement his income. Even so, the family remained constantly on the brink of poverty.[1]

Having heard of the oil strikes in Oklahoma, Pennsylvanian Tom Slick arrived in Cushing during the winter of 1911. A self-trained geologist to whom "The smell of oil sands was perfume to his nostrils," he divided his time searching for likely petroleum deposits and seeking backers to finance his drilling operations. A series of "dusters" forced Slick to turn to two Bristow, Oklahoma, bankers, B. B. Jones and M. Jones, for funds to finance his search. Slick was known for drilling wildcat wells, and many potential investors were skeptical of his idea of oil in the area, pointing out that the nearest producing well was more than twenty-six miles away.[2]

On one occasion, while exploring the region east of Cushing, an area that included the Wheeler farm and more than a thousand Creek allotments, Slick found himself miles from town late one evening. Following a barely distinguishable trail, he happened upon the Wheeler cabin and was given shelter for the night. Visitors were infrequent in the area, but he was made comfortable and given one of the makeshift beds of the family. Several oil prospectors had passed through the vicinity before but because of the distance to the nearest producing well, few had given serious thought to the possibility that the barren ground might cover hidden riches. Although somewhat hesitant, Wheeler, after being informed of Slick's reason for being in the area, agreed to lease the land. The following morning Slick departed, carrying with him the key to the incredible Cushing field, one of the richest oil strikes ever to be found in the United States.[3]

Although word spread quickly among the oilmen in the area that Slick and his associates were acquiring leases in the Cushing vicinity, they apparently paid little attention, possibly because of Slick's previous bad luck in drilling wildcat wells. Slick selected a site approximately three miles east of the Wheeler farm and, after enough land had been leased to insure adequate protection of a discovery, began drilling operations.[4]

As time passed and the well was drilled to a depth of approximately 2,000 feet without encountering oil, it appeared that Slick's bonanza was another "duster," and the Bristow bankers withdrew their financial support. Nevertheless, the sand that was brought up by the core barrels reeked of oil, and Slick remained confident that "black gold" was in the immediate vicinity. With operations halted at the well site, Slick began to study the rock samples of the abandoned well and the surface features of the surrounding land. His evaluation of the geological structure of the area convinced Slick that drilling operations should be moved closer to the Wheeler property.[5]

Slick selected a new location near a small creek that meandered through the Wheeler farm. His growing reputation for drilling dry holes, coupled with their experience with the previous well, had dampened the enthusiasm of the Bristow bankers, however, and he was unable to persuade them to finance another project. Slick scoured the area for financial backers. In an attempt to convince several citizens in nearby Cushing, he offered one-half interest in all his leases to anyone willing to advance the necessary $8,000 to drill the second well. Still, there were no takers. In desperation, he recalled an old acquaintance of the oil fields in Pennsylvania, C. B. Shaffer, then living in Chicago, Illinois. Slick borrowed $100 to make the trip and, after listening to Slick's interpretation of the geological formations underlying the Wheeler farm, Shaffer decided to risk the $8,000. Elated, the triumphant Slick returned to the Cushing area and promptly began operations on the second well.[6]

Drilling operations at the new location proceeded rapidly. By February 29, 1912, two local newspapers reported that the well had reached a depth of nearly two thousand feet, and had located "oil sand with what appeared to be an abundance of oil." Apparently B. B. Jones had second thoughts about withdrawing his financial support of the venture, for the same newspaper reported that Jones was involved in the operation. The oil sand appeared to be approximately fifty feet thick, and the journalist recorded that consideration was being given to "casing the oil off . . . and going on down with the hole to something over two thousand feet." Meanwhile, Slick had also begun operations at another location on the Thomas Maloney farm — one-half mile south and two miles east of Cushing—and newsmen reported that "a number of teams have been busy, hauling out rig material" from the town to the location.[7]

Indeed, Slick did deepen the Wheeler well and, on March 12, 1912, it blew in "with a roar of gas-driven oil that literally flooded the surrounding earth." The well was drilled to a depth of between 2,319 and 2,347 feet. The tremendous pressure of the natural gas in the well forced the crude to the surface at a rate of 400 barrels a day. The great Cushing Pool had been discovered. Within a month, Wheeler was collecting $125 daily in royalties from the barren 160 acres that had produced such a meager existence for his family only days previously. Eventually, the Cushing field was to outproduce even the famous Glenn Pool.[8]

Slick was smart. The well was immediately capped, and dirt was spread over the oil-soaked earth around the rig. Every effort was made to conceal the nature of his find. Luckily, the region was sparsely settled and there were few roads leading into the area. This, as well as a lack of telephones among the local farmers, made communications difficult and slow, and Slick made no effort to change it. He did wire Shaffer to come at once with J. K. Gano and an experienced crew of "lease getters."[9]

Word of such a strike was impossible to suppress for any great length of time, however. Although Slick had said nothing to local residents, and had no reason to seek further financial support among the area's citizens, knowledge of his extensive leasing efforts had not escaped the attention of other oilmen. Both the *Cushing Independent* and the *Cushing Democrat* made note of Slick's find in the March 21, 1912, issues. "STRUCK AN OIL GUSHER" proclaimed one headline, while the other declared "NO-LONGER DOUBT—ABSOLUTE CERTAINTY . . . Splendid Oil Find." "There had been a great find at the Wheeler well. Shaffer and Slick had struck oil, and not only struck oil but abundance of it and of fine quality," reported the newspapers. The strike was the topic of conversation among oilmen in the Tulsa area and, before long, Cushing was flooded with people scrambling to secure leases near the well.[10]

By the time Shaffer arrived, representatives of a number of the major producers were on hand. However, Slick, aware that word of the discovery would bring an onslaught of oilmen to the area, had taken the precaution of renting "every horse and rig in Cushing." In addition, Slick and his associates had even hired all the stock on nearby farms and placed the animals in a pasture under armed guard. Notaries in the town were also on "vacation," making it impossible for other oilmen to execute lease agreements that they might secure. Reportedly, there was one automobile in Cushing that had been overlooked by Slick, and the owner charged $25 in advance to haul passengers to and from the discovery site. Even so, Slick and the others must have laughed at the antics of the multitude of lease purchasers scurrying to find a means of transportation to the Wheeler farm, a number of whom "struck out" on foot for the hills east of town.[11]

At the well site, Slick and the others placed an inverted tub over the hole and secured a drill weighing several hundred pounds to the top of the tub to contain the petroleum. The pressure was so great for several days after the crude had been tapped, however, that at seventeen-minute intervals, enough force was built up to lift the tub and drill from the top of the hole and force two barrels of oil from the pipe. The overflow was trapped by a hastily constructed earthen dam, and there were soon several hundred barrels of crude on hand. Finally, they were successful in forcing the casing to such a depth that the flow of petroleum was more easily regulated.[12]

Slick's well was not the only activity being conducted in the Cushing area at this time. Another well, approximately eleven miles east of Cushing, was drilled near Slick's operation by Charles J.

The Dropright oil field. Stretching for miles along the Cimarron River, this pool was but one of several major finds in the Cushing field. *Courtesy of the Bartlesville Public Library.*

Wrightsman. Wrightsman, a Tulsa attorney, was just as secretive about his well as Slick. However, the excitement over the discovery on the Wheeler farm had already focused the attention of the oil industry on the area, and Wrightsman's construction of two large storage tanks near his location confirmed another large strike—two wells, both producers. It was a sure sign. The influx of oilmen continued to swell and the competition for leases grew more intense.[13]

Shaffer and his crew of "lease getters" opened an office above the Cushing Drug Store and immediately began to expand their holdings. Work also continued at the well site. Lumber used in constructing three large wooden storage tanks was shipped in by the Katy Railroad and hauled to the Wheeler farm. In addition, work began on a steel tank capable of holding 32,000 barrels of oil. In order that the Cushing discovery would not suffer from the lack of transportation facilities that had plagued many of the other Oklahoma discoveries, crews were already at work marking routes for pipelines to connect the new field to refinery and shipping facilities. Although Slick was suddenly called away from Cushing because of the illness of his father in Pennsylvania, Shaffer and the others continued to oversee the operations and protect their interests in the rapidly growing boomtown.[14]

Leasing activity in the Cushing area continued unabated. Money flowed in a seemingly unending stream to those fortunate enough to own land near the discovery well. Aaron Drumright, the owner of 120 acres located south of the Wheeler farm, received $20,000 for property he had purchased for $1,500 only six months previously. R. A. Fulkerson had chopped wood at $2 per cord prior to Slick's find but,

Main Street in Drumright. This community became a center of the production in the Cushing field and a well-known oil boomtown. *Courtesy of the Petroleum Publishing Company.*

after selling his small acreage just south of Drumright, was able to establish himself as a well-to-do farmer in his native Kentucky. Sara Rector, the orphaned daughter of a Creek slave, inherited her father's 160 acres of land at the age of ten and saw a well drilled on the barren land flow at the rate of three thousand barrels of oil daily. On April 12, 1912, the Prairie Oil and Gas Company paid $64,000 for 40 acres south of Slick's well, even though the land generally sold for approximately $55 an acre.[15]

By the end of April, 1912, at least six drilling rigs were in operation in the area of the Wheeler well, and the Prairie Oil and Gas Company was in the process of laying a pipeline to Slick's discovery site. Drilling was also being conducted at numerous other locations around the Cushing area. Eventually, the pool would expand to approximately thirty-two square miles, stretching from the middle of T16N-R7E northward to the lower portion of T19N-R7E, or approximately eighteen miles in length. To service the wells in the field, no fewer than five oil well supply companies opened branches in Cushing during the month of July, 1912.[16]

Within a year after the initial discovery, there were 150 producing wells in operation, with a total daily production of approximately 23,079 barrels of oil. In February, 1913, 67 wells were drilled and, of these, 64 produced oil, 2 produced natural gas, and only

Burning gas well in Dropright Pool. With no market available for the natural gas produced in the Cushing Field, in many instances it was simply allowed to burn. *Courtesy of the Bartlesville Public Library.*

"Great Lakes of oil, Drumright Field." So great was the production of the Cushing field that the output overtaxed all existing storage facilities, and the crude was held in large, open earthen excavations that literally were lakes of oil. *Courtesy of the Petroleum Publishing Company.*

one was a "duster." The natural gas associated with the oil production in the Cushing area was a boon to the oilmen, in that the crude was lifted to the surface by the gas pressure, thus eliminating the need for pumping equipment in the majority of the wells. One well, located on the Ollie Landon farm, produced an estimated ten million cubic feet of gas daily. In many instances, the gas encountered in the search for oil was simply "vented," or allowed to escape into the atmosphere, a practice that proved economically unsound, for as the gas pressure fell, the oil flow was reduced.[17]

The astounding amount of gas that was allowed to spew into the air each day often produced tragic results. This was especially true when atmospheric conditions created by damp, cloudy days caused the gas to ettle into depressions around the countryside. On such a day, W. J. Flanagan, Mr. and Mrs. Ora Lyle, and Miss Lulu Reed were touring the oil fields by automobile when the vehicle was suddenly engulfed in a huge explosion. The Lyles were killed, and the other two were severely burned by an accompanying fire ball. At the time, the episode was baffling to the citizens of the area, but shortly thereafter, two drilling contractors from Drumright were killed in a similar explosion while in the vicinity of the

Drumright oil field fire. Huge amounts of crude stored in open pits created a fire hazard, and in August, 1914, Drumright was swept by an out-of-control oil-field fire. *Courtesy of the Bartlesville Public Library.*

wells. It was theorized that the high humidity and low-hanging clouds caused the escaping gas from nearby wells, pipelines, and storage tanks to accumulate in low-lying pockets, which were ignited by a spark from the engine of the automobile or a lighted cigarette.[18]

The oil found in the Cushing Pool was "the highest grade of crude so far found in any important pool west of the Alleghenies." When refined, it was approximately thirty-five percent gasoline and benzine. The tremendous volume of high grade oil made the Cushing Pool a source of great riches to many people. The initial production of the wells in the field amounted to nearly a thousand barrels of oil per well, and it was reported that there were several wells with "a daily production greater than that of all the Pennsylvania oil fields; and at least one with an output so large as all Illinois." Drilling activity continued at a hectic pace throughout 1913 and 1914. The tremendous production of the field was, however, double-edged. It "has made millionaires of some and cut the income of others; caused a piling up of stocks and added new refineries and pipelines," wrote one journalist.[19]

Activity in the area also uncovered several minor fields in close proximity to the Cushing Pool or, more accurately, extensions of it. Situated in the central portion of the original Cushing field, the Drumright Pool, including Slick's original discovery well, was opened in March, 1912. Discovered in 1914, the Oilton field was situated on the extreme northern edge of the Cushing District. During the field's early development, two wells were drilled by the McMan

Oil Company in Section 23-T19N-R7E; one flowed at the initial rate of approximately 150 barrels of oil daily and the other was a dry hole. The Shamrock Dome was the southernmost of the three principal domes of the Cushing District and was named for the post office two miles southwest of its apex. One of the early developers of the region, the Numa Oil Company, reported the initial output of its first well at 8,000 barrels of oil daily. The Mount Pleasant Dome, named for the church located on its north slope, was situated within the same general area. In 1914, the Yale pool, opened by the Alice-Katherine Oil Company in Section 7-T19N-R6E, resulted in wells "of only moderate size," a few of which had an initial production of 1,000 barrels of oil daily. The Ripley and Ingalls pools were also found T18N-R5E, but ultimately yielded only "gas and a few small oil wells" during the years 1914 to 1916.[20]

Nonetheless, the Cushing boom added even more oil to what was rapidly becoming a glutted market. In March, 1914, the most productive well in the district was brought in with an initial flow of 6,120 barrels of oil daily. That same month, the total monthly production climbed to 1,577,168 barrels of oil. The huge amounts of crude oil dumped on the market began to have an adverse effect on the price of crude, but drilling continued. In July of that year, there were 890 wells in operation in the district. Although 32 sites had been abandoned, crews were busy drilling 182 additional wells and workmen were hastily erecting 78 additional derricks. The processing system was literally flooded by the flow of crude from the Cushing field. By May, 1914, 155 steel tanks, each capable of holding 55,000 barrels of oil, were under construction. In addition, four refineries were in operation in Cushing and four more were being built; however, their combined capacity was only 25,000 barrels daily, far short of the field's production.[21]

By April, 1914, the output of the Cushing Pool had affected the price of oil nationwide. Although the outbreak of war in Europe created a greater demand for petroleum products, World War I also disrupted the normal flow of international trade. Thus, while new markets were created, old ones disappeared. As a result, crude flowing from the wells at Cushing, Glenn Pool, the Osage Reservation, and others continued to swamp the market. A barrel of oil that sold for $1.05 in April had dropped to 40 cents by February, 1915. At the same time, the monthly production rate grew from 1,668,866 barrels of oil in April, 1914, to 8,353,020 barrels in March, 1915—from Cushing's wells alone.[22]

Meanwhile, labor problems had also reached the Cushing field. In the summer of 1914, the Boiler Makers Union, heavily involved in the construction of the steel storage tanks in the area, began to consolidate its power over workers in the oil fields. Because of a shortage of workers, there were a number of non-union men employed in the oil fields. Although the questions of wages, hours, and worker's accommodations were generally satisfactory to the union leaders, there was a problem with the funds demanded by the union for non-union members employed on tank construction. Most were recognized by the union and given "permits" in exchange for a payment to the union treasury of ten cents for every dollar earned. Nonetheless, there were many workers with no "permits," who charged that the union issued certificates to only a small percentage of those who applied, without regard to experience or ability. Finally, the construction companies announced that they would no longer participate in the "permit" policy and would employ whom they wished. At this announcement, the union men walked out on strike. But labor was plentiful, and their ranks were quickly filled.[23]

Apparently, the labor trouble had little effect on the output of the field, in that production in 1915 was more than five times greater than that in the preceding year. It was, however, becoming apparent to most of those involved that the uncontrolled expansion of the Cushing field was having a disastrous effect on the price of crude. In May, 1914, the Oklahoma Corporation Commission issued orders which began the assumption of regulatory authority over the oil industry by that branch of state government. Order No. 813 formalized an agreement among pipeline companies to share the cost of hiring a pipeline inspector and his staff. Order No. 814, although directed against the Magnolia Oil Company in the Healdton field, was an effort to fix the price of crude.[24]

The oil processing systems were swamped by the continued steady output of Cushing and other Oklahoma oil fields. By July, 1914, the daily flow of oil from the wells was an astounding 161,078 barrels, and the total amount stored in wooden and steel tanks amounted to 1,843,338 barrels. In addition, 249,418 barrels were being held in earthen storage excavations, some of which were nothing more than dammed creek beds. The daily surplus was increasing at the rate of more than 100,000 barrels, and anticipated production from wells currently being drilled

One of the many gushers in the Cushing oil field. *Courtesy of the Petroleum Publishing Company.*

was expected to add another 40,000 barrels daily within a short time. In addition, it was estimated that the untapped reservoirs within the boundaries of the pool were capable of pouring 1,500,000 barrels per month into the already glutted market and storage system. The implications were easily recognized by the area's oilmen who were in the forefront of those demanding that some action be taken to curb the gush of oil.[25]

In response to their pleas, the Oklahoma Corporation Commission issued Order No. 829 effective on July 1, 1914, which discouraged the continued overproduction of the Cushing field. The order relieved the "Texas and the Gulf Pipe Line companies of Oklahoma and the Prairie Oil and Gas companies from taking the production of any new wells brought in after the 1st of July." Exception was made for wells when lease conditions required completion by a certain date; also for existing producing offsets, short term leases complying with court orders, and several other specific conditions. In addition, the corporation commission could "determine special cases where equity or justice may fail in compliance with the rule."[26]

Order No. 829 also required the owners of "all producing properties in the Cushing district" to file a plat prior to July 10, 1914, "showing producing wells and such wells as are contemplated to be drilled under this order." Pipelines were released from the requirement of carrying oil from "any wells squibbed or shot in the Bartlesville sand in the Cushing field during the period of exemption and from taking oil from any new well that may be drilled under the exemption if it penetrated more than sixty feet of Bartlesville sand." In justifying their actions, the corporation commissioners declared that "providence has smiled so bountifully upon the producers," and believed that "they can well afford to sacrifice some . . . advantages," so that the entire industry did not "fall into contempt" because of "the destructive production of crude petroleum."[27]

The corporation commission's order had resulted from a previous meeting of oil producers in Tulsa. Recognizing that continued overproduction from the Cushing field would only continue to glut the market, the oilmen agreed among themselves that "it would be necessary to curtail the production of oil in order to relieve the depression in the price resulting from overproduction." Although the corporation commission had released the pipelines from the requirement of carrying additional crude, it had not taken the step of actually rationing well production.

The Empire Refinery in the Cushing oil field. *Courtesy of Cities Service Oil Company.*

As a result, total production remained at a fairly high level, and the price continued to fall.[28]

As production continued to grow at the Cushing field, eventually reaching its peak in June, 1915, when 8,002,500 barrels poured from the field, it became apparent that there were four major sands (the Layton, the Wheeler, the Bartlesville, and the Tucker), as well as five minor sands (Musselman, Jones, Cleveland, Squirrel, and Skinner), capable of producing oil or gas in the pool. The principal producers were found in the Layton, Wheeler, and Bartlesville horizons. The Layton Sand, encountered between 1,200 and 1,500 feet, varied from 20 to 100 feet in thickness. The Wheeler Sand, generally found at 600 to 900 feet below the Layton, was approximately 50 to 100 feet thick. The deepest of all, the Bartlesville Sand, was usually found between 350 and 500 feet beneath the Wheeler formation and was approximately 200 feet thick. In 1914 there were 166 wells drilled at an approximate cost of ten thousand dollars per well to the deepest oil-producing horizon, with an average flow of 158,183 barrels of oil daily.[29]

The Cushing oilmen were, however, in a much better position than their earlier counterparts had been. Within a short time after the field opened, the area was served by three main pipelines: The Prairie Oil and Gas Company, the Texas Company, and the Gulf Pipe Line Company. As a result, there were easily accessible markets for the crude through the Prairie Company's lines to the Standard Oil refineries located in Neodesha, Kansas; Sugar Creek, Missouri; Wood River, Illinois; and Whiting, Indiana; or to the Texas Company's own refineries at Dallas, Port Arthur, and Port Neches, Texas; and the Gulf Company's plants in Fort Worth and Port Arthur, Texas. In addition, a number of Oklahoma refineries were connected to the Cushing pool by their own pipelines: the J. S. Cosden Company in West Tulsa, the Milliken Refining Company at Vinita, and the Pierce Oil Corporation of Sand Springs. Aside from these outlets there were ten refineries located in Cushing by 1917: the Chanute Refining Company, the Brown Refining Company, the Chelsea Refining Company, the Consumers' Refining Company, the Jane Oil Refining Company, The New State Refining Company, the Cosden Refining Company, the Colonial Refining Company, the Cushing Refining Company, and the International Refining Company. Their total capacity was, however, only approximately 21,000 barrels. Although there were railroad facilities available, very little of the Cushing oil was transported by rail.[30]

Previously, oil strikes in Oklahoma had been plagued by a lack of adequate means to transport the petroleum to market. With the many facilities available at Cushing, it seemed that such would not be the case there; however, the tremendous quantity of the crude that poured from the field's wells overwhelmed the system and huge amounts were wasted. Because the lighter compounds in Cushing crude evaporated rapidly when exposed to the atmosphere, the quality depreciated when the oil was stored in open reservoirs. This was a great financial loss to oilmen; however, it was not a lesson unheeded. It was the oilmen in the area who took the initiative in an effort to hold down production and remove the glut from the market.[31]

By the end of 1919, Cushing was producing about seventeen percent of all the oil marketed in the United States and approximately three percent of all oil produced in the world. The total output from 1912 through 1919 amounted to approximately 236,000,000 barrels of oil, and Cushing proved to be one of the greatest oil fields of all time with a peak production of 305,000 barrels of oil daily from 3,090 wells. Everything about Cushing was astounding. The huge scale of the discovery did much to bring maturity to the state's oil industry. With Cushing came the money needed to finance the future petroleum development in Oklahoma, as well as the realization by oilmen of the necessity to curb the abuses of overproduction.[32]

Cushing was, however, only one of two incredibly rich oil strikes in pre-World War I Oklahoma. The other was located in the rolling hill country near the town of Healdton, approximately twenty-three miles from Ardmore, on the western border of Carter County. Production at Healdton peaked in 1916 with an estimated daily output of approximately 95,000 barrels of oil. Like Cushing, Healdton was plagued by a shortage of storage facilities for the oil that poured from the earth in such great volume and, during the period of the field's heaviest production, was the scene of one of the state's greatest oil disasters.[33]

## Chapter 8.
# Healdton

ALTHOUGH THE HEALDTON POOL was not "discovered" until 1913, the first attempt to locate oil in the area came near the end of the nineteenth century. Circa 1888, a petroleum prospector named Palmer (his first name has been lost to history) selected a location in the S½ of Section 5-T4S-R3W, in what was then Pickens County, Chickasaw Nation. Using a hand-powered spring-pole, Palmer sank a shaft to an approximate depth of 400 feet, and struck a sand which "filled the hole with tarry fluid and slopped over, letting heavy black asphaltic oil creep down the slope." Apparently, Palmer was unable to secure a lease from Chickasaw officials and abandoned the well; however, the heavy oil continued to seep from the hole for several years. Reports vary on Palmer's activities after he abandoned the well. One version has it that he left the region for a period time, returning shortly before the turn of the century. He allegedly divulged the location of his previous discovery to two other oilmen, who also attempted to secure a lease on the location, but to no avail. The other version has it that Palmer left the country and was killed in an accident in South America. Regardless of what became of Palmer, the fact remains that no commercial development resulted from his early effort; therefore, the discovery of the Healdton field was postponed for a number of years.[1]

Nonetheless, the presence of oil in the Chickasaw Nation was generally acknowledged by the late 1890s, a matter underscored by the fact that many farmers in the Healdton, Kingston, Clifton, and Cumberland areas were often forced to abandon water wells when oil was found. Over a period of five or six years, reports of these instances brought the region to the attention of numerous oil prospectors. Although oil of "good grade and quantity" was there, the sites were within the boundaries of the Chickasaw Nation where oilmen were unable to secure leases.[2]

In May, 1904, after acquiring title to 1,980 acres of land, by the Atchison, Topeka and Santa Fe Railroad undertook development of the area. H. B. Goodrich, manager of the fuel oil production section of the A.T.& S.F., after examining published reports of oil seeps and conducting a personal observation of the region's geological structure, ordered a well drilled in the NE¼ of Section 21-T3S-R2W. Using a Columbia Drilling Machine, the crew started work on the location near Wheeler, approximately twenty miles northwest of Ardmore. Before much progress had been made, however, Goodrich discovered that the subsurface formations were not stable and the well was prone to cave-ins. As a result, the site was abandoned.[3]

Later, a second well was drilled and, at a depth of approximately 150 feet, gas was encountered. Although the flow was large enough to supply operating fuel for nearly two months, this well also had a cave-in, and the hole was eventually deepened to 516 feet and converted into a water well. Another well was attempted but, at adepth of approximately 500 feet, it too collapsed, losing most of the tools and equipment as well. Utilizing a heavier rig, in October, 1905, the A.T.& S.F. drilled a fourth well to a depth of 860 feetwhere gas flowed at a rate of six million cubic feet daily. In an attempt to find oil, the well was deepened until a "pay of oil" was finally located, and plans were made to place an eight-inch casing in the hole to carry off the gas while the oil was brought up through a four-inch pipe. Unfortunately, the fourth well also caved in before the casing could reach the gas-bearing sand. Then, during the following year, the drillers determined that there were "two pays, one of gas the other of oil, separated by a few feet of shale, the oil being below."[4]

Once oil was found on a paying basis, the drilling activity accelerated until there were approximately

Robert A. Hefner, a prominent early-day Oklahoma oilman, who began his petroleum exploits in the Healdton oil field of southern Oklahoma. *Courtesy of the Oklahoma Heritage Association.*

thirty wells in the Wheeler vicinity. Most were producers with a flow of between five and sixty barrels of oil daily; the gas flowed between 750,000 and 13,500,000 cubic feet per day. There was, however, no attempt made by the railroad to market the oil produced. Nonetheless, when reports of the A.T.& S.F. discovery reached Ardmore, several individuals and companies negotiated leases in the region. Because oil was twenty-five percent cheaper than coal, the railroad gave some thought to the use of oil as fuel and of constructing a refinery near the discovery, but no action was ever taken.[5]

Sporadic exploration continued throughout the south-central portion of Oklahoma during the first years of the twentieth century. In the summer of 1906, a paying well was completed on the Arbuckle farm in the SW¼ of Section 25-T5S-R5E. Soon, additional wells were spudded "in every direction from the pool," and the oil was of such a high grade that it purportedly would burn in kerosene lamps. On March 22, 1909, the largest well of the field came in with a flow rumored to be "one thousand barrels, natural, per day;" however, it later appeared "that 400 barrels would have been a liberal estimate of its capacity." Regardless of the amount of the flow, the well excited many area oilmen, and prospectors and lease buyers from throughout Oklahoma, as well as Dallas, Denison, and Fort Worth, Texas, were soon on the scene. Although the field was viewed as a potential source of petroleum by many observers, several unavailing efforts were made to determine the limits of the new strike. It appeared that "the prolific territory" did not extend "far beyond the limits of the Arbuckle farm," and by April, 1909, there were only "four or five producing wells" in the area.[6]

Although more than a score of years had passed since Palmer had drilled his well in the Chickasaw Nation, the knowledge of his efforts had not been lost. Roy M. Johnson, the owner and editor of the Ardmore *Statesman*, listened to the stories told of Palmer and his ill-fated effort and, in 1912, decided to drill a well near Palmer's old discovery. Joining with Edward Galt and A. T. McGhee, Johnson formed the Plains Development Company and, inasmuch as government regulations dealing with tribal leases had eased somewhat, announced plans to seek a source of "grease" for the region's farming machinery.[7]

The new company found it difficult to secure leases to the property where traces of oil had polluted several water wells near Palmer's old well. The land in question was owned by two Ardmore attorneys, Wirt Franklin and Sam A. Apple, who wanted a share in the enterprise. As a result, a new corporation was formed. In addition to Franklin and Apple, the new firm, the Crystal Oil Company, included Johnson, Galt, and McGhee, who quickly secured leases on approximately six thousand acres of land. Thus far, the oilmen had managed to finance their operations with borrowed capital, and they searched for a knowledgeable driller to join the project for a share of the anticipated discovery. J. M. Critchlow, from Titusville, Pennsylvania, agreed to drill the well if lease holdings of the company were divided into blocks, in which he would receive a "checkerboard" lease comprised of alternating parcels of all the leases held by the company. In exchange for his share in the holdings, Critchlow was to use his own equipment to drill three wells.[8]

According to the story, the discovery well of the rich Healdton pool was not drilled at the original location selected, but was inadvertently moved one-half mile away, near a bog in the road approximately twenty miles northwest of Ardmore. A teamster was employed to deliver heavy timbers to the location for use in construction of the rig; however, his wagon became stuck in the mud approximately one-half mile from the stake. The loaded wagon proved too much for a single team of horses to pull from the bog, and the teamster, faced with the unpleasant task of unloading the wagon, freeing it from the mud, and reloading, decided to leave the load on the spot. The next day, a construction crew found the supply of timbers and immediately began building the rig at the altered location.[9]

Drilled in the NE¼ of Section 8-T4S-R3E, the discovery well was completed at a total depth of 920 feet in only eighteen days. It had an initial flow of between twenty five and one hundred barrels daily. Critchlow had organized the Red River Oil Company from his share of the lease holdings and, after the success of the first well, begun work on another well located on a farm owned by Mary McClure approximately one-half mile northwest. Situated near Palmer's original discovery, a "sixty-foot gusher" was brought in with an initial flow reported to be three hundred barrels of oil per day.[10]

The McClure well touched off a rush to secure leases in the area. Hundreds of oilmen from Oklahoma and Texas converged on the scene and, within two weeks, enough material had been purchased in nearby Ardmore to construct five drilling rigs. Officials of the Magnolia Oil Company were given an opportunity to examine the discovery well. They announced that it was capable of producing a large daily flow. This, of course, only added to the excitement, and when the officials of the Corsicana Oil Company inspected a sample of Healdton's crude, they also declared it to be high grade oil. In October, 1913, the Twin State Oil Company, a subsidiary of Sun Oil Company, purchased a thousand acres of lease land from the Crystal Oil Company, and plans were made to connect the field with refineries located in Port Arthur, Texas. By the end of 1913, there were fourteen wells in the field, and the output for December of that year was estimated at twenty thousand barrels of oil.[11]

In 1914 during a period of rapid expansion,

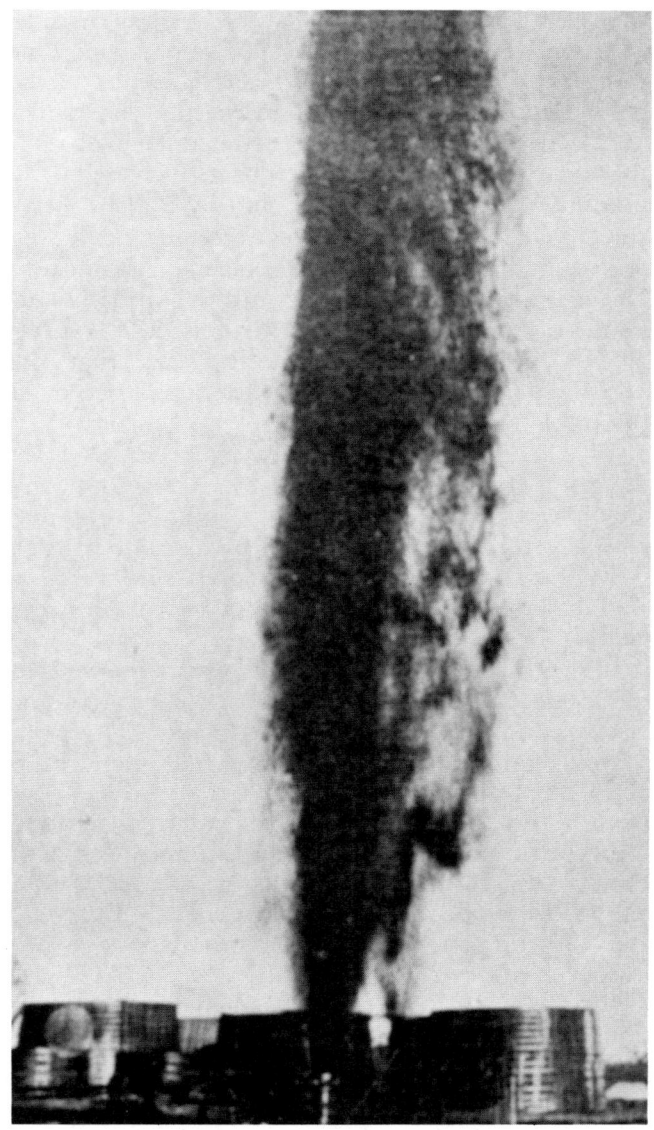

Healdton's largest producer. The Watchorn well of the Healdton field had a daily production of 5,200 barrels. *Courtesy of the Petroleum Publishing Company.*

Healdton came into its own as a major oil field. The majority of wells drilled in the pool produced approximately 100 barrels daily; however, a few flowed as high as 5,000 barrels or more per day. An ideal field for area oilmen, Healdton produced oil at such a shallow depth that the cost of drilling was almost negligible, seldom in excess of four thousand dollars, in comparison to the ten thousand to twenty thousand dollars cost per well at the Cushing field. In addition, by March, 1914, the Magnolia Pipe Line Company had completed a six-inch pipeline to the field to provide a means of marketing the crude. With these incentives, drilling continued unabated. In January, 1914, ten new producers increased daily production by 1,600 barrels, and by March, forty-three wells were in the process of being drilled by fourteen different companies operating in the area. In October, the Ardmore Refining Company announced plans to construct a 1,500 to 2,000-barrel facility in Ardmore, and it appeared that the Healdton field was destined to be one of the most profitable oil fields in the state.[12]

Their optimism was shattered, however, when on March 3, 1914, the Magnolia Pipe Line Company announced that they would pay only 70 cents per barrel for crude with less than 32° gravity. Previously, D. C. Stewart, with Magnolia's pipeline division, had told several area operators that his company would pay the current price of $1.03 per barrel for the field's petroleum. Because the majority of the wells in the Healdton field produced oil with a gravity range of between 29.27° and 33.93°, dangerously close to the cut-off figure, the figure announced meant a drop of 33 cents per barrel of oil. This cooled the excitement noticeably and the drilling activity slowed somewhat. Nonetheless, by the summer of 1914, there were 214 producing wells in the field with a daily flow of approximately forty thousand barrels, and 120 companies were in the process of securing leases for additional wells.[13]

In an attempt to solve the problem of price reduction, a number of Healdton oilmen met to form the Ardmore Independent Oil Producers Association twelve days after the Magnolia announcement. Selected as president, Wirt Franklin announced that the objectives of the organization were to implement a self-imposed reduction of crude output in order to prevent additional price restrictions. At a hearing of the Oklahoma Corporation Commission in Oklahoma City on March 13, 1914, Commissioner George A. Henshaw attempted to determine why a barrel of "Pennsylvania oil sells for $2.50 a barrel, and Oklahoma oil for less than half of that amount," while officials of Magnolia defended the actions of their company. Little was accomplished, however.[14]

Oilmen were quick to recognize the problem. By April, 1914, there were ninety-eight operating wells in the Healdton field, ninety-three others were being drilled and work was under way on thirty-nine additional drilling rigs. The pool's total output was approximately seventeen thousand barrels daily and, when compared to Magnolia's pipeline capacity of approximately ten thousand barrels daily, this left a surplus of seven thousand barrels. There was simply too much oil. On April 14 of that year, the members of the Ardmore Independent Oil Producers Association voted to cease all drilling activities until July 1, in an effort to curb further production. On April 23 a number of oilmen in the Healdton area attended the organizational meeting of the Independent Development League in Oklahoma City and urged some sort of control to prevent the continued drop of prices. These efforts were to no avail, however, for on March 23, the Magnolia Company gave notice that its pipeline would accept only four thousand barrels of Healdton crude daily. Three days later, it announced a price of seventy cents per barrel for all grades of Healdton oil. This action was followed by a reduction of an additional ten cents on April 13 and again on April 20—setting the price at fifty cents per barrel.[15]

It soon became apparent that voluntary reduction was not coping with the problem. In an attempt to alleviate the situation, the Oklahoma Corporation Commission, on May 7, 1914, issued Order No. 814, which instructed Magnolia to increase its acceptance of Healdton crude to eight thousand barrels daily by the date of May 14, then increase it to twelve thousand barrels daily by July 1 and maintain that total until April 1, 1915. In exchange, the pipeline company was released from any obligation to connect with new wells for six months, with certain exceptions. To oversee the implementation of the order, Wirt Franklin was appointed "Umpire" and Vern Calvert named "Field Inspector" for the Healdton field. Their salaries were paid jointly by the producers and the pipeline company. Although it corrected some of the problems, Order No. 814 did not set a base cost for crude, and the decline of crude markets caused by the outbreak of World War I, as well as the vast production of a higher grade at Cushing, continued to plague efforts to secure a higher price at Healdton. In the meantime, the surplus mounted.[16]

The low prices created difficult times for the oil field workers in the region. There was, however, a feeling of sympathy for the workers who put in such long hours under hazardous conditions. Ward Merrick, Sr., recalled that while he was loading a tank car with casinghead gas at Healdton on one occasion, a highwayman approached him and demanded his money. All Merrick had was fifty cents and a gold watch. Apparently, the bandit empathized with the

Healdton field was important not only for its huge volume of production but also for the beginning of one of the country's most famous oil-field service companies—Halliburton—whose name became known worldwide for its innovations in cementing and pumps. Here are the young company's blacksmith and its truck sheds, carpenter shops, and warehouse in Duncan. *Courtesy of Halliburton Services.*

young oil field worker, for he returned the fifty cents in order that Merrick might purchase a bowl of soup for his evening meal.[17]

By August, 1914, 242 wells were reported producing a daily output of 58,431 barrels of oil, far in excess of that which could be transported to markets, even if available. This huge accumulation of petroleum stored in the Healdton field led to one of the state's greatest oil disasters: the Healdton fire. On August 27, 1914, the Healdton area was struck by a violent electrical storm between three and four o'clock in the morning. During a deluge of rain, lightning struck a number of the tanks in the storage section and ignited a fire that eventually would destroy between 375,000 and 400,000 barrels of crude in a raging inferno visible "for many miles."[18]

This was not the first fire caused by lightning at the Healdton field. Two weeks earlier, on August 12, an electrical storm had ignited three 1,600-barrel and four 250-barrel storage tanks of the Crystal Oil Company and two large tanks owned by Dundee Petroleum, and although "it looked very dangerous for a time," the fire was brought under control. It was, however, an omen of what was to come, and many people recognized the danger inherent in the huge concentration of crude. Nevertheless, apparently few precautions were taken.[19]

On August 27, however, the sudden storm set off a

conflagration that sent "lurid flames of fire . . . leaping high into the air and black columns of smoke [to] envelop the Healdton fields." Witnesses reported that "there was flash after flash of lightning" during the storm, "and each stroke was followed by flames that seemed to reach back to the clouds from which the lightning came." A general alarm was sounded soon after the fire broke out, and those who "lay quietly sleeping a few moments before" rushed to the scene in an effort to extinguish the flames. Everyone "turned out" to fight the inferno and "render assistance to the field workers." Although a heavy rain accompanied the storm, the flames leaped from tank to tank during the early morning hours. Very little could be done, except to attempt to pump the oil from endangered or burning tanks with available equipment.[20]

James W. Hannah, superintendent for the Geneva-Pearl Oil Company, was in the field at the time of the electrical storm. Awakened from his sleep by the thunder, he was looking out his window when lightning struck the first large storage tank. "The blaze seemed to mount fully three hundred feet into the air," he reported. A dense column of smoke rose from the site, partially hiding the hungry flames as they reached toward the clouds. Lightning played everywhere. "One stroke after another came, and each stroke left a blaze of fire," Hannah recalled. It appeared as though "the whole face of the earth would be destroyed." Although Hannah stated that he was not frightened, the reporter to whom the oilman told his story described Hannah as "pale about the ears," with a voice that "trembled slightly."[21]

In the field, communications were difficult to maintain during the holocaust, and early reports of the disaster were sketchy. The storm dumped nearly six inches of rain on the area, and the muddy roads hampered relief efforts. The huge storage tanks burned for almost twelve hours before the heat became great enough to cause the oil to boil over the sides. When this occurred, the flaming crude poured down the sides of the tanks "in great volumes," and spread the conflagration "in every direction." By the evening of August 27, flames had engulfed three miles of the field, and thirteen different fires were raging. Although the tanks were surrounded by embankments that should have prevented the flaming oil from spreading, the intense heat caused many of the containers to explode, thereby breaching the dikes with blazing oil.[22]

By the end of the first day, flames had spread to two 1,600-barrel tanks owned by the Coline Company, one 55,000-barrel tank owned by the Corsicana Company, two 55,000-barrel tanks owned by the Magnolia Company, one 55,000-barrel tank owned by Twin State Company, one 55,000-barrel tank owned by the Crosbie Company, two 1,600-barrel tanks owned by the Red River Company, and four 1,600-barrel tanks and one 500-barrel tank owned by the Gunsburg-Foreman Company. Moreover, several earthen storage excavations were also aflame, including one holding 35,000 barrels of oil owned by the Rex Oil Company, as well as one of an undetermined size owned by the Red River Company. However, it was not just the crude that was consumed by the inferno. Several gas wells in the field also were on fire, including one, owned by the Crystal Oil Company, producing approximately forty million cubic feet of gas a day, and the roar of the gas fires could be "heard a long distance."[23]

One portion of the field, Section Four, was reported to have been "swept clean" by the spreading flames, which destroyed not only the oil storage facilities, but drilling rigs and powerhouses as well. In addition, several homes in the area were threatened by the flaming oil after restraining embankments surrounding the storage tanks gave way. Luckily, the occupants were forewarned and, when they observed the approaching flames, fled to places of safety in the rain. The scene was described as one of confusion. "Men, women and children were running in all directions, and the stoutest-hearted suffered from fright when a mass of burning oil, shooting flames high into the air, threatened to submerge the entire field." Those men away from home when the fire broke out hurried to rescue their families from danger.[24]

Although efforts to fight the flames were futile in most instances, steam engines were used to smother the flames of the flaming crude flowing through low places. This method saved a vast amount of property from being ravaged by fire. Most of the time, however, the oil was simply allowed to continue its course until creeks in the area were full of the escaping crude. In addition, heavy rains accompanying the storm weakened or breached several earthen tanks, allowing the crude stored there to escape and join the river of oil.[25]

Descriptions of the fire left little doubt that the spectacle was awe-inspiring. Judge J. T. Dickerson from Edmond, Oklahoma, at the field when the fire reached its height, declared that in all his years as a world traveler, "the Healdton fire eclipsed all other scenes he had witnessed," including many active volcanoes. When a 55,000-barrel storage tank owned by Corsicana exploded, "the fire rose to a height of five hundred feet," and the "heated gas at the bottom of the tank rose with a fury that scattered burning oil in every direction . . . [as] the gas ascended toward the heavens, while it was being consumed by the angry flames." Several other witnesses described the "red flames [that] rolled over the sides of tha tank and went in every direction, covering a large area in flames." In viewing the spectacle, Judge Dickerson "thought the writers of the Bible had selected the most awful thing they could describe as being hell for the punishment of wickedness."[26]

Several veteran oilmen on the scene expressed amazement at the violence of the electrical storm that had touched off the fire. C. B. Goddard and R. W. Coe of the Gypsy Oil Company were in the northern section of the field when the storm swept through the area. "Each rig could be seen as plainly as if it had been day," they declared. The low clouds of the storm merged with the black columns of smoke from the field and as "the light shone into the clouds, . . . this vortex was surrounded on the outside by Stygian darkness between the flashes of lightning." "Lightning played about the steel tanks with a seemingly fiendish delight," the two oilmen recalled, and the "peals of thunder were deafening as one stroke followed the other in quick succession." Moreover, the rain fell in torrents and "the soot that went up in the smoke was filtered down by the rain, and roofs were painted black . . . It was a scene that would make the strongest heart quiver."[27]

Throughout Thursday, August 27, the men in the field labored to subdue the flames. Many individuals paid "big prices" to hire men and their teams to dig fire ditches around equipment in order to divert the flow of burning petroleum. Not only were many of the oil tanks lost, but several water tanks were also destroyed. The field operations were "very much demoralized," and there were only two wells drilling in the pool at the end of the day. That evening, a second storm appeared on the horizon accompanied by lightning, and "a large number of people living in the center of the field left on account of being afraid that another storm was coming." When the fire broke out, Magnolia joined other companies in suspending operations in order that its employees could join the ranks in attempting to stem the fire. Working in relays, the men took little time to sleep as the fight continued throughout the day and into the night of August 27–28.[28]

By August 28 a report from the field stated "that the situation is pretty well in hand" and, although

Two 55,000-barrel storage tanks ablaze during the great Healdton oil-field fire of 1914. *Courtesy of the Petroleum Publishing Company.*

the fire "is still burning . . . the oil workers believe they have it under control." At 6:20 P.M., a fire devouring the third 55,000-barrel storage tank owned by Magnolia reached its height. The tank that had blazed for nearly eighteen hours exploded. Fortunately, the adjoining tank did not ignite; however, the intense heat generated by the explosion of Magnolia's tank "reduced the oil to soot and ashes."[29]

On August 29, the Magnolia Pipe Line station resumed operations, as the "fear of the further spreading of the fire now is past." At the same time, the total amount of the destruction was being totaled: in addition to the twenty storage tanks of oil consumed, several derricks had also burned and much equipment was destroyed. It was estimated that the losses would amount to more than $500,000. Most of the creeks were also flooded by oil, and the total amount of petroleum lost was estimated to be between 375,000 and 400,000 barrels.[30]

The huge fire focused as much attention on the field as did the discovery well. Many people visited the site to view the destruction and, as one journalist reported, "no one can describe what they saw." "It was a sight once seen never to be forgotten." At times, the flames were "mountains-high, and lit up all the country," and the heat from the fire was felt in

The Whittington Hotel office in Ardmore, Oklahoma. During the oil boom this was the scene of much excitement as lease traders sought the most likely drilling sites. *Courtesy of the William A. ("Mac") McGalliard Collection.*

Healdton "a mile or more away." Many tanks of rain water were full of black liquid and covered with oil. Spots of crude were found in fields more than a mile away from the wells. The conflagration convinced many industry observers that petroleum should not be pumped from the ground until means were available to carry it to market, and they argued against storing the crude on the surface. "It is better to leave it under the earth until the pipe line facilities are ready to care for it," one reporter declared.[31]

Within a very short time after the inferno, the field was again bustling with activity as the rebuilding process began. Steel was ordered for new tanks, and the debris was cleared away for reconstruction. A tremendous amount of crude had to be gathered, in that much of it had been carried away in the rain-swollen creeks. A farmer, several miles from the field, reported a large area of his land covered with oil. The catastrophe, coming at a time when prices were already depressed, was a severe financial shock to those oilmen without insurance. It was, however anticipated that those with coverage "will be able to rebuild out of the funds," and it would not be long until the wells were flowing once again. In fact, on September 25, 1914, it was reported that there were 255 wells in operation, producing an estimated 65,171 barrels of crude daily.[32]

The field continued to be plagued by overproduc-

Healdton, Oklahoma. This is the site of one of Oklahoma's greatest oil finds. *Courtesy of the William A. ("Mac") McGalliard Collection.*

tion, and prices remained unstable in spite of the efforts of the Oklahoma Corporation Commission. On September 30, the Magnolia Pipe Line Company announced a price of forty cents per barrel for Healdton crude; however, the corporation commission forbade the sale of the oil at less than fifty cents per barrel, and Magnolia did not put the new price into effect. Nonetheless, the commission's rulings and action taken by the state legislature could not overcome the downward price spiral and, at one point in August, 1916, Healdton crude dropped to forty cents a barrel. This happened at approximately the same time that the field reached the peak of its production. Soon afterward, the World War I economic expansion drove the price of crude upward, and by August, 1917, it was selling for $1 per barrel. In addition, several pipelines were completed that, coupled with increased refinery capacity, would allow Healdton production to reach an expanded market.[33]

In 1915 the United States Geological Survey mapped the subsurface of the Healdton Pool and discovered that the oil came from an anticlinal structure known as a "buried hill." The bulk of the production from the field came from a group of Pennsylvanian sands known as the Healdton Sand Zone that ranged in depth from between 800 and 1,200 feet. The "extraordinary although variable thickness of oil sands in the main Healdton field" was responsible for the long life and slow decline of crude production in the area. In addition, the Healdton region proved to be a sizable source of natural gas.[34]

Production of crude in the Healdton field continued for several years. On January 1, 1915, the average daily output stood at approximately ninety thousand barrels; however, within one year this had declined to approximately 75,000 barrels per day. Drilling activity increased in 1916. During the latter part of that year, there were more than 1,200 producing wells in the field. Two years later, in 1918, the daily production was down to approximately 40,000 or 45,000 barrels, and it continued to decline steadily until, by 1924, the annual production was less than 6,000,000 barrels from 1,915 wells.[35]

The excitement generated by the Healdton discovery prompted the rise of one of the state's most successful independents, Eason Oil Company. It started when several citizens of Marlow, Oklahoma, organized the Healdton Oil and Gas Company in November, 1913. The following year, T. T. Eason was elected secretary-treasurer and manager of the firm. Later, through a series of corporate mergers and reorganizations, he acquired control of the Oil State Petroleum Company and, in 1924, the two firms were combined to form the Eason Oil Company. In June, 1929, Eason acquired the Bolene Refining Company and, later that same year, all the various firms in which he was involved were brought under the corporate umbrella of the Eason Oil Company.[36]

Through such an arrangement, Eason was able to control all aspects of the petroleum industry and develop Eason Oil into one of the region's most active companies. Controlling the production, refining, and marketing of his petroleum products, Eason offered his goods in a wide swath across mid-America, from Texas to Wisconsin. When the Great Depression struck the country, however, he was forced to sell a portion of his holdings, and after the firm's natural gasoline plant at Crescent, Oklahoma, burned in 1938, Eason Oil divested itself of the refining and marketing aspects of its operation and concentrated only on the exploration and production phases.[37]

Throughout its development, Healdton was a "poor man's" oil field. The lease size, the low cost of drilling operations, and the short period of time required to drill a well made it possible for independent operators with smaller financial reserves to compete with the larger oil companies. As a result, several financial bases were developed by independent oilmen within the state. Moreover, at the time, the output of the pool, combined with the huge production at Cushing, helped make Oklahoma number one in the nation in petroleum production, with a total of 97,915,243 barrels valued at $56,706,133.[38]

The rapid influx of money, manpower, and materiel associated with the expansion of the petroleum industry in the early twentieth century created a "boomtown" era in Oklahoma similar to earlier mineral rushes in the American West, and this phenomenon was to have a tremendous initial impact as well as an incontrovertibly lasting influence on the development of the new state.[39]

## Chapter 9.
# Oil Boomtowns

DURING THE FIRST THREE DECADES of the twentieth century, the sustained impact of "black gold" in Oklahoma not only proved to be an important sociological, economic, industrial, and political factor in the development of the state, but had national importance as well. The lure of vast fortunes to be made in petroleum, bolstered by the inherent optimism of the wildcatter, played a major role in the transformation of picturesque, quiet farming communities into expanding, bustling oil towns. The development of the majority of Oklahoma's oil boom communities can be divided into five phases: the excitement generated by the discovery of a potentially valuable reservoir of crude; the frantic scramble to lease the most promising land; the tremendous influx of oil field workers into the community during the process of drilling and development; the mushrooming growth of small towns and communities, resulting from a steady stream of men and women from extremely diverse backgrounds intent on reaping the riches of the oil find; and in most cases, the eventual return of the boomtown to a small, tranquil farming community (even total obscurity) once the boom fever had run its course.[1]

With the possible exception of Oklahoma City, boomtowns in Oklahoma were generally characterized by their development in a sparsely populated rural setting. Large sums of investment money brought by oilmen, speculators, and promoters into an area previously dependent upon a farming economy was the most noticeable change wrought by an oil strike in the vicinity. When it is realized that petroleum extracted from the earth in Oklahoma between 1901 and 1933 approached five billion dollars in value, the vast infusion of money into the state becomes readily apparent and its staggering effect upon the economy of Oklahoma during that period can be more easily understood.[2]

Employment and the opportunity to earn high wages caused not only many rural residents to forsake the farm, but also brought in large numbers of young, healthy men from nearby communities and adjoining states. Without the stabilizing influence of their families, these men were often plunged into an astounding cross section of humanity. "Millionaires, laborers, hoboes, gamblers and prostitutes," as well as lawyers, hucksters, speculators, and men of modest means flocked to the boomtowns of Oklahoma with the hope of sharing in the proverbial pot of (black) gold. With the discovery of the prolific Cushing field in 1912, "men with surface-pipe boots and swedge-nipple pants" descended on the town of Cushing, overflowing the hotels, boarding houses, shanties, and tents. "Closemouthed prospectors, promoters, . . . scouts, scalpers, and scavengers" were everywhere. Some were "tramps" while others were "men you would not be afraid to take home to meet your wife." A large percentage of oil field workers belonged to a transient culture that followed the booms throughout the United States, flowing with the crude from one oil strike to another and enduring the nomadic life that the profession demanded of them.[3]

Most of the men who poured into Oklahoma during the oil boom period had little or no previous experience in the oil fields. J. G. Beard, founder of the Beard Oil Company, originally came to the state as a representative of the Dierks Lumber Company of Nebraska to oversee the construction of wooden derricks for which his firm supplied the lumber near Duncan, Oklahoma. Arriving at a time when excitement was running high over the discovery of the Empire field, Beard abandoned the lumber business and entered the petroleum industry. By 1921, he had brought in his first well and laid the foundation for the future of the Beard Oil Company, still in opera-

Oklahoma's oil boom was the culmination of earlier mineral rushes to the American West. *Courtesy of the Bartlesville Public Library.*

tion under the direction of his son, William M. Beard, more than half a century later.[4]

The economic opportunities offered by the new industry enticed many family men to join the rush to the state. A machinist in the Oklahoma oil fields recalled that after his marriage in 1920 and the birth of a daughter six years later, he "didn't want to be like a lot of the boys, [and] leave my family stuck off someplace while I worked in another town." As a result, they accompanied him to the Earlsboro field in central Oklahoma, where he could work for "good money . . . $1.25 an hour, with time and a half for overtime, and all the overtime a man could stand up to." This amounted to approximately $125 per week.

Once there, they found that the oil companies had purchased all the timber in the Earlsboro area, and they had considerable difficulty in securing enough lumber from a nearby community to construct a house. But, after shipping in as "much as we could afford to build us a little three-room 'shotgun' house, the kind that runs straight from front to back," the family established a home.[5]

Working "as long as I could keep my eyes open," he stated that he sometimes fell asleep "standing up against the lathe," while putting in as many as twenty hours a day. He returned to the house "long about midnight or early in the morning," and, in an attempt to avoid the ever-present threat of typhoid, "cussed till I was black in the face" about having to haul water for the family from a well "ten or fifteen miles away." Because he did not want his wife among the more unsavory characters on the streets, he also drove to a nearby grocery store for food and supplies before obtaining his badly needed rest. "When I finally did get to sleep, I'd lay there like a chunk of steel as long as I could; sometimes twenty or thirty hours at a stretch." On many occasions, however, he had "been asleep only an hour or two [when I] . . . had some driller come by and shake me awake to get me to turn him out a pipe joint." Such a schedule left little time for family life.[6]

Hours on the job were long and demanding, and roughnecks and roustabouts worked hard. Time was money in the oil business, and "time wasted was money lost." One rig-builder who traveled throughout the Oklahoma fields described the hard, tedious labor needed in the construction of wooden derricks. He recalled that "working then was pure hell . . . we had to hit a hard lick every time we raised our hands, and keep it up all day long. I've worked till my shoes would squish every step I took with the sweat that'd run down in them," and at night, because his hands were cramped from "holding a rig-hatchet or a crosscut saw all day long," he had to "take one hand and bend the fingers down to grab something small, like a match."[7]

The work was also extremely hazardous, and he related a serious accident that involved the dangerous machinery used on rigs in the early part of the century. While in the process of skidding a rig to a new location, "Me and another fellow were standing by the exhaust pipe [of the steam engine used to power the drilling equipment] knocking the sheeting off and fixing to put skids under the engine house." "Steam's always kept up on a rig," he explained, "specially if there's a job like that going on." Unaware of the work being conducted near the exhaust pipe, a driller handling the boiler "fed it to her." "That live steam blew outa there and scalded me from my waist to my heels." "I couldn't do a damned thing," he said, "I couldn't even holler, just dropped on the ground and laid there, trying my best to holler or do something to relieve myself." The steam was shut down when a co-worker spread the alarm, and the crew carried him to his room in a nearby town. But, "by the time the doctor got there, I had big blisters raised up under my thighs and the calves of my legs that looked like footballs, only bigger." The physician "gave me a shot to relieve the pain and then took out his knife and ripped the blisters open. A

Work was dangerous in the Oklahoma oil fields, as is indicated by this derrick, which collapsed under the weight of heavy drill pipe. *Courtesy of the Oklahoma Publishing Company.*

half-gallon of old blister water poured outa each one of them blisters." "The old doc smeared me with salve," and then left. In fourteen days of recuperation, he received $20.00 per week and the contractor paid the doctor's bill.[8]

The lives of the oil field crew were in constant jeopardy and courted disaster. One tool pusher recalled that the crew worked after dark by the light of black dogs—"kerosene drilling lamps that look something like bombs suspended from the derrick"—and on one occasion, the bit penetrated a pocket of gas and extinguished the lamps. He volunteered to relight the lamps while the crew continued to work as best they could under the circumstances. "I climbed up there and hooked my leg around one of them wooden derrick braces and struck a match easy as I could," he recalled, but "I didn't get but a few of 'em lighted till we hit some more gas down in the hole, and whoosh! That whole damn rig went up like a blowtorch." He was severely burned, and the rig, completely destroyed by the fire, was later abandoned.[9]

Stanley Learned, at one time chairman of the board

for Phillips Petroleum Company, started his career in the oil industry in 1924 as a surveyor for the engineering department of the company at $175 per month. According to Learned, the Cromwell field in southeastern Oklahoma had one of the highest accident records in the state because of the high concentration of natural gas in the area. Sent there to survey the location for prospective wells on one occasion, Learned recalled that he drove "twenty to thirty miles" each night to sleep in Wewoka on the advice of his supervisor because of the impending danger of escaping gas. It made for very long days and short nights but, once he had observed the dangerous situation at the Cromwell field, Learned willingly complied.[10]

Commuting to and from work was also often a chaotic experience. William M. Barber, one of the first employees of Phillips Petroleum Company, described the problem of getting to and from the oil fields near Bartlesville prior to the 1920s. The men traveled to their jobs in all manner of conveyances: by horse and buggy, or horseback, or on foot; and because of the lack of roads, they just "struck out across the prairie."[11]

Ralph L. Stewart, another early-day employee of Phillips Petroleum Company, was a truck driver in the Bartlesville area during the boom. In describing his experiences, Stewart related that it was not uncommon to work seven days a week to maintain his rigorous workload. The constant traffic of heavy machinery being moved to various locations on the oil fields kept the existing dirt roads in a deplorable state. A twenty-four mile round trip often required a full day and, on many occasions, only two trips could be completed in three days. Inclement weather often rendered the roads impassable.[12]

Those men with families were fortunate to have the luxury of food and shelter but, for the majority, these items proved difficult to obtain. One operator of an "eating house" during the Oklahoma City boom period recalled "Why, I've seen the time when this place wouldn't hold the men that wanted to eat with me, and this is big enough to feed a hunnerd men," she declared. "They'd drive for miles to get here, and come crowding in, shoving and pushing, and all of 'em hollering at Lovie [the waitress] to give 'em personal attention." "That kind of stuff would keep up for hours, with the cash register dinging like a patrol wagon bell," and by the time the noon rush was over and the cooks had prepared the food for the dinner meal, "why it'd be evening again, and time for 'em to come back." "But eat! . . . I never saw anything could eat like those men . . . ," she reminisced, "They would set down and eat a half-dozen eggs, a side of bacon apiece, four cups of coffee, and push all of that down with a loaf of bread and a couple of pieces of pie."[13]

Not all the people who followed the oil booms worked on rigs, however. Parts salesmen and tool suppliers relied upon sales of the massive oil field equipment for a living. Sam Barkley, an early-day parts salesman, recalled that it was not an easy occupation, but the money was good. In just three months at the Oklahoma City field, his sales totaled $12,000, $42,000, and $84,000 respectively. Barkley, just starting out in the business in 1912, was sent to Cushing to manage a "store," and remembered "the place was on a hell of a boom." "Everybody was hog-wild." "Every room in town was taken; there was a line a block long in front of every restaurant all day; the drilling was spreading out in every direction," he declared. "I had to pitch a tent to sleep in at the edge of town," and "I ate in a tent which an old farm couple had thrown up nearby," he continued. "They had come in from the country with lots of canned vegetables and fruit and home-cured meat" but "they spread a better meal than you could get for two dollars, even after you got a seat in a restaurant downtown, and all for thirty-five cents." Barkley offered the couple "six-bits for a meal like that," but "the old man shook his head, and said thirty-five cents was enough." Indeed, he was fortunate to find such accommodations, for later, the president of the company for which Barkley worked "couldn't get a room in the hotel at any price, and he had to sleep with me in the tent." The following morning, "somebody had thrown up a big tent to stable in" next to the restaurant, and the "air was full of manure-dust," but there was not another place to eat in town.[14]

Workers in the oil fields were pretty much on their own. "We didn't have compensation or workers' insurance in those days," recalled one pumper, "had to pay our own doctor's bills, too." When a "man got hurt it was his own tough luck," unless he was fortunate enough to have his employer pick up the bill. As a result, injuries were often treated by the workers themselves. "All we had to do was throw a rag 'round it, maybe daub it with some medicine if we had any at home, and go on back to work," the worker reminisced. Even when available, medical aid was not the best. To treat a thumb smashed with "a big S-wrench," the pumper recalled that he simply "tore me off a piece of shirt and wrapped it up, and doctored it that night when I got home." "Four or five days later it got to hurtin' pretty bad, so I went on in to town and got me a doctor to fix it up. An old Army doc, tougher'n a 20-penny nail, didn't do nuthin' but take his scissors and cut off all that proud flesh, not givin' me a thing to help the pain." "He was a good doctor," the pumper asserted, "but he'd just been in the Army too long."[15]

Most early arrivals in a boom secured lodging in local farm homes. The accommodations were, however, far from luxurious. These farmers, turned hostelers, offered beds that were nothing but a "tick stuffed with prairie grass" propped "upon store boxes." Space was money, and any available spot was utilized for improvised bedding. In describing such an arrangement, E. L. Malone complained that in 1904 the upstairs sleeping area was "cold and so cramped that he could scarcely stand up without bumping his head on the rafters." One meal, usually consisting of bread without butter, hash, "some fat-pork and some cow's liver," was also provided by the farmer's wife. The going rate for such accommodations was $5.00 per week.[16]

Opportunistic entrepreneurs reacted immediately to news of an oil strike by constructing cheaply built dwellings along a route considered most likely to be in proximity to the activity. These structures were quickly leased to individuals who offered a myriad of services to oil field workers. Restaurants were open around-the-clock. Because sleeping facilities, no matter how meager, were always in short supply, many buildings served a dual purpose. During the height of the Cushing boom, pool halls were converted to hotels after midnight in order that workers might sleep on or under the pool tables, on chairs, or anywhere space allowed at prices ranging from fifty cents to a dollar per night. Cots were rented by the night or by the week in hotels on a first come, first serve basis. It was not uncommon for three individuals to occupy one room in shifts, paying from two dollars per night to thirty dollars per week. A worker often climbed into a bed still warm from the previous occupant and, when linen supplies were exhausted, a single blanket was all the bedding available. E. A. Smith, founder of the Service Drilling Company, found people "sleeping in the cellars and basements," during the peak of the Bristow boom. "They were sleeping in any kind of shacks they could garner or get accommodations for them." Cots overflowed into the alleys, and barber chairs as well as theater seats frequently served as beds. Yet in some instances, this was not enough, and men camped on the ground.[17]

Shotgun houses in Seminole. Such houses were typical throughout the Oklahoma oil fields. *Courtesy of John W. Morris.*

Aware that their services were in great demand and commanded high wages, carpenters from surrounding areas swarmed to the site of a new boom to construct temporary shelter for the oil field workers. These dwellings were so hurriedly and haphazardly constructed, it usually required only three men working a single day to complete a one-room dwelling with dimensions of twelve by twenty feet. Batten was placed on the exterior to cover the cracks between the boards, and beaverboard, decorated with wallpaper of random design, insulated the walls of the interior. The roofs were covered with tar paper, and large wooden wheels from the oil field were sometimes added to give a dome-like effect to the simple structure. As needed, rooms were added directly to the back of the building, and, because a shot fired from the front door exited without obstruction through the back door, they were popularly known as "shotgun" houses.[18]

Many families suffered from illnesses resulting from improper sanitation while living in the crudely built dwellings, but conditions improved as development of the oil field progressed, and many oil companies provided housing for their employees. Neat rows of cottages gradually replaced the shacks and tents, and in many instances, were painted a color that symbolized a particular oil company: grey identified the homes of employees of the Gypsy Oil Company, later part of Gulf; grey trimmed with red, the property of Amerada; and gold represented the

Oil boomtowns were also characterized by impressive mansions built by fortunes made in oil. Such a structure was the W. E. Grisso home in Seminole. *Courtesy of John W. Morris.*

property of Carter Oil Company, a subsidiary of Standard Oil. As housing conditions continued to improve, residents of these oil communities landscaped the lawns with flowers and shrubs. In addition, they supplemented their income with livestock and vegetable gardens.[19]

Not all men in the fields were transient workers who followed the boom activities of new oil finds but, as described by an early-day reporter, the mass of humanity that generally came on the heels of an oil discovery was "gathered from the four corners of the earth, and while there are a lot of good law-abiding men among them, there are many desperate characters who hesitate at nothing to accomplish their purpose." Penniless, many were lured to the boomtowns under the mistaken impression "that work is abundant and that money grows on trees." Some were unable to find work, and a number turned to crime or became dependent on the public dole for food and shelter. Such individuals incurred the scorn of the community's permanent inhabitants.[20]

Those who found work were paid between $6.00 and $15.00 per day, a fairly high wage for the time period, but spent their money freely to relieve the monotony of the work and wretched living conditions. They generally resided in a separate section of the town which housed those establishments catering to the whims of the oil field workers. Prostitutes and dance halls abounded, gambling flourished, liquor flowed freely, and every known method was utilized to separate the men from their money. Such a "socially uprooted and morally apathetic environment" was a "ready breeding ground" for crime and violence.[21]

Men fought "to the finish" in the mud and dust on Tiger Creek Avenue, center of the roughnecks' district during the Drumright boom. "By merely turning one's head," declared an eyewitness, "it was an easy matter to see a dozen fights in full blast at the same moment." William B. Osborn, Jr., whose father was involved in the development of the Greater Seminole field, recalled that his mother usually took him through the back door of the bank building in Sasakwa, where his father maintained an office, to avoid the ever-present brawls on Main Street. To keep him entertained, she placed him in a window to watch the activity. "The Bowery" was the name applied to the row of saloons, brothels, and gambling dens in Kiefer, and "Bishop's Alley," a well-known district which occupied four square blocks in Seminole, was perhaps the worst of all.[22]

William T. Payne, pioneer Oklahoma oilman and founder of the Big Chief Drilling Company, recalled that he was forced to abandon a hotel room in Seminole on one occasion in order to sleep because of the women continually running through the halls. As an alternative, he built a small "bunkhouse" which he shared with his truck driver. Next door, however, was the Rainbow Dance Hall, one of the fifty-six so-called night clubs in Seminole which ran "wide open" twenty-four hours a day. What Payne called a "forty-niner dance hall," was a establishment where a man purchased a twenty-five cent ticket that entitled him to a dance with one of the girls. The girls kept a dime for each ticket they accumulated, and the house received fifteen cents.[23]

After putting in twelve to eighteen hours a day at the oil field, Payne was usually too exhausted to frequent the Seminole nightspots, but he did make one trip to the Rainbow Dance Hall. While dancing with a young woman who had caught his attention, he was astonished by her "rough" language. Later, while standing by the side of the dance floor, Payne was informed by his companions that he had been dancing with the girlfriend of the notorious gangster, "Pretty Boy" Floyd.[24]

Payne's excursion to the Rainbow Dance Hall was the exception rather than the rule. On most evenings, he retired to his bunkhouse for a few hours of sleep, provided there was no trouble at one of his wells. As a precaution, his truck driver fired a pistol into the air at night to warn anyone intent on doing them harm

Seminole's jail at the onset of that county's spectacular oil boom. *Courtesy of the Oklahoma Publishing Company.*

that they were armed. Payne said Seminole, last of the big oil boomtowns in Oklahoma, was reputed to be one of the roughest cities in the United States. After his work was completed there, he was glad to "move to paradise," the Oklahoma City field.[25]

A small creek, covered with crude which had escaped from the field's storage tanks, flowed through Kiefer, in Creek County, near one of the town's largest establishments, the Mad House Saloon. Bodies were frequently fished from its turgid depths because it was a popular site to deposit murder victims. It was reported that twelve dead men were found in the murky waters within a two-year period. Other "convenient" disposal sites for unwanted bodies included the hundreds of oil storage tanks that dotted the landscape around the boom area. A tank behind the Mad House Saloon yielded seven skeletons when drained. Violence trailed oil strikes, and killings were commonplace in some of the worst sections of the towns. A number of establishments made an attempt to curb the carnage—the North End Club of Seminole displayed a prominent sign which read "No Firearms or Knives Aloud [sic] Here"—but they had little or no effect on the situation. Life was cheap in a boomtown.[26]

Earl E. Emerson, superintendent of schools at Cromwell during the heyday of that town's oil boom, recalled stopping at a service station in 1928, while en route to assume the job. He fell into a conversation

A portion of the infamous "Bishop's Alley" in Seminole. *Courtesy of the Oklahoma Publishing Company.*

with Henry Beck, the owner about the town and what he might expect when he arrived. "Emerson," Beck declared, "I've given you credit for having good judgment until today, but now I think you're a damn fool for going into a mess like they have at Cromwell." The new superintendent had reason to give some thought to his friend's advice when, upon taking occupancy of the office, he found an agreement signed by the school's teachers promising "not to have or carry concealed weapons of any kind on or about the school grounds."[27]

Mayhem associated with oil booms in Oklahoma was not limited to the oil fields, however, and one of the most sensational crimes of the era was the abduction of Oklahoma oilman Charles F. Urschel on July 21, 1933. Having married the widow of pioneer oilman Tom Slick, Urschel was the trustee of the Slick estate when he was kidnapped from his home at 327 Northwest Eighteenth Street in Oklahoma City. Urschel and his wife were entertaining friends when George "Machine Gun" Kelly, Albert L. Bates, and Harvey Bailey burst into the house and spirited away Urschel and W. R. Jarrett, a guest.[28]

Jarrett was released shortly after the abduction by Kelly's gang with instructions to return to Oklahoma and deliver a $200,000 ransom note. Two hundred thousand dollars was quite a sum in the 1930s, and the crime shocked the state's citizens. While Jarrett made the kidnappers' demand known, Urschel was carried southward into Texas, where he was held at a farm belonging to Kelly's wife's parents. Kept in a shed, he managed to make note of the time that a commercial airliner flew overhead each day. Later, after payment of the ransom was made, Urschel was released unharmed near Oklahoma City and relayed this information to the authorities, who used it to locate the outlaw's hideout and apprehend the criminals. Although Kelly and his gang were captured, the kidnapping of Urschel touched off a wave of apprehension among the oilmen of the state and brought about a demand for an end to the lawlessness associated with the oil fields.[29]

Although Article I of the Oklahoma Constitution, adopted in 1907, specifically prohibited the "manufacture, sale, barter, giving away, or otherwise furnishing . . . intoxicating liquors," and the Eighteenth Amendment of the United States Constitution, adopted in 1919, expanded the regulation of alcoholic beverages to a national scale, these laws had little effect on the liquor that flowed freely in the boomtowns of Oklahoma. The problem of enforcement simply overwhelmed what law officers were available, and bootleggers plied their trade with little difficulty. In Healdton, for example, fifteen officers were charged with overseeing twenty thousand people, with generally only two deputies available at any one time for duty in the oil fields. Corrupt officials often hindered the effort by closing their eyes to illegal activities. Liquor was, in many instances, considered a necessary part of doing business. "You can't go out in the field and talk to a man for half an hour, and sell the stuff," Sam Barkley stated, "you've got to get chummy with him, throw him a party in the hotel with plenty of Scotch and plenty of girls . . . some way to get to be his friend." "If he likes you," the parts salesman continued, "he'll come through with a swell order from your [supply] house."[30]

Despite their notoriety, the lawless element of an oil field community was in the minority and a majority of the workers were law-abiding citizens, who only occasionally violated the law. One "Godfearing" individual discussed his reluctance to move to Seminole during the boom: "Our preacher had been to Seminole, and he told us of the lust and drunkardness and murders," he recalled. "It was not a fit place to bring your family." "I felt the temptations in an oil field would be too strong for . . . [my children]" inasmuch as "youngsters are weak in resisting the ways of the flesh," But "we came to Seminole and continued leading Christian lives." "You can guess my gladness of heart," he continued, "when I found thousands of other folks living in God's grace." In addition, most oil producers "were all prohibitionists" and realized that the wide-open atmosphere of the boomtowns hampered their worker's ability to do a good job. A man who appeared at work in a drunken condition could cause a costly or dangerous situation around the drilling rig. Oil producers in the Cushing and Drumright region jointly raised five thousand dollars to provide for rigid law enforcement. The efforts of the oil producers, working with businessmen and those individuals who wished to establish a law-abiding atmosphere in which to raise their families, ultimately ended the worst of the violence and crime in the boomtowns.[31]

To further offset the more unsavory character of the community, churches in nearby towns established congregations in the boom area and, if enough volunteers were located, constructed a building in which to hold services. There were two rural churches near Three Sands, a field just west of Ponca City, attended by many workers; but by 1923, the Holiness Mission had also been established, followed shortly by the Methodist Church. Some of those individuals engaged in religious work were as rough as the flock to which they ministered. A. L. Snyder, a one-time cowboy, was one of the earliest preachers to arrive in the area of Three Sands. A burly man, Snyder spent his first night in the area locked inside his car with a revolver, ready to discourage any would-be hijackers. He obtained employment in a restaurant, but soon purchased a general store, in which he also held religious services. After collecting nearly five hundred dollars, he purchased one of the worst nightspots in town and converted it into a mission catering to the oil field workers. Performing his ministry in the heart of the lawless district at Three Sands, he was not opposed to physically ejecting anyone who attempted to disrupt the services and encountered little opposition after throwing out a few "rowdies."[32]

"Scottie the Baptist" conducted services in a downtown building in Seminole until he had raised enough money to construct a church building, and "Sky Pilot," a Methodist minister, attempted to cleanse the town of Drumright of vice with the aid of his revolver. A number of lay ministers worked on the rigs during the week and held services on Sundays. J. M. Critchlow, the discoverer of the Healdton field, was a preacher and always halted work on his drilling rigs each Sunday. Wells were sunk on church property in some instances, and often the churches themselves became involved in the drilling efforts. One driller acquired a lease to two acres of church property in Tonkawa but, when he began to drill in the church's cemetery, was dissuaded from his efforts by armed church members. Many oil companies contributed financially to the construction of church buildings, and church members often provided meals for the oil field workers to raise funds for religious activities.[33]

As the churches became more firmly established with restoration of law and order, rowdies were jailed or driven from the towns. The problem was compounded in some instances, however, by the lack of suitable facilities to incarcerate the undesirables. At

Company housing such as these structures in South Ponca City marked an end to the boomtown atmosphere in Oklahoma's oil communities. *Courtesy of Allan Muchmore.*

Three Sands, prisoners were often chained to telephone poles because no other facilities were available; a special wire stockade was constructed at Newkirk on one occasion to hold the results of a raid by law officers; and a hastily erected wooden stockade, the "Bowlegs bastille," was easily torn down by those inmates sober enough to escape the town. Jake Simms, the police chief at Seminole, compensated for a lack of facilities by utilizing an empty boxcar to hold the drunks, gamblers, and toughs who were arrested nightly in the community. The boxcar proved insecure, unfortunately, and was rebuilt to more exacting standards with provisions made to secure the men on one side and women on the other. First Simms fastened leg irons to the prisoners, and then he connected them to a long iron bar attached to the exterior of the boxcar.[34]

Once the more reputable portions of society eliminated the gambling dens, brothels, and saloons, the character of the boomtowns underwent a drastic metamorphosis, and schools and other "civilizing" social aspects appeared. At first, the education of children was neglected or they attended the few rural schools available in the surrounding countryside. The original school at Three Sands consisted of a single room and one teacher. When oil was discovered, however, there was such an influx of people into the area that the enrollment rose from 14 to 106, with 3 students occupying the same desk. The tax burden to support the schools was borne at first by the area's

Bowlegs, Oklahoma, one of the many oil boomtowns of the Greater Seminole field. *Courtesy of Cities Service Oil Company.*

Seminole, Oklahoma, one of the most famous oil boomtowns. *Courtesy of Cities Service Oil Company.*

farmers, but eventually, the oil companies paid their share and compensated for the increase in enrollment resulting from the children of their workers. Nevertheless, the three percent production tax levied on the petroleum companies needed to be supplemented by outright donations in many instances. The situation gradually improved and more educational facilities became available. The establishment of an adequate school system did much to bring a stabilizing social influence to the boomtowns, for it attracted more workers with families to the strikes.[35]

Once the process had begun, many oil companies attempted to cultivate a more stable environment in order to make their communities more attractive. In addition to providing housing, the companies often offered organized sporting events, card clubs, and dances for the employees. Newspapers also sprang up in the oil boomtowns which, in many cases, acted as the conscience of the community. As the business section of the town grew and became more diversified, streets were improved, municipal services were expanded, and the rougher elements either went "their way or reformed." Conditions that had discouraged many oil field workers from bringing their families with them began to disappear and, in their place, were the institutions necessary for social stabilization. The end of the boomtown was imminent.[36]

Boomtowns in Oklahoma can be classified into two categories: those dependent upon a single pool of petroleum, such as the Kiefer field, near Supulpa, or the Three Sands field, near Ponca City, and those dominated by several pools and taking an active role in the processing and shipping of crude, such as at Tulsa or Cushing. The former generally disappeared once the boom was over, while the latter often became thriving communities. If the field was small, most of the residents left when drilling activity ceased, and the town faded into obscurity. If the find was sufficiently large, however, those who made the initial discovery were quickly replaced by stable family men.[37]

A product of one of the greatest mineral rushes to the American West, the boomtowns of Oklahoma bore all the distinguishing characteristics of previous scrambles for wealth in America: a special attraction to those individuals in search of riches; unlimited potential for the ones lucky enough to stumble upon a find; rapid population shifts; settlements which were born and died as a result of the quest for wealth; hard work; a hoard of camp followers preying on the oil field workers; hardships; and the gradual replacement of the less stable early arrivals by men with

Mud—the curse of all boomtowns.
*Courtesy of Cities Service Oil Company.*

families and an established social order. The oil boom in Oklahoma at the turn of the century lacked nothing: glamour, excitement, adventure, violence, and riches. It gave birth to a special breed of men willing to risk all on the turn of a bit. Fortunes were made, lost, and, in some instances, made again. In the wake of the gold rush to California and the subsequent stampede for silver in Colorado, the quest for oil in Oklahoma ranks among the most romantic and flamboyant eras in American history.[38]

## Chapter 10.

# The Roaring Twenties

By the end of World War I, the petroleum industry in Oklahoma was considered "big business." Crude from the Glenn Pool, as well as the Cushing and Healdton fields, had pushed the state into the forefront of national oil production. The conflict in Europe had ended, and the American petroleum industry was entering a new phase of development spurred by the growing importance of the automobile in American society, the need for a higher-quality gasoline to fuel the newer engines, and an ever-increasing output of crude from the nation's oil fields. The automobile and its need for improved oil products provided a "dynamic dimension to the reshaping of the American economy" in the 1920s. Although drilling activity had languished to a degree during the war years, it took a dramatic upturn once the hostilities had ceased, and in 1920 alone, 34,000 new wells were drilled in the United States. Oklahoma continued to provide much of the crude utilized during the decade, and between 1920 and 1929, it was estimated that the state produced 1,843,477,000 barrels of petroleum and 2,378,014,200,000 cubic feet of natural gas, at a total estimated value of $3,587,139,500.[1]

In addition, between 1920 and 1929, the number of automobiles registered in the United States increased from 9,231,941 to 26,501,443, and petroleum products shipped to foreign markets rose from 77,279,000 to 152,771,000 barrels. During this time span, new pools were opened over the entire state. So great was the production from the more than forty producing fields within the Sooner State that, in 1929, a total of 251,269,400 barrels of oil flowed from Oklahoma's wells at a daily average of 688,400 barrels—24.01 percent of the nation's total oil production in 1930.[2]

One of the first regions in Oklahoma to undergo a resurgence of drilling activity was the old Osage Reservation, now Osage County. The initial discoveries of oil in the area had been made in the eastern part of the county, and it was not until December 9, 1918, that the western portion, west of Range 8, was opened to public leasing. A number of oil companies, including Gypsy, Guffey-Gillespie, Marland, and Carter, secured drilling rights to several scattered tracts in ranges 5, 6, and 7. Soon, both Gypsy and Guffey-Gillespie began drilling south of the community of Foraker, in the northwestern sector of the county; however, the test wells proved to be a disappointment. Nonetheless, the Gypsy Oil Company drilled another well north of the Foraker community on the White Horn allotment where oil was found at 2,880 feet. Although other tests were drilled in the western portion of the county, most did not find commercial quantities of oil, and many oilmen became reluctant to attempt further drilling operations in the area.[3]

Unperturbed by the previous failures, Frank Fisher with the Carter Oil Company undertook a test well in Section 9-T26N-R6E which, although plagued by a series of mechanical difficulties, eventually became a "gasser." Fisher then began an offset well which was brought in as a a small oil producer in September, 1920. In May of that year, however, the Marland Oil Company, under the direction of Ernest W. Marland, brought in a producing well on the Bertha Hickman farm located in the SE¼ of Section 36-T27N-5E. They found crude in what they first thought was the Burgess Sand, but which later proved to be the Bartlesville Sand, at a depth of between 2,949 and 3,001 feet. The well flowed 150 barrels of oil per day. Later that same year, the National Exploration Company drilled a well in the SE¼ SW¼ of Section 4-T26N-R6E that flowed 3,450 barrels of oil during the first twenty-four hours of production.[4]

E. W. Marland—United States congressman, governor of Oklahoma, and pioneer oilman in the state. *Courtesy of the Oklahoma Heritage Association.*

The Burbank field became the center of attention of the oil industry in the north-central portion of the state. Consequently, many large companies moved into the area and proceeded to extend the pool in a methodical manner. At first, geologists believed that the Marland and Fisher wells had discovered two separate pools, but it soon became apparent that they were in the same field, and the distance between the two wells, Fisher's in Section 9-T26N-R6E and Marland's in Section 36-T27N-R5E, indicated a large accumulation. Drilling operations spread in every direction from Marland's discovery well in an effort to determine the boundaries of the pool; however, major activity was centered to the north.[5]

Under the direction of the Department of the Interior, thirteen lease auctions were held in Pawhuska to dispose of the Osage oil holdings. By the time the sales were concluded in March, 1927, a total of 532,876.75 acres of land had been leased. The per capita payments made to the Osage Indians were astounding. In 1920, a total of $18,032,610 was paid to the Indians in oil royalties or bonuses, a headright payment of $8,090 for each tribal member. Three years later, the per capita payment was a staggering $12,400.[6]

Because of inclement weather, one of the auctions conducted by "Colonel" E. E. Walters on February 3, 1920, was held at a Pawhuska theater instead of under the famous "Million Dollar Elm." As reported by H. L. Wood in the *National Petroleum News* and *Oil Weekly*, Walters handled the sale with a "booming voice, expressive gestures, and entertaining humor."[7] At this auction alone, Phillips Petroleum Company, the largest purchaser, paid $495,000 for seven 160-acre leases. Tract No. 75, the NE¼ of Section 20-T22N-R10E in the Wildhorse Creek District, attracted a winning bid of $222,000. At the same auction, the Carter Oil Company paid $154,000 for one quarter-section lease, and other 160-acre tracts went for as much as $70,000. By the end of the day, the total sales amounted to $3,102,700. As excitement grew over the Burbank discovery, a total of 102,192 acres of land were leased during a series of three auctions from December, 1921, to June, 1922. One 160-acre tract, located in the NW¼ of Section 14-T27N-R5E, drew a winning bid of $1,990,000 from the Marland Oil Company. In March, 1924, the Midland Oil Company paid $12,437.50 per acre for the NW¼ of Section 14-T27N-R5E in the Burbank Pool—the largest bonus paid for acreage in Osage County. Paul Endacott, who became president of Phillips Petroleum Company almost half a century later, recalled the excitement generated by the sale of the Burbank leases. Just out of college and newly employed by Phillips, Endacott constructed oil storage tanks at the time. Usually, only top men from the various oil companies attended the sales which were described by Endacott as "spirited poker games with very high stakes." Because of the substantial amount of money involved in the transactions, decisions necessarily came from

Skelly and Oklahoma. Skelly Oil Company traced its tremendous success in the oil industry to the Roaring Twenties and the fabulously rich strikes in north-central Oklahoma. *Courtesy of Getty Refining and Marketing Company.*

the top. Before the Osage auctions, Frank Phillips usually sent survey crews into the region to mark those leases on which he intended to bid in order that there would be no delay in starting operations.[8]

Early in the morning before each sale, workers and their wagons loaded with equipment assembled at a Phillips warehouse near the location of the leases to await the results of the sale. In the meantime, Phillips was bidding on leases in Pawhuska. If he obtained the lease he wanted, an associate telephoned the warehouse (a signal to spring into action) and within a matter of minutes, a string of mules and wagons was winding its way toward the location in order to begin work as quickly as possible.[9]

Many prominent firms participated in the development of the Burbank field; however, in September, 1921, the field's largest producer was the Carter Oil Company. Involved in twenty-three of the pool's sixty-three wells, Carter's wells produced more than 5,445 barrels of the field's daily production of 23,570 barrels of oil. In 1920, crude flowing from the Burbank field jumped from 134,408 barrels to 4,986,340 barrels in 1921, and reached an annual production of 24,230,563 barrels of oil in 1922. Peaking in 1923, the annual production was 26,206,741 barrels of oil. Hoping to eliminate some of the problems faced by earlier oil fields in Oklahoma, the companies involved in the Burbank field agreed to drill only one well on each ten-acre tract.[10]

When excitement over the Burbank field was at its peak, several companies, such as Phillips, Skelly, Comar, and Gypsy, purchased acreage west of the Osage leases in Kay County, Oklahoma, at fairly low prices. On January 16, 1923, the Sapulpa Refining Company brought in a wildcat test in the SW¼ of Section 15-T26N-R5E, with a flow of 500 barrels of oil per day. Characterized by a large amount of natural gas and high oil saturation, the new region in Kay County eventually proved to be the richest portion of the field, and its additional output pushed the daily flow of the Burbank Pool to 121,700 barrels of oil on July 21, 1923, its largest daily production figure. Shortly thereafter, the field entered a period of decline. Production dropped slowly at first; in 1924, the annual production was 21,994,479 barrels of oil, by 1926, it had fallen to an annual output of 13,352,917 barrels.[11]

Much of the natural gas encountered by the early oil operators in Oklahoma was wasted, however, simply because there was no readily available method of transporting it to market. As a result, it was often looked upon as a "stepchild" of the petroleum industry and, regardless of the large accumulations of natural gas found in the process of drilling for oil, most was allowed to escape into the air. Nevertheless, there were some farsighted oil men who sought to tap this natural resource. Two such individuals were Henry L. Doherty, founder of Cities Service Company, and Frank Phillips, originator of Phillips Petroleum Company.[12]

After graduation from the University of Kansas, Stanley Learned joined Phillips Petroleum Company in June, 1924, and was involved in much of the firm's early-day development of natural gas. Learned recalled that Phillips was very concerned with the development and utilization of natural gas during this period. As Phillips' involvement in gas fields in Oklahoma and surrounding states grew, Frank Phillips directed more and more of the company's research efforts toward natural gas. Eventually, the firm was to reap huge gains from the farsightedness of its founder, and natural gas became one of the major aspects of the operation of Phillips Petroleum Company.[13]

At the same time, Doherty was channeling the resources of Cities Service in the same direction. The efforts of the two men were rewarded when an accidental discovery in 1927 created a petrochemical industry based on natural gas. During that year, a temporary shortage of dry gas resulted in the substitution of casinghead gas at Empire Gas and Fuel Company's natural gasoline plant near the Burbank field. This caused some problems for the company, soon to become a subsidiary of Cities Service, for the Casinghead gas contained moisture and air concentrations of up to four percent, which resulted in serious corrosion problems. In an effort to counteract the corrosion, engineers attempted to remove the "contaminants" in the gas by combustion, utilizing aluminum phosphate and small amounts of metal oxides as catalysts to promote the reaction. This experiment was attempted under a pressure of 100 to 300 pounds per square inch at a temperature of between 800° and 900° Fahrenheit; however, the controlled oxidation did not proceed completely to carbon dioxide as anticipated. Instead, the "partial oxidation of the light hydrocarbons resulted in the formation of methanol, formaldehyde, acetaldehyde, dimethylacetate, and higher alcohols from traces of propane and butane in the gas stream." The potential of the discovery was quickly recognized, and with continued experimentation, the basis for Oklahoma's petroleum-chemical industry was established.[14]

Many prominent Oklahoma companies were in the forefront of the rush to the western part of the Osage Nation. By 1930, two thousand wells were in operation in that region, but production had dropped to 43,000 barrels of oil daily. Six producing horizons were present: Hoover, Suitcase, Layton, Oswego, Burbank, and Wilcox. The Burbank Sand varied from fifty to eighty feet in thickness and ranged in depth from 2,800 to 3,200 feet. Eventually, the field covered approximately 132 quarter-sections (33 square miles) and ultimately produced about 160,541,000 barrels of oil. Furthermore, the field returned approximately $285,762,980 to the producers, who had spent nearly $78,000,000 developing the field. As for the Osage Indians, they received more than $45,000,000 in royalties.[15]

Burbank was not the only field opened on the old Osage Reservation during the 1920s. The first year of the new decade also saw the location of the Frankfort Pool in T29N-R6E with a discovery well that initially flowed 10 barrels of oil and one million to three million cubic feet of gas per day. The pool was developed chiefly by the Tidal Oil Company. That same year, the Madalene and Prue fields were also discovered in T21N-R10E; the latter produced some of the best gas wells in Osage County and had an initial flow in several wells ranging from twenty million and sixty-five million cubic feet of gas per day. In February, 1926, the Devonian Oil Company found oil in Section 27-T21N-R10E of the Prue field, with the initial well flowing 27 barrels daily. Exploration continued in the Osage region, and in 1922, the Foraker field discovery was made in T28N-R7E, with gas flowing from three horizons and oil flowing between 10 and 100 barrels daily from the Mississippian Sand at a depth of slightly more than 2,700 feet. Two years later, in 1924, the Atlantic Oil Producing Company's discovery well in the Atlantic Pool had "an initial flow of 3,000 barrels from a siliceous sand." The well was drilled approximately seven miles west of Pawhuska in Section 27-T25N-R8E, and touched off some active drilling efforts in the area. Located in T23N-R9E, the Manion field discovery also showed promise, with an initial flow of between 12 and 400 barrels of oil recorded from several different horizons. These finds, coupled with production from older fields, combined to maintain a high annual production from Osage County wells outside the Burbank field. The crude from these wells rose from 17,077,348 barrels in 1920 to a substantial 41,810,178 barrels in 1923.[16]

The southern portion of Oklahoma also witnessed a surge of drilling activity after the location and development of the Hewitt field, approximately fifteen miles west and two miles north of Ardmore in Carter County. In October, 1916, William J. Millard had mapped the pool's structure as a "buried hill" covered by an anticline. Located only four miles east of the Healdton Pool, the formation occupied Sections 9, 10, 15, 16, 21, 22, 23, 27, and 28 of T4S-R2W. On June 5, 1919, the Texas Company brought in the No. 1 A. E. Denny, located in the NE¼ NW¼ NW¼ NW¼ of Section 27-T4S-R2W, which flowed at a rate of 410 barrels daily from a major producing sand found between 2,100 and 3,134 feet.[17]

The well was a source of excitement while it was being completed in that it "got beyond control" twice and sprayed oil over the drilling rig. Storage tanks were hastily constructed on the drilling site, and men worked overtime to set up pumping equipment capable of transferring the crude to the Texas Company's tank farm, a mile and a half to the southeast. Some oilmen proclaimed the well to be "the best ever drilled in Carter County," and with 35.5° gravity, it was hoped that the crude would sell at approximately $2.25 a barrel. Although later disappointed, officials of the Texas Company expressed the belief that the newly completed well would, without a doubt, open "another field as large, if not larger than Healdton."[18]

Optimism concerning the future of the field created considerable interest among oilmen in the state. It also resulted in a great increase in lease speculation in the region. Several oil companies dispatched representatives to the locality in order to "scout the land" before speculation drove the prices too high. The businessmen in nearby Ardmore were understandably elated with the prospect of another oil boom in the region and did little to discourage the efforts. By March 20, 1920, there were 20 wells in operation, producing 11,805 barrels of crude. Drilling activity continued to increase, and by the summer of 1921, there were 570 wells in operation. The peak of production was attained in September of that year, with a total monthly production of 43,902 barrels of oil. The year 1921 was also the top production year for the Hewitt field, with a yearly output of 13,095,000 barrels of oil from 605 wells. This encouraged further exploration, and in June, 1922, the Baker and Strawn Oil Company completed a well in the NW¼ of Section 22-T4S-R2W that made 450 barrels of oil per day from a lime at a depth of 2,863 to 2,878 feet. The most spectacular find in the pool was made by the Carter Oil Company in May, 1924, when its No. 33 Noble, located in the SW¼ NE¼ NE¼ of Section 21-T4S-R2W, was brought in at a depth of 2,491 to 2,492 feet with "an initial flow of 12,800 barrels of oil," one of the largest producers in the state.[19]

While the Hewitt field did not live up to expectations, it did contribute substantially to the state's petroleum output during its heyday. At the close of 1927, there were 3,050 productive acres in the pool, with an average per acre yield of 19,786 barrels. While this did not match the huge output of Healdton or Cushing, the field's total production for the decade of the 1920s helped Oklahoma maintain its standing among the nation's oil-producing states.[20]

Many Oklahomans who had served in World War I returned home to the oil boom at the onset of the 1920s. One such man was T. H. McCasland, Sr., founder of Mack Oil Company. Although he had been involved with the Three-in-One Oil and Gas Company prior to the war, McCasland returned to Ardmore after the conflict and began trading leases in the Hewitt Pool. This was not an uncommon practice. Many prominent oilmen began as lease traders who, after accumulating enough capital, entered the production phase of the industry. The life was not easy, however, and McCasland recalled that the hours were long. Because most oil business was conducted on a personal basis, it was not unusual for McCasland to leave Ardmore before sunrise and drive to Enid, Ponca City, Bartlesville, and Tulsa, before returning home after midnight. According to McCasland, such an itinerary, conducted in early-day automobiles and over a rudimentary road system, required an understanding wife.[21]

Because of the nature of the business, lease trading also had obvious advantages. It was often possible to "turn a profit" out of a dry hole. After acquiring a block of leases, the broker sold acreage to several production companies for more than enough capital to drill a well, then contracted a driller whose salary was paid from a portion of the lease money. In most instances, the lease broker was assured of a profit regardless of the success of the operation. Further-

The Lew Wentz Oil Company in Three Sands, Oklahoma. *Courtesy of Allan Muchmore.*

more, it enabled several production companies to participate in an exploratory well without sinking a large amount of capital into the venture.[22]

One of the major fields developed during the 1920s was located in Kay County, west of Osage County on the border of Oklahoma and Kansas. Originally part of the old Cherokee Outlet, the new field centered around Tonkawa, located approximately thirty miles west of Burbank. The area had a history of early oil exploration. As early as 1883, a scouting report was prepared by a "Professor Bowers of the Department of the Interior"; however, it was not until the twentieth century that the region was extensively explored by the Marland Oil Company (later Continental Oil Company) under the direction of its founder, Ernest W. Marland.[23]

In 1917, Marland had organized the Marland Refining Company and laid the groundwork for one of the state's most prosperous petroleum firms. Previously involved in the oil industry in both Pennsylvania and West Virginia, he had "already enjoyed modest success producing oil and gas." By 1907, he had located a sizeable gas field near the boundary of West Virginia and Ohio, but financial reverses suffered during the Panic of 1907 dashed his hopes, and in December, 1908, he came to Ponca City, Oklahoma, where he hoped to recoup his losses and revive his oil ventures. Educated as a lawyer, Marland, through a self-study of geology, was convinced that he would be able to locate profitable supplies of petroleum in the north-central area of the state.[24]

A few miles south of Ponca City, Marland located what he believed to be a "prominent outcropping or rock," and "was certain that the formation revealed a perfect oil bearing structure." After further examination of the terrain, he obtained a lease from the Miller brothers, Joe, George, and Zack, owners of the 101 Ranch. The 101 Ranch Oil Company was organized from this early contact, and the new company subsequently secured leases on nearby Ponca Indian land. The company's early attempts to find oil were conducted under exhausting and adverse conditions and the initial results were most discouraging. Because horses were not available, oxen were used to haul "timbers for the rigs, tools, boilers, and casing material." Workers were in short supply, and Marland, dressed in "Knickerbockers, Norfolk jacket, and oftentimes spats," often drove the teams himself. Financing was a constant problem as well, especially after the first well was a dry hole. Moreover, the next seven wells produced only a small amount of gas.[25]

Determined, Marland refused to give up and selected his next drilling site, "an elongated and isolated hill," near Bois d'Arc Creek. Unfortunately, the site was on "sacred ground for the bodies of dead Ponca Indian warriors," and only after prolonged conferences with the Chief of the Poncas, White Eagle, was he given permission to drill a well. Drilling on the slope of the hill, away from the crest, Marland and the Miller brothers soon found oil, the first petroleum discovered in the Ponca field, and over the next ten years, Marland aggressively developed the production in the area. Furthermore, this success launched Marland on his rapid rise in the petroleum industry, and during the next few years, he expanded his operations throughout northern Oklahoma. By 1917, Marland had absorbed the 101 Ranch Oil Company into the newly organized Marland Refining Company.[26]

Nearly forty years before Marland's entry into the petroleum industry, Isaac Elder Blake had experienced his initial involvement in the oil business in Pennsylvania during the late 1860s. By 1869, he was earning up to a thousand dollars per day selling crude oil, and in one year alone, cleared a hundred thousand dollars; however, like Marland, he suffered a financial setback in Pennsylvania and came west. In 1874, Blake traveled to California, then later to Ogden, Utah, where he hoped to enter the marketing sector of the oil industry. In March, 1875, he and four associates petitioned the council of Ogden to establish a "Depot for the reception of their Oils, and from which it could be distributed in the Territories of Utah, Idaho, Montana, and Nevada." After some delay, Blake organized the Continental Oil and Transportation Company on November 25, 1875, with the objectives of "producing, purchasing, refining, transporting, selling and delivering and dealing in petroleum, and in all its products." Although incorporated in Council Bluffs, Iowa, the firm's first business was located in Ogden.[27]

Basically, Blake's plan was to transport oil products westward to market in Ogden. Later, after moving the company's marketing center to Denver, Colorado, he expanded westward into the oil fields of California. During the next few years, Continental Oil and Transportation Company and its affiliates continued to increase their interests throughout the West to the point that it eventually challenged the Standard Oil Company. Finally, in 1884, Blake accepted an offer to become an affiliate of Standard Oil and, on January 1, 1885, Continental Oil and Transportation Company was reincorporated. Under the arrangement, "The Standard Oil Company and the Continental Oil and Transportation Company . . . consolidated their business in Colorado, New Mexico, Wyoming, Montana, and Utah, under the style of the Continental Oil Company." Soon afterward, the firm expanded into Arizona Territory, Idaho, Nevada, the Pacific Coast, Alaska, and the northwest part of Canada. The firm continued to grow and expand in its relationship with Standard Oil throughout the remainder of the nineteenth century and into the early twentieth century; however, the United States Supreme Court ordered Standard Oil to dissolve some of its holdings in 1911, and in 1913, the Continental Oil Company again became a separate business.[28]

Throughout the oil boom era of the early twentieth century, both Continental and Marland oil companies continued to develop. Marland was involved in Blackwell, Garber, Billings, Burbank, Tonkawa, and a number of other plays throughout Oklahoma. At the same time, Continental was expanding its products, marketing, refining, and exploration efforts from the Mississippi River to the Pacific Coast, and from the Gulf of Mexico to the Great Lakes. Nonetheless, in 1927, the oil business in general, and Marland's enterprises in particular, fell upon hard times. Marland found it more and more difficult to finance his operations. In less than two years, he was to lose a corporate fight for control of the company and resign "his chairmanship even though he was almost broke and needed the income." Dan Moran assumed the presidency of Marland's company and organized an aggressive campaign to place the firm "back on its financial feet." During the same time, Continental had continued to increase its activity, and "The situation was now opportune for the merger of Continental and Marland." After months of negotiation, the executive committee of Marland's company "Approved [the] plan of reorganization and agreement with Continental Oil Co." on April 30, 1929. The agreement called for the two firms to enter into a plan of reorganization, in which Marland's company agreed to purchase "all the assets and prop-

Women in the oil fields. The 1920s was the era of flappers and suffragettes, and in the oil fields women were also making a place for themselves. Two examples were Mrs. W. L. Umburn, *left*, who worked on several drilling rigs in the state, and Mable C. Orr, *right*, who was with the firm of Dunham & Orr in Blackwell. *Courtesy of Mrs. Claude V. Barrow.*

erties of Continental and to deliver Marland Oil Company stock in exchange." Technically, Marland Oils was acquiring Continental, but the latter's name was to be retained. On July 18, 1929, Marland's board of directors approved the acquisition and "authorized the name of the newly-formed corporation to be Continental Oil Company." Marland Oils passed from the scene. The new firm made its headquarters in Ponca City and became a major factor in the growth of Oklahoma's petroleum industry.[29]

Marland's zealous efforts had resulted in the location of the Newkirk-Mervine field in 1913. That year, on the basis of information published by the Oklahoma Geological Survey, a well was drilled on the Murdock farm located in the SE¼ of Section 2-T27N-R3E. Flowing at an initial rate of a hundred barrels a day, the discovery well touched off a flurry of activity that failed to establish significant production, and interest in the area quickly subsided. In late 1916, however, Marland opened a gas well in Section 17-T28N-R3E with a flow of 28,000,000 cubic feet of gas per day, and a new round of exploration began which resulted in a series of producers during 1918 and 1919, greatly intensifying the development of the field.[30]

The next locale of interest to oilmen in Kay County was the Blackwell area. Marland had already developed the north end of the field by 1912, and as a result, twenty-one gas wells were brought in; nevertheless, new interest in the pool was spawned after B. B. Jones drilled a well in 1914 that had an initial flow of 600 barrels of oil per day. Shortly thereafter, the Duluth-Oklahoma Company brought in a producer in Section 1-T28N-R1W with a flow of 2,000 barrels of oil per day. Southwest of Kay County, in Garfield County, the Sinclair Oil and Gas Company opened the Garber field in September, 1914, with a discovery well located on the Hoy farm in the NE¼ of Section 25-T22N-R4W. After the discovery, the pool was developed fairly rapidly by the Roxana Petroleum Corporation, the Healdton Oil and Gas Company, the Cosden Oil and Gas Company, the Marland Refining Company, and the Atlantic Petroleum Company. One well, brought in by Sinclair in Section 18, flowed "27,000 barrels," the largest initial potential of any Oklahoma well to that date. Noble County, directly south of Kay County, was the site of the Billings field discovery well in February, 1917. Drilled by the Midco Petroleum Company in Section 22-T23N-R2W, it produced 250 barrels of oil per day. Fifteen miles from the nearest production, the Midco well confirmed a potentially productive area south of Kay County. West of Kay County, the Deer Creek Pool was discovered in July, 1920, in Grant County. Drilled by the Western States Land and Development Company in Section 22-T27N-R3W, the discovery well initially produced eighteen million cubic feet of gas daily, and when deepened, it flowed at the rate of 10 barrels of oil per day. Later, the same company brought in the first oil well in the field, located in Section 15, with an initial flow of 450 barrels a day.[31]

The discovery of the Garber Pool in 1914 led to the founding of one of Oklahoma's most successful oil companies by Herbert H. Champlin, an Enid banker. The George Beggs farm was located in the midst of the new find and, although it had originally been leased by the Sinclair Refining Company, the firm had allowed the lease agreement to expire. Beggs, seeking someone to pick up the lease offered it to Champlin for $11,500. Champlin had never invested in petroleum property and was hesitant, but his wife, Ary, persuaded him to take advantage of the opportunity, and he acquired the acreage. The property proved to be one of the most prolific producers of oil in the entire field, and Champlin, by now greatly interested in the oil business, began to organize a company that eventually would span the spectrum of the petroleum industry.[32]

During this time, Victor Bolene had begun construction of a refinery, or "skimming plant" as they were sometimes called, in Enid to handle crude flowing from the nearby wells. Champlin, recognizing the advantages of controlling not only production but processing as well, purchased the refinery before Bolene had finished construction and organized the Champlin Refining Company. Realizing the need for access to a reliable facility to transport the crude from the wells to the refinery, he also built a pipeline from the Garber Pool to the refinery in Enid. This line was later expanded as the Cimarron Pipe Line Company into other parts of Oklahoma, Texas, Kansas, Nebraska, and Iowa.[33]

When the Goodwell Oil Company placed a string of bulk plants and service stations on the market in 1920, Champlin quickly purchased those facilities as well. With this acquisition, his oil ventures became a completely integrated petroleum company involving production, refining, and sales. Champlin continued to reinvest profits from his company back into the business until his death in 1955, and by that time, the firm "was the world's largest family-owned company engaged in every phase of the [petroleum] business." The solidarity of Champlin's financial base prompted a vice-president with Standard Oil Company in New Jersey to remark on one occasion, "I wish you would look at this. No one in the Standard offices has ever seen anything to equal it. Here is man who owns all the stock in his company. He has his own oil wells, crude pipelines, refinery, products lines, bulk plants and service stations; even his own bank and we cannot find where he owes a dollar to anyone."[34]

Although production resulting from drilling activity in and around Kay County had been somewhat of a disappointment to oilmen up to this point, the situation was soon to change. In 1920, Marland recognized his geological staff, and E. Park "Spot" Geyer persuaded him that the Tonkawa anticline held great prospect. Several "dusters" had already been drilled in the area, however, and many oilmen questioned the opinion of Marland's men. Nevertheless, the Humphreys Petroleum Company was persuaded to grant Marland half interest in a 320-acre lease, and Marland began raising the capital needed to drill a well. The Prairie Oil and Gas Company advanced $2,500 in "dry hole money" in exchange for geological information that might be of use on their leases in the vicinity. In addition, the Cosden Oil Company was sold a half interest in the 640-acre section, and the Kay County Gas Company, a Marland subsidiary, contributed $5,000 to the venture. Nevertheless, Marland remained nearly $16,000 short of the funds necessary to drill the wildcat well. After selling Southwestern Oil Company 160 acres of the property received from Humphreys Petroleum Company on a "farm-out basis," he was finally ready to begin drilling operations.[35]

Once adequate financing had been assured, Marland embarked on a ten-well exploratory campaign to test the most promising locations. The first eight wells were disappointing and seemed to confirm the

The Champlin Petroleum Company building in Enid, Oklahoma. Founded by Herbert H. Champlin, the firm has been among the most successful Oklahoma-based operations. *Courtesy of Milton B. Garber.*

generally accepted theory that the region contained no profitable deposits of petroleum; however, in the spring of 1920, Marland began work on the ninth well approximately eight miles south of Tonkawa, just across the county line in Noble County. The undertaking was hampered by delays in securing needed equipment from producing areas some distance away. Nonetheless, work continued on the J. H. Smith School Land Well No. 1, located in the NW¼ of Section 16-T24N-R1W, and by fall of that year, the actual task of drilling had commenced.[36]

As usual, Marland's determination paid dividends. The School Land well encountered gas at a shallow depth, and the formations indicated that the location was on an oil-bearing structure. Traces of crude were found shortly thereafter, although not in sufficient quantities to be commercial. Delays in securing equipment continued to plague the operation, and it was not until shortly before midnight on June 30, 1921, that the well came in. It was a huge success. Adquate storage facilities were not available in the beginning, and it was impossible to gauge the well's flow for several weeks; however, when measured, it was flowing a thousand barrels of oil per day that tested between 41° and 43.9° gravity.[37]

Excitement over the well's production was understandable when compared to the average flow of less than 200 barrels per day from other wells in the Mid-Continent area. Proclaiming the discovery in bold headlines, the *Tonkawa News* declared that "No doubt now remains that Tonkawa is located on, over, nearby or adjacent to an oil pool of the first magnitude." A witness present the Friday evening the well came in described his joy as "he stood upon the brink of the pool of oil that soon formed," and "gazed upon its velvety greenness." So excited that he let out a "prolonged shout of joy," he "cupped his hands and dipped them into the pool and bathed his smiling countenance with oil."[38]

Because there were inadequate storage facilities immediately available at the drilling site, the oil "came gurgling and bubbling up through half a mile of pipe and spread out over the slush pond." It soon became apparent that the well was producing a strong flow and "to give satisfactory evidence that it was not a free strike," it was shut off to prevent continued waste. Although a local journalist declared the oil of "a high grade, said to be the best west of the Pennsylvania field," and proclaimed it a major find, there was apparently no immediate rush to the region by other producers. This oversight was to cost the larger companies dearly, for later, they were "forced the pay big money to get into the Tonkawa for drilling purposes." "Sixteen months after the initial well was drilled," a journalist reported that Comar Oil Company paid $1,300,000 "for the lease on four hundred acres in the northern portion of the field to the Alcorn Oil Company, another Marland concern," and the Blackwell Oil and Gas Company sold "a quarter-section in the northern Tonkawa for $2,000,000." Only the larger firms could afford such high prices for leases and, two years after Marland's find, "approximately all the acreage in the field," was held by the "big companies."[39]

Marland did not delay in constructing the necessary storage tanks at the well site, and they were quickly filled. In addition, plans were made to lay a pipeline from the Smith lease to Marland's refinery in Ponca City, in order to handle the field's production. Some concern was expressed over the fact that the low price of crude might delay the development of the field, but Marland began work on offset wells almost immediately. Marland negotiated an agreement with the Roxana Petroleum Company and the Comar Oil Company, which he had organized and taken the name from "companies of Marland," to develop the region. Under the arrangement, Marland agreed to turn over his holdings in the Tonkawa field to the Comar Company and, in exchange, Roxana was to advance a million dollars for the pool's immediate development and another million dollars when needed. Roxana was to receive fifty-one percent of Comar, and headquarters for the concern were to be based in Ponca City.[40]

The Tonkawa field was "a money-maker from the start," and by April 1, 1922, Marland had already recovered the cost of drilling the discovery well. As the region's potential became apparent, oil scouts and lease brokers arrived in droves; however, they soon discovered that most of the land within the pool's boundaries had been leased. When oil and gas leases were not available, attention was then focused on the landowner's royalty. Generally, the landowner was granted a one-eighth royalty, but this was increased if a particular tract held more promise. Even so, the "standard" one-eighth royalty was generally in great demand. "Enormous prices have been paid for forty-acre, eighty-acres and quarter-section royalties," one jounalist reported, and related the instance of another farmer, Sam McKee, who refused an offer of two million dollars for the "royalty on a quarter-section farm." So great was the quest for crude that one lease included one-half acre of the graveyard of the Prairie View Church in the northwestern portion of the field. The driller purportedly had made preparations to drill a well thirty-two feet from a grave, when horrified trustees of the church, fearful that saltwater would seep into the graves, sought a court order to stop the drilling. The fields surrounding the church had, according to one newspaperman, been "transformed into a forest of derricks" which "crept nearer and nearer to the place of worship." The church members won the legal battle, however, and "the shadow of the spire often fell across the graveyard as though attempting to assure the dead that they would not be reached by the advancing . . . derricks." Nonetheless, wells were drilled wherever leases could be acquired, including a one-acre schoolyard lease.[41]

Comar quickly added to its original 580 acres of holdings a large block of leases to the northwest of the discovery well. This proved to be a very shrewd move. The discovery well was in the southern portion of the field, and it was soon determined that the major producing area was to the north, "along the township line between 24 and 25-1 west." Develop-

The Three Sands oil field. *Courtesy of Allan Muchmore.*

ment of the field proceeded at a rapid rate during the latter half of 1922 and into the first portion of 1923, and an entirely new area was opened on April 10, 1924, when the Tom B. Slick Company's Slick No. 1A, located in the sw¼ of Section 35-T25N-R1W, found production in the Wilcox Sand at a depth of between 4,065 and 4,085 feet. Total Wilcox Sand production in the Tonkawa field as of January 1, 1926, amounted to 28,305,000 barrels of oil, compared with 49,258,000 barrels from shallower sands as of the same date.[42]

The Tonkawa Pool was the wonder of the oil world at the time of its discovery and development. Oilmen were amazed to find so many producing horizons of oil and gas present in a single field. Beginning at a depth of 1,150 feet, the horizons: Hoy, Hotson, Newkirk, Upper Hoover, Middle Hoover, Lower Hoover, Carmichael, Endicott, Tonkawa, Layton, Oswego, Mississippi, Chattanooga, and Wilcox, extended to 4,330 feet, with thicknesses ranging from 6 feet for the Endicott to as much as 150 feet for the Chattanooga. Four sands were major producers of crude: the Middle Hoover at a depth of 1,850 feet with an initial flow of 25 to 375 barrels per day; the Carmichael at 2,050 feet with between 15 and 1,225 barrels per day; the Endicott at 2,100 feet with 50 barrels per day; and the Wilcox at 4,330 feet with between 40 and 10,000 barrels per day. Three were major producers of a combination of oil and gas: the Newkirk at 1,450 feet with an initial flow of between 20 and 75 barrels of crude and between one million and fifteen million cubic feet of gas per day; the Lower Hoover at 2,000 feet producing 35 to 300 barrels of oil and between one million and ten million cubic feet of gas per day; and the Tonkawa at 2,500 feet with a flow of between 50 and 3,000 barrels of oil and between one million and six million cubic feet of gas per day. Two horizons were major gas producers: the Hoy at 1,150 feet and the Upper Hoover at 1,800 feet, which flowed between one million and forty million cubic feet and between three million and sixteen million cubic feet of gas respectively.[43]

With so many producing horizons, it was no surprise that the Tonkawa field, despite its comparatively small size, became a major supplier in the state. Although the field covered only approximately eight square miles, as many as five derricks might be clustered around one location producing oil from several different horizons. Such prolific production resulted in the completion of no fewer than fifteen pipelines into the field by May 1, 1923, with eleven others under construction. The high gravity of the crude made it desirable, and the price remained fairly high. In addition, the demand led to the development of adequate storage and transportation facilities to handle the field's flow. But it was not only oil that made Tonkawa famous; gas wells were also a major factor, and in April, 1921, twenty gas wells had a daily open flow of 408,250,000 cubic feet. Crude remained king in the eyes of the oilmen, however, as well it should, for between February 27 and April 23 of 1923, Tonkawa produced oil valued at $208,000 daily for a sixty-day total of $12,480,000. The highest daily production for the field was 112,112 barrels of oil, and by the end of 1927, the field had produced an average of 30,000 barrels of oil per acre.[44]

A significant sidelight to the tremendous impact the Tonkawa field had on the oil output of Oklahoma was the technological advances made in conjunction with the pool's development. In an effort to solve the problem of the interference of natural gas while drilling for oil and the need to constantly "mud off" during the drilling process, the Southwestern Petroleum Company first adopted rotary drilling rigs which revolutionized drilling procedures at Tonkawa. Marland, who had been one of the leaders in the application of geology by oilmen and had relied on the science in his early search for oil at Tonkawa, also used the diamond core drill at the pool, and its success led to increased use throughout the industry.[45]

During the 1920s, great strides were also made in the marketing of the state's oil and natural gas production. On November 19, 1927, Phillips Petroleum Company undertook an operation which would, in many respects, have a great impact on Oklahoma's petroleum industry. The firm opened its first retail gasoline outlet for motor fuel in Wichita,

*The Pioneer Woman.* This statue by Bryant Baker was E. W. Marland's gift to the people of the state. *Courtesy of the Interstate Oil Compact Commission.*

Kansas. The company's researchers had successfully stabilized natural gasoline in order that it might blend with straight-run gasoline, and this procedure gave Phillips a big lead over its competitors in production of a higher-gravity gasoline with easier-starting characteristics, better performance, and controlled volatility. Before it could be profitable, however, a retail market had to be established. V. D. Peters, a longtime supervisor in the marketing department of Phillips Petroleum Company and an attendant at the firm's first service station, recalled the excitement generated over the station's opening. Great emphasis was placed on customer service. The facility was the first in Wichita equipped with air and water available to customers on the service island. They were supplied for the automobiles of all customers who came through the driveway, and service people also cleaned their windshields. As a further enticement on opening day, each individual who purchased a full tank of gas received a coupon good for five free gallons of gasoline on the next visit to the station. This strategic move by Phillips greatly increased the market potential for Oklahoma petroleum products.[46]

Such innovations and opportunities resulted in the founding of a number of successful Oklahoma-based oil companies during the roaring twenties, and among these was Apco. L. H. Prichard and J. Steve Anderson had legal backgrounds and had been involved in the oil business in Oklahoma for several years; in 1918, Prichard had been one of the organizers of Caddo Petroleum Company, and two years later, Anderson joined the firm as secretary-treasurer. Between the two, they controlled sixty-three percent of the corporation's stock. Forming the partnership of Anderson and Prichard in May, 1919, they operated in Oklahoma's Cement field, as well as in Louisiana and Texas. After acquiring the bankrupt facilities of the Cyril Refining Company in 1920, they organized the firm of Anderson-Prichard and Company; however, the business failed and went into receivership.[47]

Undaunted, they made a comeback. By October 6 of that year, Anderson-Prichard and Company was incorporated under the laws of Delaware and agreed to assume the previous contract with the Cyril Refining Company. Once again, however, the firm was forced to liquidate its assets in January, 1922, and faded from the scene. Displaying the perseverance characteristic of many pioneer oilmen, Anderson and Prichard then formed the Apco Refining Company on June 21, 1922, with headquarters in the Colcord Building in downtown Oklahoma City. The refinery at Cyril was again acquired, and in February, 1923, Anderson-Prichard Oil Corporation accumulated all Apco stock. The corporation continued to grow during the next several years, expanding into nearby Texas fields. Eventually, Apco entered the marketing phase of the petroleum industry, and expanded its marketing region from the Gulf of Mexico to the Canadian border and from the Rocky Mountains to the Mississippi River. For more than fifty years, Apco played an important role in the development of the Oklahoma oil industry before the company was sold in 1977.[48]

Following Tonkawa, several other pools were opened in the same general region: the Thomas and Braman pools in 1924, and the Hubbard and Vernon fields in 1925. In 1923, however, the attention of the oil world was shifted to a different section of Oklahoma, embracing a vast region in the central portion of the state. Exploration and development of the old Seminole Nation, now Seminole County, Oklahoma, led to the discovery of "one of the leading producing areas of the United States," the Greater Seminole field.[49]

## Chapter 11.

# The Greater Seminole Field

THAT THE INCREDIBLE EXPERIENCES of Red Fork, Glenn Pool, Cushing, Healdton, and Tonkawa could be repeated during the latter portion of the 1920s must have seemed impossible to oilmen, most of whom believed that Oklahoma's major fields had already been discovered. Nonetheless, this thinking was quickly cast aside when the tremendous potential of the Greater Seminole field became known. Not a single field, but a series of pools underlying the east-central portion of Oklahoma, the Greater Seminole discovery and feverish development thrust the state once again into the national spotlight of the petroleum industry. So great was the flow from wells in the area that, in 1927, the added production placed Oklahoma at the head of the oil-producing states in America. The subsequent threat to glut the petroleum market, however, led to the proration of well production in the state and, as a result, the Greater Seminole field was the last region in Oklahoma where wells were allowed to produce at full capacity. Afterward governmental regulations made this virtually impossible.[1]

The Greater Seminole field encompassed a number of individual pools, found primarily in Seminole County, but locally extending into the surrounding counties of Pottawatomie, Okfuskee, Hughes, and Pontotoc. There were thirty-nine pools within the Seminole District, and of these, twenty-two were located in Seminole County proper: Keokuk, Bowlegs, Searight, North Searight, Bethel, North Bethel, Seminole City, East Seminole, East Earlsboro, Carr City, Mission, Little River, East Little River, Rosanna, Wewoka, Grayson, Dora, Fish, Sacred Heart, Wetley, Konawa, and Traugh. In four instances, the main body of the production was within the boundaries of Seminole County, while a smaller portion was located in an adjacent county: Earlsboro, which extended into Pottawatomie County; Cromwell, partially under Okfuskee County; Wewoka, a portion of which was in Hughes County; and Francis, on the border of Seminole and Pontotoc counties. To the south, the Beebe pool, the Holdenville pool, and most of the Allen field were located in Pontotoc County. To the east of Seminole County, Okfuskee County contained the Dill, East Cromwell, and Olympic pools. To the west, Pottawatomie County was the locale for the Shawnee, West Earlsboro, Maud, St. Louis, North St. Louis, Asher, and Gray oil fields.[2]

There had been oil discoveries in the region of the Greater Seminole field prior to the boom of the 1920s. The Allen Pool, situated along the border of Seminole and Hughes counties, was opened in 1913, and although encouraging, production was not large until the wells were deepened in 1926. West of the Allen Pool, the discovery well for the Francis field produced as much as fifteen oillion cubic feet of gas and 100 barrels of oil per day in 1916. The Holdenville Pool, opened the same year, yielded "an initial flow of as much as 4,000 barrels." Maud field, also found in 1916 by the Prairie Oil and Gas Company, produced only a "small amount of gas." It was not until the Mid-Kansas Oil and Gas Company's wildcat well came in with an initial production of 250 barrels of crude daily, that development of the area began in earnest. Nevertheless, these finds were generally overshadowed by the contemporaneous events at Cushing, Healdton, and other larger discoveries, and as a result, were overlooked by oilmen of the state for a number of years.[3]

The boom at the Greater Seminole field was set off when R. H. Smith's No. 1 Betsy Foster was brought in on March 16, 1923, near Wewoka. This was the culmination of a twenty-year search for crude that had started in the area as early as 1902, when the Wewoka Trading Company drilled a well on the

Seminole, Oklahoma, in 1926. *Courtesy of Cities Service Oil Company.*

B. F. Davis farm. Five years later, another well drilled in the same area was brought in as a gas well, but when an attempt was made to deepen the well, the hole was lost. Undaunted, the drillers sank another well approximately sixty feet northwest of the first location, and in July, 1907, encountered the "Wewoka Sand," which brought in a large producer. "The original or flush production from this well was tremendous," and flooded nearby streams with escaping crude before "arrangements could be made to take care of the production." Eventually, the well flowed at the rate of approximately 150 barrels daily; however, during this year, the price of oil and the lack of adequate pipeline facilities prevented the area from being further developed. Later, an attempt was made to deepen the well but the hole was lost. In 1921, another well was started in the same general region by F. E. Montee and several others, but they were using a "small and antiquated rotary" drilling rig, and it was not completed before Smith's well was brought in two years later.[4]

Located two miles south of the corporate limits of Wewoka in the NW¼ SW¼ NW¼ of Section 33-T8N-R8E, the Smith well had reached a sand at 3,150 feet in early March, 1923, but problems with the casing resulted in a two-week delay. After drilling resumed, the bit penetrated the sand "only a few inches" when the well blew in, flowing "twenty million feet of gas and spraying oil." Quickly suspending operations, Smith immediately began work on storage facilities to hold the crude. At the time of discovery, the initial flow was estimated at approximately 100 barrels; however, the following day, it had increased to between 200 and 300 barrels daily and, after the well was deepened, reached an astounding 3,500 barrels per day.[5]

The rapid development of the Greater Seminole field proved to be a great boon to many area residents. Many residents of Seminole County, formerly the Seminole Nation, were apprehensive of outsiders, but two Sasakwa businessmen who understood the Indians' customs developed a lucrative practice of lease-buying in the area. Because William B. Osborn, Sr., and Jim Fleet were well-known in the region, they encountered little trouble and, after accumulating enough capital, formed a partnership

Thousands of oil field workers, such as this pipeline gang, flocked to the Greater Seminole field. *Courtesy of Cities Service Oil Company.*

and entered the production phase of the petroleum business. In the early 1930s, Fleetborn Oil Company was organized and eventually expanded throughout Oklahoma and Texas.[6]

Concurrent with the Smith discovery, the Gardner Oil Company brought in a well in the NE¼ NE¼ of Section 34-T11N-R8E which found a productive sand at 1,930 feet. Located in the northern portion of Seminole County, the Gardner well had an initial flow of "two hundred barrels of high grade oil, and getting better." The simultaneous discovery of two producing wells in the same general location sent oilmen scurrying to the area, and the price of land quickly rose. Within a month, one-quarter royalty on a forty-acre lease was bringing as much as fifteen thousand dollars and the price was climbing.[7]

Improvements were quickly undertaken at the Smith well. In addition to a large earthen tank, three 1,600-barrel wooden tanks, four 250-barrel metal tanks, and a 210-barrel flow tank were constructed to

Mud, the nemesis of the Greater Seminole field. *Courtesy of Cities Service Company.*

The Seminole rail yard became the shipping center for material in the Greater Seminole oil field. *Courtesy of Cities Service Oil Company.*

handle the No. 1 Betsy Foster's production. The Prairie Pipe Line Company immediately began extending a two-inch pipeline from Wetumka to the new pool to provide an outlet for the crude. Wewoka went wild. One local bank reported an increase in business of more than $125,000 by the end of April, and the town's facilities were strained by the tremendous influx of oilmen. As described by one reporter, the scene was one of "numberless big automobiles constantly going to and fro; . . . crowds of strangers on the streets; . . . big 8-Wheeled wagons and four-up and six-up teams of draft horses; . . . powerful trucks and trailers; . . . switch sidings filled with cars of timbers, . . . casing, riggings, and timber; . . . stirring and ever-increasing excitement, all evidence the opening of a great oil field."[8]

The frenzy of drilling set off by the two wells soon led to additional discoveries in the region. Before the end of April, 1923, the Wewoka Oil and Gas Company, the Cosden Oil Company, the Indian Territory Illuminating Oil Company, the T. B. Slick Oil Company, the Independent Oil and Gas Company, and many others had equipment in the field. The Magnolia Petroleum Company tapped the Sykes Sand about fifty feet below the Smith Sand in October, 1923, opening a new horizon, and two years later, the Dixie Oil Company opened the Hunton Limestone. By the spring of 1927, these three formations, Smith, Sykes, and Hunton, had cumulatively produced 15,000,000 barrels of crude. Eventually, eight producing horizons were located in the Wewoka Pool. As the boundaries of the Wewoka field were located, it was found to cover an area of approximately 1,700 acres of land, located in the southwestern corner of T8N-R8E and the northwestern corner of T7N-R8E. Peak production from the Smith and Sykes formations was reached on June 12, 1925, when 134 wells averaged 19,860 barrels of oil; the height of the Seminole Sand production was reached the week of August 20, 1926, with an average daily yield of 29,023 barrels of oil.[9]

On the heels of the Wewoka find, another discovery jolted the state's oilmen. At Cromwell, which had experienced some previous drilling activity, the Cosden Oil and Gas Company brought in the No. 1 Jim Wallace in November, 1922. Located in the SE¼ SE¼ NW¼ of Section 15-T10N-R8E, it had an initial flow of thirty million cubic feet of gas per day from the Cromwell Sand at a depth of 3,456 to 3,466 feet; however, it was not until October 2, 1923, that the pool's first oil well was brought in by the company. Located in the NW¼ NE¼ SW¼ of Section 15-

Okemah, Oklahoma, one of many neighboring communities that flourished because of the Greater Seminole strike. *Courtesy of Alma Marsh.*

T10N-R8E, approximately fifteen miles northeast of Seminole near the Seminole-Okfuskee county line, the No. 1 Bruner had an initial production of 312 barrels of oil daily and quickly captured the attention of oilmen throughout the state. Soon, additional wells were under way.[10]

The exploration activity triggered by the Cosden well resulted in an additional discovery in March, 1924, when the Frensland Oil Company brought in its No. 1 Bruner in the NE¼ SE¼ NE¼ of Section 16-T10N-R8E. Although it was impossible to get an accurate gauge of production, the well quickly overflowed the three-thousand-barrel storage facilities available. On March 25, the hole was deepened four feet, resulting in a flow of five thousand barrels per day. Reported to be "by far the biggest-producing well in the field today and possibly the biggest well in the state at this time," the Frensland well set off "a scramble for royalty and lease interests in that territory never before seen in Seminole County."[11]

At the same time, the McMan Oil Company completed a well in the SE¼ SW¼ NW¼ of Section 22-T10N-R8E with a flow of nine hundred barrels daily. The opinion of a number of oilmen that the well's production would increase if it were deepened generated more excitement for the so-called "north field." Many producers believed it would prove to "be the biggest oil field in the state—bigger in production per well, and a larger territory covered." Other companies were soon drilling in every direction from the initial discoveries as "Materials are being rushed to the north field as fast as possible." Within a short time, the Amerada, Mid-Kansas, and Livingston companies were drilling wells in the region, and by August, 1924, seventy-five wells were producing in the pool.[12]

Ultimately, the field was found to include 3,560 acres located in the northeastern corner of Seminole County and 140 acres of land in adjacent Okfuskee County. Peak production of the pool was reached

Walter Helmerich, *left*, and William T. Payne, *right*, two of Oklahoma's pioneer oilmen, who made the rush to the Greater Seminole field. *Courtesy of the Oklahoma Heritage Association.*

within a relatively short period of time after its discovery and, during the week ending August 20, 1924, seventy-five wells averaged 62,391 barrels of oil per day. As development continued, most wells averaged between 500 and 2,500 barrels daily; however, daily flows as high as 5,600 barrels were reported. By November 22, 1927, there were 393 wells in the pool, and as of that day, production for the week had decreased to an average of 10,823 barrels daily per well. Nonetheless, the total production still remained a respectable 4,253,424 barrels.[13]

The year 1924 ended with the discovery of the small Bethel field in the northwestern part of T9N-R8E, approximately seven miles north of the Wewoka Pool and approximately five miles southwest of the Cromwell field. The Independent Oil and Gas Company completed a discovery well on December 9, 1924, in the NE¼ SE¼ NE¼ of Section 7, from a depth of 3,275 to 3,302 feet, and although the well was not of the magnitude of the Wewoka or Cromwell discoveries, it had an initial flow of 200 barrels per day which assured the continued search for crude in the Greater Seminole area. This was the only major discovery in the region for more than a year. Limited to approximately five hundred acres, the field's production peaked during the week ending April 12, 1927, with a daily average of 2,786 barrels per day. A peculiar aspect of the Bethel Pool was the fact that the Cromwell and Simpson formations (so productive in other portions of the Greater Seminole area) were dry, and most of the crude came from the Booch Sand at a depth of between 3,200 and 3,300 feet.[14]

Drilling activity in the Greater Seminole field continued at a rapid pace during 1925 and early 1926, and, although the existing fields were expanded to their limits both in area and depth, there were no new major discoveries. Activity in 1925 was expected to center along the South Canadian River, which meandered through Seminole, Pottawatomie, and Pontotoc counties, and the area was extensively leased. Expectations were also high for the northern portion of the Cromwell Pool because of good shows in a number of wells in that area. The boom was still going on by the end of the year, and leases were still bringing high prices. According to observers, "Royalty interests have exchanged hands at a rapid rate . . . while most anywhere in a reasonable distance from . . . [a] big gusher, fancy prices are being paid." Nonetheless, the lack of significant new finds did affect a number of oilmen, who left the region to search elsewhere. However, this "off" period in the development of the area was short-lived in that a rapid series of new strikes overshadowed the previous finds.[15]

The lull abruptly ended in the spring of 1926 when, within a sixty-day span, three fields were brought in: Earlsboro, Seminole City, and Searight. In less than two years, a total of 673 wells located in these three pools, the oldest of which had been producing for less than eighteen months, yielded 105,326,194 barrels of crude as of January 1, 1928. Covering a total of 7,130 producing acres in the northwest quadrant of Seminole County, the wells averaged 46,790 barrels of oil per acre up to January 1, 1928, rekindling the oil rush to the region.[16]

The events began approximately eight miles northwest of Seminole, near Earlsboro, just across the border in Pottawatomie County, when the Morgan and Flynn Oil Company brought in the No. 1 Ingram on March 1, 1926. Drilled in the NW¼ NW¼ SW¼ of Section 10-T9N-R5E, the well produced an "initial flow of 200 barrels of oil" from a depth of between 3,560 and 3,580 feet in what was called Earlsboro Sand. As activity in the region increased, an even larger discovery was made in late 1926, when the Gypsy Oil Company completed its No. 1 States in the northeast corner of Section 16 at a depth of between 4,275 and 4,291 feet. From the Seminole Sand, the well flowed 500 barrels daily, and when deepened slightly, its daily output increased to 3,000 barrels. When the well was completed on December 3, the flow was a remarkable 8,050 barrels of crude daily.[17]

There was no longer any doubt of the productivity of the new field, and oilmen poured into the area in search of potentially productive leases. Most of the wells in the Earlsboro Pool flowed with an initial production of between 1,000 and 4,000 barrels daily; however, reports were received of 14,000-barrel-per-day producers. Rapidly developed, there were approximately 130 wells in the field by August, 1927, with a weekly output in excess of 192,000 barrels of oil. On the first day of 1928, 286 wells were pumping crude, and the total output since its discovery was pegged at 37,775,722 barrels of oil. The Earlsboro Pool's limits were eventually expanded and, by the end of the decade, included 4,170 acres, with an average-per-acre yield of 22,234 barrels.[18]

Following the discovery well at Earlsboro by a scant six days, ITIO's No. 1 James was completed on March 7, 1926, in the SE¼ of Section 24-T9N-R6E. The region was not without a history of drilling efforts, for only three years previously, W. B. Pine had drilled a 3,685 foot well in Section 22; however, it was abandoned after the hole was lost. The James well came in on March 1, a cold, overcast day with drizzling rain. While huddled inside the rig's doghouse for protection, the crew members suddenly heard a steady downpour on the roof. Expecting to see rain, they were astonished to discover that the noise was caused by oil spewing from what proved to be an eleven-hundred-barrel-per-day producer.[19]

Heralded in the local newspapers as the "First Good Well of Much-Talked-of-Seminole Dist.," the location was just two and one-half miles east of the town, and the pool was named the Seminole City field. Its activity touched off a wave of renewed drilling efforts in the vicinity, with many "leases and royalties . . . changing hands at a high figure."

Work was started immediately on adequate oil storage facilities, and the Prairie Oil and Gas Pipe Line Company rapidly constructed a line to the new field. Activity was feverish. As one journalist reported, "Tents are going up and material . . . is on the ground. Many workmen, truckmen and teamsters are busy in the field during the day and some at night . . . and the work getting ready to start after the liquid gold is progressing rapidly."[20]

William B. Osborn, whose father was a lease-trader in the region, recalled that oil scouts from throughout the world rushed to the Greater Seminole field, a number of whom had trouble adapting to the peculiarities of the Oklahoma oil boom. The great influx of people searching for prospective leases placed a great strain on the area's transportation facilities, and automobiles were frequently rented at outrageous prices. At the time, geologists with Shell Oil Company were controlled from The Hague in the Netherlands and, as was the customary procedure, Shell's men submitted their expenses to the company's offices in The Hague for reimbursement. The home office apparently had difficulty justifying the high cost of automobile rentals, however, for the company's geologists in Seminole received a shipment of bicycles by return freight.[21]

Within a week of the completion of the discovery well, it was reported that "anywhere from 20 to 30 new locations," were active in the new field. ITIO began three offset wells from the No. 1 James, which, after being deepened one foot ten days after it was brought in, had a flow increase of from 960 to 995 barrels of oil per day. Development continued at a rapid pace and geologists were busy throughout the area "making new locations for their companies on their leases of close-in acreage." It was reported that "many oilmen from all over the country are here to investigate the new field," and that within "sixty or ninety days many new wells will be in and will materially increase the production of the new field." Their optimism proved to be well-founded.[22]

The excitement generated by the ITIO well was given greater emphasis when, about one quarter mile east of ITIO's discovery, the Amerada Petroleum Company completed a sixty-barrel-a-day oil well in the Seminole Sand at a depth of between 4,258 and 4,277 feet. When the R. F. Garland Independent Oil and Gas Company brought in its No. 1 Fixico in the NW¼ NW¼ SW¼ of Section 26-T9N-R6E, one mile east of Seminole: the rush to the area began in earnest. Although only four feet of the Hunton Limestone was found by the drillers, the well was

Seminole became a one-of-a-kind boomtown with as many hucksters flocking to the crude as honest workers. Here is a typical sidewalk show catering to oil-field workers. *Courtesy of the Oklahoma Publishing Company.*

deepened to the Seminole Sand formation at between 4,065 and 4,073 feet and completed on July 16, 1926, with an initial flow of an "amazing 6,120 barrels" of oil daily. Described as "the largest in the state," on July 29, the well was reported to be producing "255 barrels of high grade oil per hour, after drilling one foot deeper into the sand." By early August of that year, fourteen new wells had been started in the Seminole City field within a week's time, and fifty derricks were drilling deeper holes to reach the high productive Seminole Sand. Among those most pleased with the activity touched off by the Fixico gusher were Garland and his wife, who, after placing most of their possessions in the effort, watched their gamble pay far greater rewards than either had hoped. Like many of the early-day Oklahoma oil pioneers, they were willing to risk almost everything in the emerging industry.[23]

After the Fixico discovery, development of the Seminole City Pool proceeded at an almost frantic pace. Unprecedented drilling activity continued the remainder of the year. Although hampered by snow, rain, and severe cold which lashed the area during the winter, the work continued and the field's production steadily climbed, reaching a peak the week ending on February 28, 1927, with an output of 253,192 barrels of oil daily from 211 wells. Expansion continued, but production dropped off until by

The loading racks for oil at Seminole. *Courtesy of Cities Service Oil Company.*

November 1, 1927, 326 wells were averaging only 64,351 barrels weekly. Nonetheless, up to that date a total of 51,250,633 barrels of crude had flowed from the pool. Most of the wells in the area had initial flows that ranged between 1,000 and 3,500 barrels daily; however, a flow as high as 9,000 barrels per day was reported. The high production generally resulted from "shooting" the wells or the application of air lift. Production dropped fairly rapidly after February, 1927, even though the number of wells in the field continued to rise. By the end of the 1920s, the limits of the Seminole City field had been fairly well defined, covering approximately 3,600 acres of land: to the east, northeast, and southeast of Seminole proper.[24]

Within two months after the Seminole City discovery, oilmen in the Greater Seminole region witnessed another discovery to the north and northwest. F. J. Searight had been working on the No. 1 Youngblood in the SE¼ NE¼ SE¼ of Section 33-T10N-R6E for nearly two years before it started producing gas. Greatly encouraged, Searight continued his operations and, on April 21, 1926, the well, located four miles north of Seminole, was completed as a 312-barrel-a-day producer from the Hunton Limestone at a depth of between 4,090 and 4,157 feet. Although it was at first thought that the No. 1 Youngblood simply marked an extension of the Seminole City Pool, it later proved to be an entirely new pool and was subsequently named for its discoverer: Searight.[25]

Shortly thereafter, the well slowed to a steady 200 barrels daily, from a hole drilled six feet into the Hunton formation. Searight then shut down drilling operations to allow for the construction of a pipeline to carry the crude, but acreage around the drilling site was already "changing hands at from five hundred dollars to eight hundred dollars per acre."

Excitement over the discovery ran high and reached a fever pitch within six months when Searight brought in the No. 3 Youngblood in the NW¼ NE¼ SE¼ of Section 33 as a "gusher, making around 7,000 barrels daily." The bit had penetrated the Seminole Sand formation a scant two feet before the crude rushed to the surface and overtaxed the capacity of the four-inch pipeline carrying the petroleum to storage tanks. Within one twenty-three minute period during the night of October 13, the well produced a total of 117 barrels of oil.[26]

Northeast of the Seminole City Pool and east of the Earlsboro field, the Searight discovery was not so prolific a producer as the previous two wells; however, it maintained a high state of excitement. The initial production of wells in the pool generally flowed between 500 and 2,000 barrels daily, but ranged as high as 4,500 and 5,000 barrels per day of oil possessing a gravity of between 39° and 42°. Maximum daily production was reached the week ending June 21, 1927, which showed an average output of 39,857 barrels of oil per day from forty-two wells. Developed from both the Seminole Sand and Hunton Limestone, the field was extended to approximately seven hundred producing acres of land and, on November 1, 1927, the production from sixty-two wells was 25,663 barrels of oil. A fairly large number of the wells, between 20 and 25 percent, showed signs of water. By November of the following year, the pool had produced a total of 11,282,174 barrels of crude.[27]

The rapidity with which new oil-producing fields were found and developed continued unabated through the first half of 1926. The discovery of the Earlsboro, Seminole City, and Searight fields had reignited the interest of the state's oilmen in the region, and renewed drilling activity resulted in the discovery of even more pools in quick succession. The St. Louis field, southwest of Seminole, situated on the Seminole-Pottawatomie county line, was the next big find. The Darby Petroleum Company and the Independent Oil and Gas Company opened the pool in the SE¼ NE¼ of Section 15-T7N-R4E on July 20, 1926. Initially, the well flowed 125 barrels of oil a day from the Hunton Limestone which was topped at 3,652 feet. Described as "flowing by heads every three hours . . . it is showing heavy gas and had placed a considerable quantity of oil in the earthen tanks which were prepared immediately after the first flow was made." Area oilmen looked "upon the new discovery with optimism and are predicting that the well has possibilities of opening up for Pottawatomie County its first real pool and accompanying oil activity." Their hopes were eventually justified, for although the well proved dry when deepened to the Wilcox Sand, another well completed in the NW¼ SE¼ of Section 25-T7N-R4E flowed at an initial rate of 2,007 barrels of oil on March 20, 1928. That same year, on July 26, the Magnolia Oil Company completed a nearby well for 6,189 barrels per day, and it was estimated that the yield from the more productive leases could exceed 100,000 barrels per acre.[28]

On July 1, 1927, the Little River field was opened by the Indian Territory Illuminating Oil Company's No. 1 House located in the northwest corner of Section 1-T7N-R6E in the south central portion of Seminole County. Flowing from the Seminole Sand at a depth of between 4,017 and 4,028 feet, the well had "an astonishing initial production of 13,541 barrels of oil." With such an apparent bonanza awaiting them, oilmen flooded the region and, by November 1, 1927, fifty-nine wells were in operation producing 45,361 barrels of oil per day. Within six months, July 1, 1927 to January 1, 1928, the Little River field averaged 5,519 barrels per acre; however, one-eighth of the wells drilled prior to November 1, 1927, showed signs of water, suggesting the pool possessed an oil-water contact. Moreover, the Little River which flowed through the field flooded several times during the early development period, and as a result, the field was not exploited as rapidly as the previous finds in the Greater Seminole Pool.[29]

In the fall of 1928, an effort was made by the Sinclair Oil and Gas Company to reexamine the vicinity of the Allen Dome in the southeastern corner of Seminole County. Located in the SE¼ NE¼ SW¼ of Section 7-T5N-R8E, the Sinclair Company's No. 2 Amos-B reached the Wilcox Sand at a depth of 4,158 feet on September 16, 1928, to produce forty barrels of oil per hour. As additional wells were drilled, the pool was developed as a part of the greater Seminole field.[30]

Encompassing such a large number of individual pools within an extensive area, the Greater Seminole field marked an unusual era in the growth of the Oklahoma petroleum industry. In the course of its development, begun in 1923 and reaching its peak of production four years later, major companies, established independents, and new producers strove to obtain the crude from beneath their individual leases before their competitors did. This "law of capture" resulted in an almost reckless abandonment of any form of regulation, exhausting the field's reserves very quickly and swamping the oil market. When such a crisis had faced oil producers in the past, they generally were able to reach an agreement among themselves to slow down production, and the oilmen of the Greater Seminole field made a similar effort. Operators imposed voluntary proration and created a committee for compliance as early as 1926. Unfortunately, the period during which the Greater Seminole field was producing such a huge amount of crude coincided with the development of several other highly productive pools in West Texas, the Texas Panhandle and California, and as a result, the action of the Oklahoma oilmen did very little to decrease the glut of oil on the market or raise prices.[31]

In early 1927, the price of oil plummeted to fifteen cents a barrel for crude testing 28° to 28.9° gravity and thirty-nine cents for oil with 40° gravity or more. The low prices were not the only financial drawback to oilmen in the Greater Seminole field, however. The rush to obtain crude from the ground before another producer drilled a nearby well had resulted in the drilling of many unnecessary wells. Oilmen in the Seminole City Pool alone drilled a total of 3,500 test wells at an average cost of $60,000 per well, with some wells costing as much as $65,000, for a total drilling cost of nearly $210,000,000.[32]

Nonetheless, the sheer magnitude of the crude produced from the conglomerate of pools comprising the Greater Seminole field was startling. During the Seminole boom, pioneer oil man Tom Slick ordered that all of his wells not producing at least 400 barrels of oil per day be shot with nitroglycerine to increase production. Any well incapable of maintaining the required 400 barrels daily was considered a "stinker" and "not worth very much." The peak production for the entire field was reached as early as July 30, 1927, when a total of 527,400 barrels of oil flowed from its wells in one twenty-four hour period. By the first day of September, 1929, the Greater Seminole field was America's foremost source of high-gravity petroleum, 250,000,000 barrels of oil having been produced since July 16, 1926. Based on the average price of $1.50 per barrel at the time, the field generated in excess of $300,000,000 for producers and royalty holders of the area. Within a decade, the wells had produced an estimated $1,009,996,749 worth of crude.[33]

Accompanying the unprecedented growth of the Oklahoma petroleum industry during the 1920s was the need for a greater market outlet in the form of product pipelines, especially gasoline carriers. Harry A. Trower, an early-day vice-president for marketing

Seminole, Oklahoma, at the height of the Greater Seminole boom. *Courtesy of Cities Service Oil Company.*

for Phillips Petroleum Company, recalled that the pre-World War I gasoline industry within Oklahoma presented "a difficult problem" from a marketing viewpoint. Refineries were the only outlet available. During the 1920s, the gross margin of refineries in the Mid-Continent region spiraled downward. With the tremendous production of the Greater Seminole field, as well as East Texas, the decline became even greater after 1926. By 1930–31, crude in the region had dipped to thirty-five cents per barrel, a decrease of forty-six percent in five years, making the construction of a far-reaching interconnecting pipeline system linking Oklahoma production to marketing outlets across the nation a vital necessity.[34]

The result was the formation of the Great Lakes Pipeline Company by six firms: the Pure Oil Company, the Mid-Continent Petroleum Corporation, the Continental Oil Company, the Barnsdall Corporation, Skelly Oil, and Phillips Petroleum Company. Several of the firms already had existing pipelines bringing crude from their holdings in Oklahoma and surrounding areas to their "crude-oriented plants; however, with only a few exceptions, these refineries were located within the Mid-Continent area and had no easy access to expanded markets." Pooling their resources, the six Oklahoma-based companies secured the needed $16,500,000 necessary to finance the Great Lakes line. As originally constructed in 1931, the pipeline was capable of pumping thirty thousand barrels of gasoline per day through a system stretching 1,500 miles: from Okmulgee, Oklahoma, north to Minneapolis, Minnesota, then east to Chicago, Illinois.[35]

Although it owned five percent of the Great Lakes project, Phillips Petroleum Company began work on its own major pipeline in 1930. Running from the Phillips refinery at Borger, in the Texas Panhandle, the gasoline pipeline terminated at East St. Louis, Illinois, 740 miles to the east. This gave the firm another outlet by way of barge traffic on the Mississippi River and greatly increased the market potential for Phillips products. A tariff agreement was also arranged whereby Phillips gasoline could be shifted from the East St. Louis line to the Great Lakes system, expanding the Great Lakes system significantly.[36]

## Chapter 12.
# Oklahoma City

THE BOOM-BUST ERA of the 1920s culminated in one of the most spectacular oil discoveries in the state's history: the Oklahoma City field. Previously, the majority of oil had been discovered in rural areas of the state, near small communities that, in a number of instances, grew into larger towns and cities after the discovery of crude was made in the vicinity. Never before, however, had a large metropolitan area undergone the excitement of a major oil boom. Because of the location of the oil pool (one well was directionally drilled beneath the state capitol and another on the grounds of the executive mansion) it received an unusual amount of attention from journalists and involved an extraordinary number of individuals. The nation's second largest single oil field by 1935, the field provided one of the most spectacular events in Oklahoma's oil heritage, the Wild Mary Sudik, but what is perhaps more important, it also marked the beginning of a period of decline for the state's petroleum production that did not end until World War II.[1]

Literally "born grown" following the opening of the Unassigned Lands on April 22, 1889, Oklahoma City sprang into existence overnight and suffered almost immediately from a shortage of water. In an attempt to avoid the high prices charged by a few individuals who owned water wells in the area, a group of the town's citizens employed "a man versed in boring for oil" to drill a water well. After the unnamed wildcatter had successfully completed the well, he then announced his intention to drill an oil well on "the edge of our thriving city," near what is now the intersection of Northwest 4th Street and the Santa Fe Railroad tracks. An appropriate dedication ceremoney with accompanying speeches inaugurated the project, followed by an afternoon picnic for the citizens. Several weeks later, at a depth of approximately six hundred feet, the driller struck "clear, cool water," which was nevertheless, "as welcome to the townsmen as oil."[2]

As persistent reports of possible locations of petroleum continued to attract oilmen to the region, several other attempts were made to locate oil in the Oklahoma City area before Oklahoma statehood. Claude V. Barrow, a longtime oil journalist of the city, recalled that "most of them [were] honest ventures," but "others [were] purely promotions." C. B. Ames, Marvin Armstrong, John Shartel, and a number of other residents drilled a test well approximately one mile east of Putnam City, but abandoned the hole at a depth of 1,000 feet. In addition, E. J. Streeter, an Oklahoma City hardware merchant, organized a company to drill a well approximately one-fourth mile south of Spencer, a few miles northeast of Oklahoma City. By 1903, the well—contracted by L. C. Hivick—had purportedly reached a depth of 2,002 feet, having found traces of crude "at 1,630 feet in a 20-foot sand and at 1,982 feet in a 9-foot sand."[3]

Another attempt was made to locate oil northwest of Newalla, in southeastern Oklahoma County, but the hole was abandoned after a depth of approximately 1,000 feet was reached. In October, 1913, the Merchants Oil and Gas Company began a well on the G. M. Housh farm in Section 5-T11N-R2W, about five miles southeast of Oklahoma City. Completed in June, 1914, at a depth of 3,001 feet, the well showed only a trace of oil from a shale formation at approximately 2,161 feet. A possible "show" of oil was also found at the City Park, located in the NW¼ NW¼ of Section 13-T12N-R3W, in a "fifteen-foot sand" at a depth of 2,400 feet; however, the well was abandoned at a total depth of 2,786 feet, without locating oil. A dry hole was drilled by the Mutual Oil and Gas Company in the NW¼ NW¼ of Section 7-12N-R3W, and in the northeastern portion of the county, near Luther, the Luther Oil and Gas Company drilled a dry hole in Section 34-T14N-R2E. Although this activity was non-productive, the search continued.[4]

By 1917, however, science was beginning to play an important role in the state's petroleum industry, and geologists were given the opportunity to examine the Oklahoma City area. Some confusion exists over the exact date that George D. Morgan and Jerry B. Newby, early-day geologists, conducted their study. Some years later, Morgan stated that "In either 1917 or 1919, and I cannot remember which year, I worked out the structure at Oklahoma City." Although Everett Carpenter, apparently involved in Morgan's work, stated, "I am of the opinion it was about 1917," Newby emphatically declared that "My work north and northeast of Oklahoma City was done in the early part of 1919." Regardless of the date, these men were involved in the earliest attempt to provide an accurate geological view of the Oklahoma City region. Their efforts were supported by L. E. Trout, who made a reconnaissance of the area and mapped a portion of southern Oklahoma and northern Cleveland counties "Late in 1920 or early in 1921." Accompanying Trout were S. H. Woods, Claude Dalley, and L. R. Trout, who prepared what "seems to have been the first structure map prepared of the region." L. E. Trout was so impressed with the potential of the area that he drilled a well near the Oklahoma-Cleveland county line in Section 36-T11N-R2W in 1925. Although "several minor shows of oil were reported," the well was drilled to a depth of 4,480 feet and abandoned.[5]

The same year, John R. Bunn worked out a "surface high north of the State Capitol" for a "deep test." The well was drilled by the Cromwell Oil and Gas Company in the SE¼ NE¼ of Section 15-T12N-R2W in 1926. Again, "several minor shows in shallower horizons" were found, but the well was not commercial. That same year, E. A. Paschal, employed by the Coline Oil Company, traced the Hennessey-Garber formation south of Oklahoma City, where he recognized "the presence of a fold." In evaluating the findings, Coline Oil Company's chief geologist, C. T. Moore, recommended that three leases be purchased in the area of Paschal's prospect, which resulted in several successful wells.[6]

All of these early geological studies indicated the presence of potential oil-bearing formations to the north, northeast, and southeast of Oklahoma City. As a result, geologists were well aware of the possibilities of the region prior to the drilling of the discovery well of the Oklahoma City field. Because of this knowledge, oilmen demonstrated considerable interest in the area, and in April, 1926, the Rock Island Railroad instigated an overnight "Oil Special" between Amarillo, Texas, and Oklahoma City in order to accommodate them.[7]

However, much credit for the eventual discovery of the prolific Oklahoma City field belongs to Dr. G. E. Anderson, of the University of Oklahoma. Temporarily employed by the Indian Territory Illuminating Oil Company in 1927, Anderson was engaged in examining the Oklahoma City region. Reporting "a southward projecting nose running through Oklahoma City," he "outlined this nose on the Garber-Hennessey contact." Apparently stimulated by Anderson's find, the geology department of ITIO initiated detailed structural mapping of townships 10 through 12 north, ranges 2 and 3 west, in February, 1928. C. L. Wagner, under the direction of J. H. Derden and C. W. Roop, did the work, which was later approved by the chief geologist at ITIO, R. J. Riggs. Once the project was completed, ITIO leased in excess of six thousand acres of land in the Oklahoma City area.[8]

Throughout this period of geological investigation, actual drilling operations continued. The Cromwell Oil and Gas Company drilled a well to the south of the state capitol, in the SE¼ SE¼ SE¼ of Section 15-T12N-R3W. A "fast hole was made for the first 4,000 feet," and the wildcat well was drilled to the contract depth within a hundred days. It was nearly six months before drilling resumed in June, 1926, and although the operation was plagued by water (eight water-bearing sands were found) the bit pierced two "shows" of gas at approximately 2,900 and 3,400 feet; however, they were small and cased off. Drilling continued, and, on November 30, 1926, it was reported that oil brought up by the bailer was clearly visible in the well's slush pit.[9]

The Cromwell Oil and Gas Company's effort "created considerable excitement, . . . and hundreds of cars parked back of the Jodlowski store so that Oklahoma City citizens might get a look at the derrick and see the black sluice in the pit." The hole had been drilled on the Frank Martin farm, and the prospect of finding crude in paying quantities touched off a round of lease speculation in the area. "While many deals are reported and rumors fly thick and fast," one journalist recorded, "John Jodlowski is the only man to verify sales of lease or royalty." Owning fifty-nine acres that adjoined the Martin farm, Jodlowski sold one-half of the royalty on the land for two hundred dollars an acre. In response to the crowd "that has been milling about the location and asking questions," J. W. Lawrence, superintendent of the Cromwell Oil and Gas Company, and M. V. Davis, superintendent of the Reece Drilling Company, which contracted for the well, replied, "See the slush pit . . . That's oil and we believe there [sic] more of it in the hole." Indeed, the well did turn out to be a producer, but not in sufficient quantity to set off a rush to the area. Producing intermittently for several years, it furnished fuel for drilling on more productive sites to the south; however, nearly two years remained before the No. 1 Oklahoma City would touch off the boom.[10]

After leasing a ten-thousand-acre block on the structure for an average cost of only five dollars per acre, ITIO joined the Foster Petroleum Corporation in a joint drilling effort on land acquired from Mrs. Celia Hall in the C SE¼ SE¼ of Section 24-T11N-R3W. Prior to beginning the well, ITIO geologists examined the log of a well drilled by the Firestone Oil Company south of the site, in Section 36-T11N-R3W, that had been abandoned at a depth of about 4,480 feet "after getting several good showings." Work on the ITIO-Foster site, approximately six and one-half miles from the center of Oklahoma City, progressed rapidly. The Oklahoma City No. 1 was spudded in on June 12. "with a gallery of less than a dozen drillers" present.[11]

The drilling operations continued at a fairly steady rate into the fall of that year. Several impressive gas zones were penetrated: a gas flow of 25,000,000 cubic feet per day was recorded between 3,997 and 4,012 feet, and a flow rate of 47,000,000 cubic feet of gas was encountered between 4,805 and 4,816 feet. In late November, 1926, the Arbuckle Limestone saturated with crude was penetrated and casing was set. In the process of drilling the plug with cable tools, pressure from natural gas forced the equipment back up the hole and created a "bridge" by lodging the tools at around 2,500 feet. After weighting the bridge with water as a precaution, the crew began fishing in an effort to clear the hole on November 29.[12]

It was not until December 4, 1928, that the crew was able to move the bridge, and shortly after 3:00 P.M., the pressure of the gas forced the water out of the hole, followed by "the tangled mass of concrete, drilling line and mud." A "few moments later, the tools were thrown up [through] the derrick and . . . dropped on the rig floor." Pushed to nearly a hundred feet above the derrick by the force of the gas, the first "gush" was caught by a stiff north wind and "whipped toward the south." In addition to leaving a film on visitor's automobiles parked on a road to the south of the drilling site, the oil fell in such quantities that it "settled in holes, then filled the ditches of the road to the derrick and created a stream to the ditches along the county road."[13]

"When the first bubbles began to creep up the casing," the workers and onlookers were startled, but as "the white fumes of gas began to form around the top," and the water flowed over the derrick floor," those on the rig "ran like rabbits in all directions, hunting for cover." "Sixty yards away [from the rig] and still running, they glanced over their shoulders to see a white stream of water shoot up half way over the derrick top, and then with a roar like thunder, change darker as the black gold flowed." Caught by the wind, the crude blew to "the south . . . like a fast-pacing horse's tail."[14]

Photographers present to record the event were "spattered," and their lenses "smeared" by the crude. Oil scouts were covered with petroleum, and "reporters felt it dripping down their backs, as they raced madly across an open field for a telephone." Along the section-line near the location, several onlookers had gathered behind a wire installed a safe distance from the rig and had a good view of the proceedings. Any type of flame was prohibited near the well-head due to the danger of fire, and clustered along the section line road, reporters and photographers sat in their "cars, smoking the cigarettes forbidden any closer."[15]

Hurriedly setting up their cameras, several photographers recorded the event, while others less concerned with recording history were "tearing down the road for the automobiles," shouting "Water's pouring out, and gas . . . Move those cars, if you want to save them." Others scampered to report the event to their newspapers. Speeding down the road at "30 miles an hour," two newsmen rushed to the home of Frank Anderson, located about a quarter-mile from the drilling site, to use the telphone and call in the story. Alvin Rucker, a photographer and roving correspondent for the *Daily Oklahoman* reported the scene to his boss:

> Tuesday afternoon was cold, cloudy and windy. There were only two visitor's cars at the well . . . The Press car was occupied by Morris Moore, *Times* reporter, O. A. Huffman, one of the morning tower drillers, and myself. It was time for the *Times* reporter to quit and for me to begin. I drove over to Mrs. Frank Anderson's farm residence so that the *Times* reporter could make his final report of the day. When we

returned to the parking place about 1,000 feet south of the well, I began to relate one of my famous oil field stories . . . I had talked for a half-hour. It was then five minutes of three.

Huffman said: "Get your camera ready. The well's going to clean itself; I can tell by the way the drilling crew is running to turn out the fires under the boilers."

I got out of the car and strapped on the camera. Moore started the engine for the dual purpose of getting the engine warmed up and for making a quick getaway to the Anderson phone . . . .

Huffman then said, "They are running back to the rig; I guess it isn't going to clean itself just yet, but keep the engine running and keep your camera ready."

I got back in the car and resumed . . . the joke . . . when Huffman said, "Stop! she'll be out in a few minutes."

I got out of the car, and Moore took the wheel and with Huffman backed out into the road; leaving me to make the pictures, they went to the telephone.

When they got to the 'phone, Mrs. Anderson cleared the 'phone line for them and they got hooked up with [Claude V.] Barrow, the oil editor. Moore said to Barrow,

"Hold the 'phone; it's coming in in a few minutes."

Huffman was standing on the porch and watching the well. He said,

"Tell them it is in; it's gone over the top for 5,000 barrels!"

I snapped five pictures and ran to the Anderson home and got the car, and said that I would take the films to town while Moore remained to keep covering the story.

Huffman said, "If you are coming right back, I want to go with you . . ."

He and I then drove as rapidly as possible to the *Oklahoman* building.[16]

The well flowed out of control for an hour and fifteen minutes before the gate valves were turned at 4:30 P.M., and the crude was diverted into eight nearby storage tanks. The actual gauge on the first hour's production was 205 barrels, and the first 1,000-barrel storage tank was filled within five hours. At the end of six hours, the well flowed at a rate of 230 barrels of oil per hour. Described as "light green and . . . very hot when it first came out," the crude tested at 42° gravity. The initial twenty-four hour flow was reported to be 4,909 barrels, and it had not yet reached the Wilcox Sand, which was thought to be another hundred feet below in the Simpson formation.[17]

Soon after the well came in, officials of the Indian Territory Illuminating Oil Company ordered that drilling operations be resumed and the hole deepened to the Wilcox Sand. However, because of the lack of proper storage facilities and the chaotic conditions at the drilling site, W. P. Simpson, district production superintendent, postponed the operation. In the meantime, orders were placed for six additional one-thousand-barrel tanks, which would provide a total of fourteen tanks at the well site. The new storage facilities were shipped to the No. 1 Oklahoma City well from the Greater Seminole field, and work crews labored around the clock. In addition, engineers were brought into the area from the Seminole pool to begin a preliminary survey for a pipeline to carry the crude to market. ITIO was also rushing construction of "its proposed camp for employees" in the city area. Within a short time after the completion of the well, one dwelling was nearly finished and several others were planned "in the shadows of the derrick of the new well." In an effort undoubtedly designed to allow the workers an opportunity to clean up the area and keep sightseers out of the way, ITIO barred visitors from the location and required them to remain "only on the edge of the lease." Automobiles were required to stop "far back on the road," to allow the work crew room to operate, and as a further precaution, smoking was prohibited anywhere on the lease.[18]

The excitement generated by the No. 1 Oklahoma City spread rapidly, not only throughout the community and state but nationwide. Proclaiming Oklahoma City an "oil town" and Oklahoma an "oil state," local newspapers urged the citizens to support the petroleum business, in that much depended on the "good will" shown the industry. Thousands of people from surrounding communities flooded the metropolitan area, and "shanty towns" sprang up near the drilling site. Bodine City was established to the south of the well, and Emerson City to the west. Radio announcers spread word of the discovery over the air waves, as oilmen of all types: promoters, lease hounds, workers, brokers, speculators, and scouts, poured into the city, overburdening the hotels and crowding the streets. Never before in Oklahoma had a major oil field been discovered within a large metropolitan area, and the area was a mass of confusion, albeit optimistic and affluent.[19]

By December 16, 1928, the No. 1 Oklahoma City was flowing at a rate of 6,380 barrels per day, and during the first twenty-seven days of its production, 110,496 barrels of crude were brought to the surface. Eventually the hole was deepened to 6,624 feet and produced high-gravity API oil. Within a relatively short time, several other wells were being drilled in the region; however, the ITIO-Foster combine held the majority of leases within a two-mile radius of the discovery well. The only exceptions were some acreage controlled by Tom Slick, Sinclair Oil and Gas

The Oklahoma City oil field. The map shows the location of the Indian Territory Illuminating Oil Company's discovery well, as well as the drilling sites of several other companies. *Courtesy of Mrs. Claude V. Barrow.*

An enclosed derrick in the Oklahoma City oil field. *Courtesy of the Oklahoma Publishing Company.*

Company, the Coline Oil Company, Roxana Petroleum Company, W. R. Ramsey, and Wirt Franklin. In a short period of time, other wells were flowing in the field. By June 20, 1929, the Sinclair Oil and Gas Company brought in its No. 1 Stamper, and seven days later, the Coline Oil Company completed its No. 1 Olds, with a 5,400-barrel flow per day. It was rapidly becoming apparent to even the most skeptical of oilmen that Oklahoma City was situated on top of a major oil field.[20]

On December 6, 1928, the representatives of those companies holding lease agreements in the vicinity of the No. 1 Oklahoma City met in the office of the Wirt Franklin Petroleum Corporation in the Franklin Building on West Second Street in downtown Oklahoma City "to discuss plans of operation." Although Wirt Franklin was evasive and declared that he "did not know whether matters of proration drilling would be taken up," most of the local oilmen were in favor of continuing a previous agreement that limited wells to one to each forty-acre tract. Because of the depth at which the crude was found, drilling in the Oklahoma City Pool was expensive, and such an arrangement would prove advantageous to oilmen. In addition, the continued spread of the drilling activity toward the residential areas of the community posed a number of problems never before encountered, and in an effort to cooperate with local officials, ITIO employed a planning expert from Cleveland, Ohio: Herbert Standley.[21]

By December 12, the Empire Pipeline Company had completed a line from the drilling area to a railroad loading rack about five miles away. With the means available to market the crude, the field expanded even more rapidly northward toward the city limits. In an effort to exert control over the expanding oil field, the Oklahoma City Council enacted a business zoning ordinance to establish limited drilling locations and prevent haphazard expansion. Created on May 10, 1929, the first zone, U-7, covered the area north of Southwest Twenty-second Street, on the town's eastern border. This ordinance was expanded on December 3 of that year to include a second zone, stretching northward from Grand Boulevard and bounded on the west by North High Avenue. Despite the zoning ordinance, few actual restrictions were placed on oilmen and the attendent paraphernalia of drilling operations—rigs, slush pits, wooden storage tanks, steel holding tanks, and earthen excavations—spread throughout the community's industrial and residential areas. In the fever of searching for crude, drilling rigs "reared their crown blocks in school yards, slush pits were dug on playgrounds, and storage tanks were built in alleys." Dean A. McGee, chief executive officer of Kerr-McGee Corporation, was employed as chief geologist for Phillips Petroleum Company during the Oklahoma City boom and unraveled the geological mysteries of the pool so that the field could be expanded to its full potential. McGee recalled that wells were being drilled everywhere. In one instance, Phillips drilled two wells in a local refuse dump located in the Walnut Grove region of the city, and appropriately enough, they were dubbed the Ash Can No. 1 and Ash Can No. 2.[22]

The U-7 Zone and a strip of territory bordering the city on the south were opened to oil development on March 4, 1930. At the same time, the Oklahoma City Council created an ordinance designed to prohibit drilling within the corporate limits of the city,

Preparing a drilling site in front of the governor's mansion in Oklahoma City. *Courtesy of the Oklahoma Publishing Company.*

with the exception of certain exempted areas. Within the areas where drilling was permitted, wells were limited to one per block, and royalties were paid to residents of such blocks. However, the regulations were practically meaningless in that it was difficult for officials to determine which block was being drained of crude. Within five months, the area from Southwest 15th to Southeast 29th streets between North Byers Avenue on the east and the Santa Fe Railroad on the west was opened. This was followed on November 25, 1930, with the area north of Grand Boulevard between the Santa Fe Railroad and Santa Fe Avenue. Eventually, when the land surrounding the state capitol was opened to petroleum development, a bitter controversy developed between city and state officials over who would control the drilling operations.[23]

The boom period of the Oklahoma City field resulted in the establishment of several of the state's oil companies, and one in particular, Harper Oil Company, grew directly from the rush for leases in the community. The founders of the firm, F. E. Harper and Roy J. Turner, both had been associated with the land boom in Florida. Upon learning of the pool's opening, they returned to Oklahoma City. Having previously known one another, they formed the firm of Harper and Turner "in order to better exploit the opportunities that the discovery . . . presented." Recognizing that the field would eventually encompass the community, the two partners immediately acquired as many leases as possible in potential producing areas.[24]

With the wells limited to one per block, Harper and Turner quickly became experts at acquiring the twenty to thirty different lease agreements necessary to control a single block. The competition was keen. With several companies bidding on the same block, it was often quite a task for one company to secure an agreement from all the property owners in question. Hold-outs were common, driving the price of a lease to a high level before they signed. Nonetheless, the two men were able to piece together enough acreage to generate sufficient capital to enter the production phase of the industry. Harper and Turner, usually a

Oklahoma City firemen battling a wild gas well near the state capitol. *Courtesy of the Oklahoma Publishing Company.*

non-operating partner with other oil companies, invested in several wells in the Oklahoma City region. Enjoying a considerable amount of success, the firm continued to grow until it was recognized as one of the community's most successful enterprises.[25]

In 1939, the company entered the operating phase of the oil business by drilling and completing several wells in Oklahoma, Lincoln, Pottawatomie, and Cleveland counties. During World War II, Harper and Turner were involved in the West Edmond field, and Turner, having entered politics in the interim, was elected Governor of Oklahoma in 1945. Reorganized in 1948, the corporation continued its exploration activities and discovered the Peavine-Wilcox Pool in 1950. Four years later the partnership was dissolved, but the firm continued to operate as the Harper Oil Company, organized in 1955.[26]

The most exciting event associated with the Oklahoma City field was the completion of the highly publicized "Wild Mary Sudik." An ITIO well located in the southern portion of the pool in the NE¼ NE¼ SW¼ of Section 31, T11N-R2W, the Wild Mary was also the discovery well of the prolific Wilcox Sand-producing horizon. The pool's first producing well in 1928 had given a good indication of the strong gas pressure associated with the field, and in March, 1930, the Kinter No. 5 drilled by the Sinclair Oil and Amerada Petroleum companies blew out and caught fire. Consuming an estimated 1,200,000,000 cubic feet of gas before it was brought under control, the conflagration continued for twelve days; however, this was nothing when compared to the Mary Sudik, the wildest well in Oklahoma's history. The No. 1 Mary Sudik was brought in near daybreak on March 26, 1930, when the bit penetrated the top of the Wilcox horizon. The crew had neglected to keep mud in the hole while they withdrew the drill pipe and, with a thundering roar, a column of "almost invisible" gas rose "high above the forest of 110 feet [sic] derricks." Gradually tinted with oil, the gas "turned a decided brown." It was estimated that the gas flow was as high as 200,000,000 cubic feet daily with twenty thousand barrels of crude, and this tremendous pressure threw the petroleum so high into the air that the north wind carried a film of oil "as far south as Norman, eleven miles away."[27]

The crew struggled to "connect to the top of the casing the master gate that will choke the rebel into submission," but their efforts failed, and the wild well continued to spew gas and crude. "Men in slickers and steel helmets, wearing goggles, with cotton stuffed in their ears to keep out the deafening roar," did their best, but to no avail. The wind shifted, and oil was reported falling to the ground at Nicoma Park, some "10 or 12 miles distant," as the oilmen labored to save what crude they could. By March 31 it was estimated that more than 10,000 barrels of oil had been trapped by earthen dikes around the Mary Sudik, with 6,346 barrels being recovered "from the cellar and pits at the well" in one twenty-four hour period. Another well one half mile north of the Mary Sudik was brought in by the Skelly Oil Company, and for a time, it appeared that there might be two runaway gushers; however, the Skelly crew managed to bring their well under control. On March 30, it was feared that the "Wild Mary" would burst into flames after a haystack a half-mile from the drilling site caught fire, but the burning haystack was extinguished within the hour, and the fire did not spread.[28]

Although oilmen were optimistic that the Sudik well would also be quickly brought under control, their hopes proved futile. After five days of struggling with the gusher, they finally placed a master-gate on the well. However, the pipe fittings were

Wells were drilled throughout the residential areas of Oklahoma City. Here Mrs. G. K. Sutton, 1109 Northeast Seventeenth Street, is cleaning a container of lemonade that was suddenly streaked with oil from a runaway well. *Courtesy of the Oklahoma Publishing Company.*

weakened by the cutting action of the sand forced through the well by the tremendous gas pressure, and it did not hold. Within a few hours, the Mary Sudik was wild again. Fearful that the escaping gas would ignite, ITIO officials asked that all fires in the southern portion of the field be extinguished. The other companies cooperated fully, and fire-fighting crews were gathered at strategic points throughout the region. In addition, only "necessary traffic was permitted on the roads in the fields," and "aviators working out of the local airports were advised to fly high in passing over the field," after workers had been forced to "scramble away" on several occasions because of low-flying planes. Such actions were intended to diminish the chances the gas would be

Wells were drilled even on the grounds of the capitol building during the scramble for black gold in the Oklahoma City field. *Courtesy of Helmerich and Payne International Drilling Company.*

ignited from sparks from the engines of the aircraft.[29]

Not only were the citizens of Oklahoma City fascinated by the antics of the "Wild Mary," but the exciting broadcasts of Floyd Gibbons on NBC describing the runaway well captivated imaginations throughout the radio world. In compliance with a request of producers in the Oklahoma City Pool, C. R. O'Neal, the state fire marshal, created a fire hazard zone in the extreme southern end of the field. The regulations were put into effect immediately, and all roads in Section 31, the eastern halves of Sections 30 and 36, and the western half of Section 32-T11N-R2W were closed. Armed guards "charged with refusing admission to all persons other than those holding permits or on company business" turned away casual visitors and sightseers. Small fires near the well continued to plague workers and caused much excitement. On April 2, three fires were spotted in the area. The worst, a lake of blazing waste oil and a brush fire east of Moore, were in dangerous proximity to a ravine in which escaping crude from the Wild Mary was trapped. However, they were extinguished without the feared inferno.[30]

In an effort to bring the well under control, a special "bonnet" was prepared by the American Iron and Machine Company of Oklahoma City. Weigh-

The Wild Mary Sudik. Note the men in the oil-covered clothes working to cap the well. *Courtesy of Cities Service Oil Company.*

ing three thousand pounds, it was designed to "be stabbed over [the] top of the casing and anchored in place." Because it was impossible to screw a valve in place, packing was designed to insure a close fit around the top of the casing, and pipelines could then be connected to the bonnet. By April 4 all was in readiness, and more than a hundred workmen were assembled for the task of taming the No. 1 Mary Sudik. Officials were worried that the gas and sand might "cut away the casing at the top and make it impossible to cap the well," and as the time neared, H. V. Foster, president of the Indian Territory Illuminating Oil Company, remained in constant touch with the events in Oklahoma City via private telegraph and telphone wires from his office in Bartlesville.[31]

The men first fastened a "Shaffer rant" to the top of the well's casing. The tremendous pressure and the roar of escaping gas made the task very difficult. Some of the workmen wore "handkerchiefs around their faces to ward off the oil mist, most of them wore slicker suits, and a few on the derrick floor had on army helmets as a protection against falling rock or fragments of the steel derrick." "More than once it seemed the terrific pressure of the gas would fling the rant clear of the derrick," but the men succeeded in fastening it to the casing. At 6:15 P.M., "weary workmen turned a ponderous valve with grab hooks . . . [and] the roaring column of oil and gas subsided almost in a flash and an epic battle was at an end." Conquering the well was only the beginning, however, for rigs on adjoining leases had to be cleansed of a thick coating of crude, and the oil-drenched ground around the Sudik well plowed under.[32]

Although the runaway well had been tamed, the state fire marshal continued the strict blockade of the immediate area until the oil around the well was removed. However, drilling and other petroleum related activities were allowed to resume shortly after the well was tamed. The wind quickly dissipated the dangerous "low-lying white banks of gas that . . . [had] been hanging about the well," and when the flow of oil stopped, the derrick became visible. The twisted drilling tools that were thrown from the well appeared as "a gnarled, knotted mass of iron and steel"; however, the "100-foot steel derrick, dripping with its satin sheen, was intact except for battered crown block railings and a shattered crow's nest." During its rampage, homes "five or six miles around were showered" with crude and thousands of acres of nearby land were soaked with oil. An estimate of between $100,000 and $1,000,000 was lost by the owners of the well. Although there were other wild wells in the Oklahoma City field (one turned the North Canadian River into a blazing stream of fire and threatened to turn the city into an inferno) none would ever capture the imagination of the public as did the Wild Mary Sudik.[33]

The Oklahoma City field continued to expand until by 1935 it had become the second largest American oil field, covering an area of 11,000 producing acres with a total cumulative production of 290,730,062 barrels of crude. Drilling continued, and by 1940, the field covered 13,325 producing acres of land with a cumulative production of 475,640,053 barrels of oil and more than 800,000,000,000 cubic feet of gas; but paradoxically, production was on the decline. Even so, as recently as 1968, the field had produced a total of 733,706,000 barrels of oil and 1,700,000,000,000 cubic feet of gas, to remain among the ten largest oil fields in the United States. The output of many of the wells in the Oklahoma City field was tremendous. J. Sam Williams, a longtime supervisor in the land and geological department of Phillips Petroleum Company, recalled that the gauges on some wells on the Walnut Grove lease (which Phillips helped develop) indicated that if the valves were opened to allow full production they were capable of producing in excess of 100,000 barrels of crude daily. The huge production of the Oklahoma City and Greater Seminole fields, however, as well as other oil-producing areas of Oklahoma and new locations elsewhere in the United States, had a catastrophic effect on the price of crude. This, combined with the problems of controlling an oil field located in a large metropolitan area, played a major role in the regulation of the petroleum industry within the state of Oklahoma.[34]

## Chapter 13.
# Regulation and Conservation, 1910-31

DURING THE FIRST FEW YEARS after statehood, Oklahoma officials played a limited role in the regulation of the oil industry within the state. Several factors accounted for this inactivity: the office of the chief mine inspector was understaffed, and the limited personnel hindered the discharge of the duties of the office; legislative activity was not concerned with the oil companies within the state; those laws that were enacted were often quickly struck down by federal courts; and the frontier heritage and spirit of Oklahoma often made legislators reluctant to interfere in what many believed to be private business practices. Most of the action taken by state officials was directed toward controlling the "monopolistic tendencies" of the oil industry and in assuring state citizens the availability of natural gas. Indeed it was the independent oil producers of the state who took the lead in calling for an end to the "intolerable conditions" in the oil fields.[1]

The demands of the independent producers in conjunction with the huge production of the newly discovered oil fields successfully instigated more state control over the industry immediately prior to America's entry into World War I. At the same time, the Oklahoma Corporation Commission began to exert greater authority in dealing with the problems of the state's oil production. By the spring of 1913, the waste of natural gas in Oklahoma's oil fields was approaching scandalous proportions. It was estimated that more than twenty thousand dollars' worth of gas was lost daily in the Cushing area alone. Another federal report issued that same year contended that the daily waste of natural gas in Oklahoma was equivalent to ten thousand tons of coal, and that eighty percent of this waste was preventable. Many of the state's citizens viewed these reports as the beginning of an effort by the federal government to extend its authority over Oklahoma's oil fields. This resulted in a push for preventive legislation on the state level to offset any such attempt.[2]

When the Fourth Oklahoma Legislature convened in early 1913, the independent oil producers, along with other individuals and groups interested in state regulation of the petroleum industry, were successful in securing passage of two regulatory acts. "House Bill No. 723," which was approved on March 26, 1913, declared gas pipeline companies to be "common carriers." Required to take natural gas on a ratable basis "without discrimination," they were prohibited "from taking more than twenty-five (25) percent of the daily natural flow of any gas well or wells unless for good cause shown." "Senate Bill No. 130" provided additional strength for the state's regulation of ratable production of surface lease-holders from the common reservoir of natural gas. Stating that "In case other parties, having the right to drill into the common reservoir of gas, drill a well or wells into the same, then the amount of gas each owner may take therefrom shall be proportionate to the natural flow of his well or wells to the natural flow of the well or wells of such other owners of the same common source of supply of gas," it provided "that not more than twenty-five percent of the natural flow of any well shall be taken." However, perhaps the most important aspect of the Senate act was its label of any violations as "grand larceny"—the first such provision enacted into law, and one which gave some teeth to the legislation. In addition, the Oklahoma Corporation Commission (OCC) was designated the regulatory agency for the enforcement of the bill.[3]

Shortly after the passage of these acts, approximately 250 independent oil men convened at Oklahoma City on April 23, 1914, to form the Independent Producers League (IPL), which advocated regulation of the petroleum industry. With C. F. Colcord serving as president, M. C. French as vice-

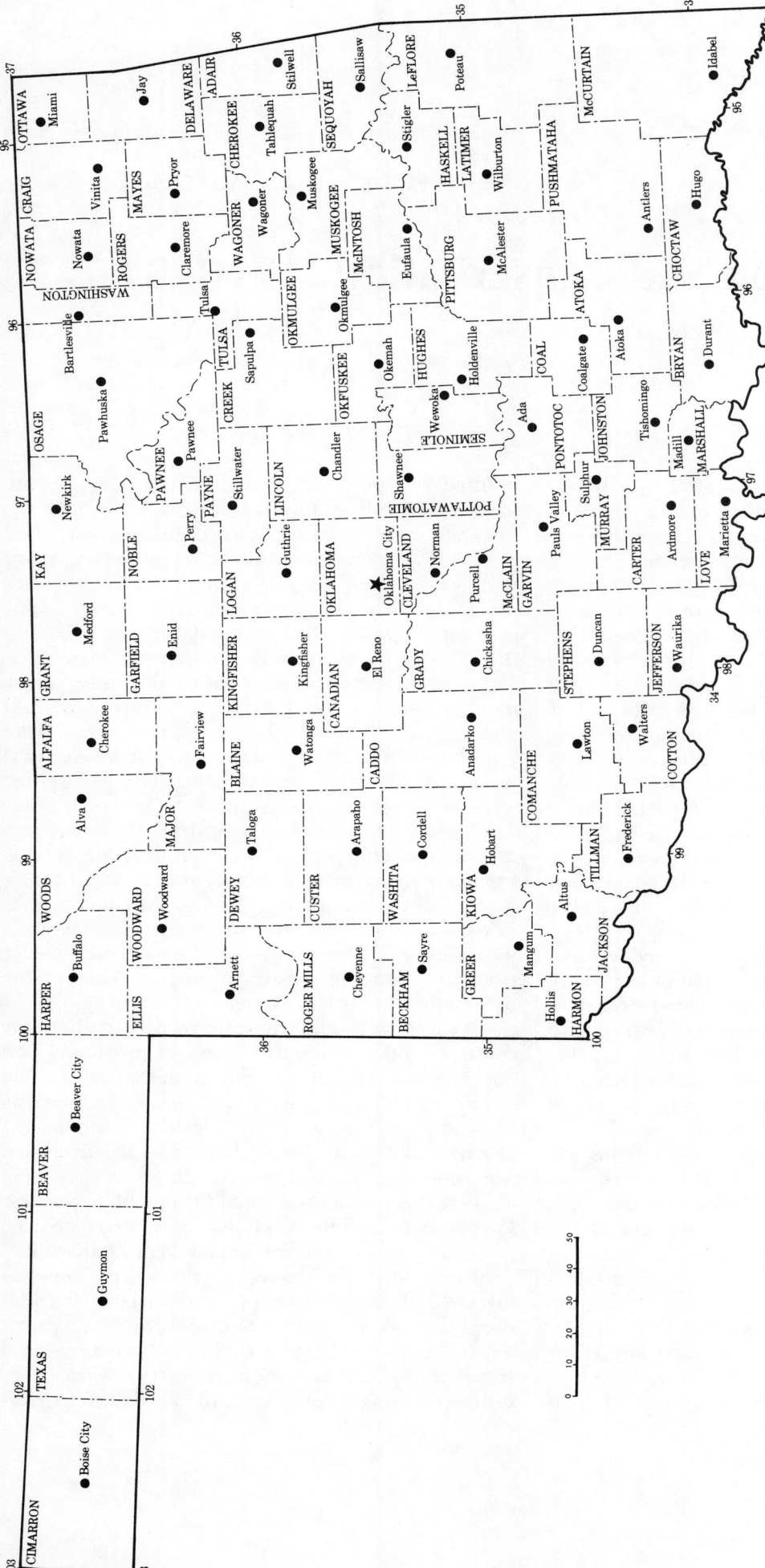

Oklahoma's seventy-seven counties. Shortly after Oklahoma became a state, the original seventy-five counties were expanded to seventy-seven. Petroleum was produced in huge quantities in a wide band stretching from the rich Osage Nation (now Osage County on the Kansas border) southward to the rich finds around Ardmore near the Red River. *From Morris et al., Historical Atlas of Oklahoma.*

president, E. E. Brown as secretary, and Robert Galbreath and Wirt Franklin elected as directors, the group's purpose was "to discuss the control of prices of crude and its products and the inadequate pipeline facilities of the state." The problem facing the oilmen was the tremendous output of crude from the Oklahoma oil fields, and the depressing effect this was having on the price of petroleum products. Drafting a series of resolutions in an effort to ease the situation, the IPL requested that all interstate pipelines be made common carriers and brought under the control of the Interstate Commerce Commission; that pipeline companies be prohibited from production operations; that the federal government construct a pipeline from Oklahoma to the Gulf of Mexico to provide Indian wards a means of disposing of their oil holdings "at reasonable prices"; and that "the monopolistic pipeline companies be compelled to carry and transport oil at a reasonable price." As time progressed, the IPL and its officers became active proponents of fair and just treatment, as well as conservation of petroleum resources.[4]

Following close on the heels of the IPL meeting was the issuance of two orders by the Oklahoma Corporation Commission that marked the beginning of an effective regulation of the petroleum industry by the state. Issued on May 7, 1914, "Order No. 813" and "Order No. 814" were "America's first proration" decrees designed to end discrimination in the Oklahoma oil fields. "Order No. 813" was expressly directed at the Cushing field, and formalized an agreement (previously reached among area pipeline companies) to share the expense incurred in hiring an inspector and two assistants. In turn, the commission was to insure that pipelines were taking production on the ratable basis, and to ascertain "the amount of oil that should be taken." All companies accepting the arrangements were not to "be considered by the commission as violating any of the . . . laws of Oklahoma with reference to discrimination or common purchasers." "Order No. 814" was the result of charges by the attorney general of Oklahoma, Charles West, regarding price discrimination in the Healdton Pool by the Magnolia Oil Company. The ruling required Magnolia to increase its takings from the field, establish a reasonable rate, and pay for some storage tanks constructed by area producers. An inspector was appointed "to prevent any discrimination," and the president of the Ardmore Producers Association was named as an umpire to settle all controversies. Although West failed to have the price of oil fixed by the OCC, these acts were initial steps of the assumption of regulatory authority by the commission.[5]

By 1915 the Fifth Oklahoma Legislature quickly passed two acts designed to implement a plan for oil and gas conservation within the state. "House Bill No. 168," approved on February 11, 1915, prohibited the "waste of crude oil or petroleum," provided for "the equitable taking of the same from the ground," and conferred authority upon "the Corporation Commission" to enforce the legislation. The act clearly stated that "whenever the full production . . . of crude oil . . . in this State can only be obtained under conditions constituting waste . . . then any person, firm or corporation . . . may take therefrom only such proportion of all crude oil . . . that may be produced therefrom without waste." Violation of the law was punishable by a five thousand dollar fine and/or thirty days in the county jail. Dealing with the waste of natural gas, "House Bill No. 395" allowed proportioning of gas production among producers, "whenever the full production from any common source of supply of natural gas in this state is in excess of the market demands." In addition, it made "every person, firm or corporation . . . engaged in the . . . purchasing and selling [of] natural gas in this state . . . a common purchaser." Violation of this law was punishable to the same degree as "House Bill No. 168." The two basic principles of the legislation were "that property rights in oil are subject to common ownership," and "that state authorities have the power to restrict output in the interest of conservation or public concern with a wasting natural resource." Such assumptions by state authorities were to have a great effect, not only on the petroleum industry within Oklahoma but nationwide as well, for it was "the first 'conservation' law enacted."[6]

The Oklahoma Corporation Commission was quick to issue new orders to comply with the new acts, and the huge production of the Healdton field presented the first opportunity to implement the 1915 legislation. In May, 1915, the OCC received a request from Healdton area producers for a set of rules to govern the region. In particular, the Ardmore Producers Association accused the Magnolia Pipeline Company, the Ardmore Refinery, and several area producers with wasting oil through overproduction. This necessitated extensive storage in open earthern tanks which increased the loss of crude through evaporation and seepage. The complainant declared that the output of the field in May, 1915, was approximately 70,000 barrels of crude daily, while at the same time the available pipelines and refineries were

Prices falling. By 1914 the tremendous output of many Oklahoma wells, such as this one in the Cushing field, was beginning to have a depressing effect on the price of petroleum products. *Courtesy of the Petroleum Publishing Company.*

capable of handling only approximately 21,000 barrels of oil per day. The difference of 49,000 barrels was held in open tanks and when the region was struck by heavy rains "much of this earthen storage was carried off by the floods . . . and greatly injured the crops and vegetation on farm lands . . . and also the water supply." Clearly wasteful, the continued operation of the field in such a manner would only contribute to future losses.[7]

As a result of the complaint, the corporation commission conducted an investigation of the conditions in the Healdton vicinity, and on June 5, 1915, issued "Order No. 920." Iterating the obligation of the OCC to the citizens of the state and the principle of conservation of oil and gas, the ruling declared that "Would it not be a spectacle for the public to assemble each morning and witness the destruction of eight thousand barrels of a commodity that is absolutely necessary for the prosperity and business of the country—a commodity which once being destroyed can never be replaced by any artificial means known to science?" Continuing, the commissioners stated that "We are now exploiting the greatest oil fields of the world. . . . This oil should be produced in order to prevent waste, not necessarily to influence the price of crude oil. It should be produced and preserved in such manner as not to be destroyed, so that the public . . . may receive the full benefits therefrom, and in order that reasonable prices for the product may be enjoyed by the public at large for a number of years." Contending that the Oklahoma Legislature "had in mind . . . two objects. . . . First, to regulate the production of oil by the operators so that the weaker or small producer would

Oil conservation. The early actions of the Oklahoma legislature were designed to prevent the waste of oil and natural gas and to conserve these natural resources for the future. In 1917 the solons required oilmen to notify state officials immediately of any "fires which destroy crude oil or natural gas." *Courtesy of the Bartlesville Public Library.*

be guaranteed his pro rata part of the oil; and second, that the oil should be produced and preserved in such a manner that the public would enjoy the full benefits thereof at reasonable prices, not only for the present time, but for years to come," the corporation commission issued ten rules conforming to these principles.[8]

Prohibiting the use of earthern reservoirs, sometimes referred to as "pond storage," in the Healdton field, the regulations also provided that "No operator . . . shall take from the potential production of the pool more than his fair and equitable proportion thereof." To insure that all would be treated equally, A. L. Walker was specifically given the power to "prohibit the raising of oil from the sands to the surface except such as can be marketed under these rules and regulations."[9]

This action was supplemented on September 1, 1915, by "Order No. 937," pertaining to the conservation of natural gas. Not only significant in its regulatory scope, this ruling by the OCC marked the first time the commissioners took the initiative in issuing a ruling without first receiving a complaint. After hundreds of pages of testimony were recorded at meetings conducted throughout the state, the final ruling concluded "That in the past eighty per cent of the gas opened up in this state has gone to waste and that the present daily waste, conservatively estimated, is 200,000,000 cubic feet." Continuing, the commissioners stated "That by proper effort upon the part of the owners and operators of oil and gas property in the state the waste of gas can largely be prevented." To encourage such conservation, the corporation commission issued twenty-eight specific rules governing Oklahoma's gas production.[10]

Specifically, Rule Number 1 required that "Natural gas shall not be produced in the State of Oklahoma in such manner and under such conditions as to constitute waste." This was a reaffirmation of the intent of "House Bill No. 395," passed earlier that year. Additional regulations spelled out in "Order No. 937," including a definition of waste to mean the "(a) escape of natural gas in commercial quantities into the open air, (b) the intentional drowning with water of a gas stratum capable of producing as in commercial quantities, (c) underground waste, (d) the permitting of any natural gas well to wastefully burn, and (e) the wasteful utilization of such gas." Also, the commission ruled that: once the production of gas was "in excess of the market demands," persons, firms, or corporations were prohibited from "securing any unfair proportion"; anyone "engaged in the business of purchasing and selling natural gas" was declared a "common purchaser"; all gas produced in the state was to be "measured by meter"; the corporation commission could "prescribe rules and regulations for the determination of the natural gas from any and all common sources of supply"; the OCC "shall regulate the taking of natural gas from any and all common sources of supply"; and before the right of eminent domain could be invoked "a proper and explicit authorized acceptance of the provisions of the law" was required.[11]

Outlawing several common practices, "Order No. 937" required a series of corrective measures to be implemented among the Oklahoma gas fields. Among these were requirements for a separate slush pit "for the reception of all pumpings from clay or soft shale formations," the plugging of dry or abandoned wells to protect sources of fresh water, the securing of proper anchorage for drilling equipment where the potential gas pressure was unknown, the prohibition of oil and gas from different strata, and the sealing off of any oil, gas, or water stratum encountered while drilling. In addition, no gas formation was allowed to be left open more than three days without "the application of mud-laden fluid" to prevent the escape of gas; all oil was to be separated from any gas deposits encountered; if necessary, producers were required to install a separating device to segregate the oil and gas; gas wells were prohibited from producing from different sands at the same time through the same casing; and vacuum pumps could not be installed without the permission of the corporation commission. Also, no casing could be pulled without first flooding the well with mud-laden fluid; the same had to be utilized "when it is thought advisable to do so in order to avoid existing underground waste, pollution or infiltration"; whenever requested, well logs and plugging records were to be furnished to OCC officials; gauge reports were to be sent to the corporation commission on the first and tenth day of each month; and gas production was limited to twenty-five percent of a well's potential capacity, based on the previous month's gauge.[12]

Coming at a time when Oklahoma was among the nation's leaders in oil production, the actions of the legislature and the corporation commission caused quite a stir throughout America and pushed Oklahoma to the forefront of the petroleum conservation question. Nonetheless, the oil fields of the state continued to be plagued with problems as production from the Mid-Continent area increased. In an attempt to alleviate the curse of overproduction, on July 16, 1917, the corporation commission issued "Order No. 1299," and its forty-one rules became "the bible" of the state's oil industry for a decade and a half.[13]

Prior to issuing "Order No. 1299" the corporation commission reviewed previously issued regulations, discussed their effectiveness and, in some cases, suggested alterations. Many of the new guidelines simply carried over the provisions of earlier orders, and the definitions of waste and commercial quantities were essentially the same. Likewise, in "Order No. 1299," there was little, if any, alteration in the requirements that natural gas had to be confined, the ratable taking of gas, the common purchaser rule, discrimination by common purchasers, the metering of gas, the right of the commission to regulate the taking of natural gas, the filing of a "proper and explicit authorized acceptance" of eminent domain, the enjoinment of conservation officers to enforce the rule concerning eminent domain, the filing of log and plugging records with the OCC, the necessity of proper anchorage, the provision for separate slush pits, restrictions from the production of oil and gas from different strata, the sealing off of strata, the utilization of mud-laden fluid in completed wells, the protection of fresh water, the installation of separating devices, the restriction on vacuum pumps, the reporting of well gauges on the first and tenth of each month, and the restriction of well production to twenty-five percent of potential capacity.[14]

However, "Order No. 1299" went further than

the earlier OCC decisions, encompassing several previously unregulated practices and expanding the amount of control of the commission's agents. Enjoining all oilmen to "use every possible precaution . . . to stop and prevent waste of oil and gas," the order prohibited allowing these valuable resources "to leak or escape from natural reservoirs, wells, tanks, containers, or pipes." Notice was required of any intention to drill, deepen, or plug a hole; the OCC had to be provided the well's exact location; and five days' notice was required prior to beginning drilling operations. The regulations regarding the plugging of dry or abandoned holes were modified and strengthened. Any plugging operation was required to be conducted under the supervision of a corporation commission agent. The wells had to be thoroughly cleaned, and the mud-laden fluid had to have a maximum density and weight. The same rules were applied to the introduction of mud-laden fluid into a well containing water. Very stringent restrictions were placed on the practice of shooting wells. No wells were to be shot with saltwater, or shot in such a manner as to "let in salt water or other foreign substance injurious to the oil or gas sand." Reports of shooting operations were to be made to the corporation commission. These were to contain information on the condition of the well "before and after shooting, [and] per cent of water in well before and after shooting." In addition, any wells sustaining "irreparable injury" were to be immediately abandoned and plugged. Oilmen were also required under the 1917 regulations to "immediately notify the Commission by telegraph or telephone and by letter of all fires which occur at oil and gas wells or oil tanks." Any tanks struck by lightning and "any other fires which destroy crude oil or natural gas, and . . . any breaks or leaks in tanks or pipelines from which oil or gas is escaping" were also to be reported.[15]

The commission extended some of its rules to pipeline companies which, under "Order No. 1299," were required to make reports to the OCC concerning various aspects of their operations, as well as daily listings of oil and gas purchased from different wells or parties. Specific conditions were prescribed concerning the coupling of pipeline companies and oil or gas wells, and certificates stating compliance with the conservation laws of Oklahoma were required. The powers of the commission's agents were strengthened. Accurate records concerning drilling, redrilling, deepening, formations, casing, and other pertinent information were to be maintained at the well sites, and the commission's agents were granted access to the lease and all records. In addition, the conservation agents were to "cooperate with and invite the cooperation of the oil and gas inspectors of the United States Bureau of Mines of the Department of the Interior," and "assist in the enforcement" of the rules and regulations of the corporation commission.[16]

Although "Order No. 1299" remained the mainstay of Oklahoma's petroleum conservation effort for nearly fifteen years, several shortcomings were to become apparent. Perhaps foremost among these was the advance of petroleum technology, which introduced previously unknown factors into the question of oil production. Nonetheless, the pattern for regulation of the oil industry by state officials had been pretty well determined and, as demand increased, offset the problem of overproduction in the fields. However, with the opening of vast new pools during the mid-1920s, both within Oklahoma and elsewhere across the nation, the specter of a glut of overproduction again reared its ugly head and increased agitation for additional regulations.[17]

Many Oklahoma oilmen supported these regulation efforts for the sake of conservation and a more rational development of petroleum reserves. Foremost among these was Henry L. Doherty, the founder of Cities Service Oil Company. Doherty's research division of Cities Service had proved that the flow of crude from wells increased when natural gas was dissolved in reservoir oil. Likewise, when the level of natural gas was diminished, the oil flow decreased. As a result, Doherty was a strong advocate of regulation that would allow unitization of oil fields, with a single operator selected to develop the region. This would permit the greatest production from a pool with a minimum number of wells and greatly decrease the waste of natural gas, thereby retaining it within the reservoir.[18]

To prove his theory regarding the role of natural gas in the production of crude oil, Doherty assigned the research to Charles E. Beecher, formerly of the Bartlesville Research Center, but then employed by the Empire Companies. Convinced that nothing would come of his research, Beecher was hesitant about taking the assignment. However, Doherty persuaded him at least to make the attempt. Beecher began work in February, 1924. His experiments were conducted in great secrecy. Beecher contended that the fluid needed to be less viscous in the formation than on the surface, and although it was generally conceded that gas pushed the oil through the underground formations, this theory could not be proved.

Oklahoma City, 1932. One problem that thrust the regulation of the oil industry into the limelight was the expansion of the Oklahoma City field into the residential and commercial sections of the city. *Courtesy of the Oklahoma Heritage Association.*

Eventually, Beecher's work led to the development of a viscosimeter which allowed him to saturate oil with natural gas to determine the viscosity. The results were astounding. The presence of natural gas in crude greatly reduced its viscosity, and the higher the gas pressure, the greater the reduction. To Beecher and Doherty, the conclusion was obvious: the practice of allowing natural gas to escape into the air actually reduced the amount of oil that could be obtained from wells, and one of the best ways to insure that enough natural gas was present to increase well production was to adopt a policy of unitization.[19]

In 1926, Doherty presented a paper to the Federal Oil Conservation Board in which he stated his arguments. He contended that the geologic waste associated with the development of American oil fields was totally unnecessary. With considerable foresight, he declared that the impending shortage of petroleum made it imperative in the national interest to stop this waste. Continuing, Doherty offered a program to accomplish such an objective by the unitization of all new discoveries. Because of his great interest in advancing the technology of the petroleum industry, he suggested the development of secondary recovery attempts in seemingly depleted oil fields. Dissatisfied with the progress of the attempt by Oklahoma oilmen to accomplish voluntary regulation, Doherty proposed mandatory reorganization of the production process and enforced unitization. Going even further, he charged that the oil industry was not operating in the national interest if it refused to adopt advancing technology to achieve superior production practices.[20]

Needless to say, Doherty's proposals were not greeted with optimism by many petroleum producers. The reaction against his ideas was severe in some instances, even to the point of almost expelling him from the American Petroleum Association. Although his criticism struck at the very foundation of the oil industry's interpretation of the peculiar "petroleum problem," Doherty continued his campaign and persisted in presenting his beliefs equating the elimination of natural gas waste with increased production of crude. Eventually, his theory was accepted, and unitization became a common practice.[21]

Another facet of the petroleum industry regulation issue was brought into the limelight when the Oklahoma City field expanded into the residential and commercial sections of the community. City officials attempted to control the expansion through zoning ordinances that described which regions could be opened and specified the type of drilling operations permitted within these areas. Several oil companies retaliated with a series of court actions that hamstrung the city's efforts to such a degree that the attempt at planned expansion was undermined. This resulted in a mad scramble to drill more wells, and as might be expected, a morass of legal problems.[22]

On May 10, 1929, the Oklahoma City Council established the U-7 Zone as a limited drilling area within the city limits. However, there was little actual control, and in the search for oil, the encroachment of derricks, slush pits, storage tanks, and other surface equipment into residential and commercial areas continued. The controversy over the invasion of the metropolitan area reached a climax on May 5, 1932, when Oklahoma Governor William H. Murray, in an effort to support the city council, declared martial law over the community and prohibited drilling operations outside of designated areas. Characterizing the drilling operations as a "danger to homes and the business section of the city," and pointing to the "possible loss of life through fire hazard," Murray authorized the adjutant general of Oklahoma, Brigadier General Charles F. Barrett, to use the National Guard, if necessary, to enforce the drilling ban.[23]

The order was intended to "prevent further trespass upon the zone." Barrett made it clear that the military would not be used "unless someone violates the provisions of the order; "however, should this occur "the national guardsmen will be used." After making his point, Murray announced on the evening of May 6 that the order would be lifted; however, he iterated that this was only to "see if the city officials will enforce the . . . ordinances against extensions." Once Oklahoma City's Mayor C. J. Blinn and City Manager Albert L. McRill complied, the martial law proclamation was withdrawn.[24]

This was the second time Murray had proclaimed martial law to exercise his will over the Oklahoma oil industry. The previous event had grown out of the disastrous plunge of the price of crude oil resulting from overproduction and the Great Depression. During the later 1920s and into the early 1930s, production had continued to mount, not only in Oklahoma's fields but in other prolific pools throughout America. Most oilmen realized that something had to be done to curb the outlandish rate at which the crude poured from the wellheads. Various attempts were made, but nothing seemed to work. In one instance in the fall of 1926, producers, pipeline companies, and the Oklahoma Corporation Commission met to formulate a plan for the ratable taking of crude production,

Martial law. Governor William H. ("Alfalfa Bill") Murray, shown here inspecting troops of the Oklahoma National Guard during the dispute over the Red River bridges, was determined to bring order to the state's oil fields even if it required martial law. *Courtesy of the Forty-fifth Division Infantry Museum.*

drilling restrictions, tank storage rules, and pipeline regulations. Eventually, a program was agreed upon, a referee was appointed, and the system was implemented. Called the Seminole Plan, it basically provided for prorated pipeline runs for individual leases in accordance with the purchasers' demands and restrictions on the drilling of new wells with the exceptions of offsets. Nonetheless, the price continued to spiral downward. A meeting of the governors, or their representatives from the states in the Mid-Continent area was called to search for a solution, but still the price fell until by mid-1931, Oklahoma crude was priced at twenty-five cents a barrel. It was becoming clear that voluntary action was not succeeding; nor was the corporation commission able to enforce its orders effectively.[25]

As the crisis continued to mount, it was obvious that drastic action would be necessary to restore order to the petroleum industry. In Oklahoma, the corporation commission was in the process of defending its rulings in a series of legal suits brought by dissatisfied oil companies and, although its actions would later be upheld by the courts, the uncertainty of the situation contributed to the chaotic conditions in the fields. Longtime Oklahoma oilman Edward A. Smith remembered that, during this period of drastic price reductions, he received a royalty check issued on July 20, 1931, from the Mid-Kansas Oil and Gas Company for $.01. As another example of the low to which petroleum and petroleum products had dropped during this period, Oscar L. Cordell, a longtime supervisor in the marketing department of Phillips Petroleum, recalled that as a part of a promotional stunt in conjunction with the opening of one of the company's new gasoline stations he suggested to Frank Phillips that the station give away five gallons of gasoline to all customers who filled their car tanks. In reply, Phillips stated, "Why that is perfectly alright—it isn't worth anything anyhow, it isn't worth as much as water, give them all you want."[26]

By this time, most oilmen within the state realized that the only means of raising the price of crude was to decrease production until demand caught up with the output. In following this thinking, Oklahoma producers convened in Tulsa in July, 1931, in what was described as a "general relief" meeting. Many of the more than one thousand oilmen present "harangued those who would not vote for a shutdown of wells," as feelings ran high. Although some producers met with Governor Murray to urge him to close Oklahoma's wells, the governor pointed out that it would be useless to do so unless other producing states, such as Texas, also complied. Voluntary proration was having some effect on the output of crude, but not enough. Eventually, Murray apparently realized that the only solution to the problem was an enforced closure of the state's wells.[27]

Governor Murray indicated his intention to impose martial law and close the state's fields if the price of oil was not raised. On July 28, 1931, he went so far as to issue an ultimatum to the large companies: a dollar a barrel or be shut down. The warning was ignored, and on August 4, 1931, the National Guard was ordered into the oil fields in an unprecedented exercise of executive authority. The order was made public late in the afternoon. The guards were called to their stations, then dispatched to the oil fields as quickly as possible. The military encountered little opposition, and Murray expected the entire state to be shut down "within three or four days." Nonetheless, preparations were made to keep the guardsmen

Curtailing production. In an attempt to raise the price of crude, Governor William H. ("Alfalfa Bill") Murray ordered the Oklahoma National Guard into the state's oil fields to enforce a cutback in production. *Courtesy of the Oklahoma Publishing Company.*

Cities Service Oil Company's Bartlesville Experimental Station. This is where Henry L. Doherty and Charles E. Beecher worked out the theory of unitization. *Courtesy of Cities Service Oil Company*.

in the fields for an extended length of time should a lengthy operation be necessary. On several leases, the oilmen demanded to be shown the executive order closing the wells, but in general, most independent producers believed "that the governor's order would accomplish the desired end."[28]

Whatever the outcome of Murray's actions, the imposition of martial law marked the "de facto" end of the early attempts by Oklahoma officials to regulate the oil industry. Although it undoubtedly appeared to many that these early efforts had ended in failure, the actions of those involved in charting a fair regulatory policy had in many instances blazed the trail. By the early 1930s, however, it was clear that the 1915 legislation was outdated, and in 1933 the Oklahoma Legislature enacted a new petroleum conservation law which listed in detail the power granted to the corporation commission to enforce its rules and regulations, and thus ushered in a new era in the regulation of the state's oil industry.[29]

## Chapter 14.
# The International Petroleum Exposition and Congress

ONE OF THE HIGH POINTS in the history of Oklahoma's petroleum industry was the creation of the International Petroleum Exposition (IPE) and Congress in the mid-1920s. The idea for such an exposition was initially expressed in an editorial which appeared in the March 3, 1923, edition of Tulsa's *American Saturday Night*, in which Earl Sneed, a local attorney, concluded that the city should host an International Petroleum Congress. In presenting a lengthy outline of his proposal, Sneed pointed out that "An International Petroleum Exposition and Congress with all its sideshow features would give thrills to the young people, knowledge to the oil fraternity, opportunity to make world-wide acquaintances, renew friendships, and firmly establish Tulsa for all time to come as the oil center of the entire world." In conclusion, Sneed invited "all manufacturers of oil field equipment and refinery supplies, and all those interested in production and distribution [to] form such an organization." Such an occurrence would, he pointed out, "be the first time in the history of petroleum that an exposition, international in character and designed and planned for the oil industry and those directly and indirectly dependent upon it, has ever been held." Comprising "all branches of the industry," it would allow oilmen to get "together, with the spirit of rivalry absent, to discuss problems which affect the good of the entire fabric of oil, and allied industries."[1]

Sneed's editorial stirred the interest of the Oklahoma petroleum community, and with the backing of such prominent oilmen as W. G. Skelly, the founder of Skelly Oil, plans developed rapidly. The Tulsa Chamber of Commerce presented the proposal to its convention committee, which in turn returned a favorable recommendation, and the International Petroleum Exposition and Congress was soon incorporated. Edward F. McIntyre was appointed general manager of the organization, headquartered at 212 South Boston in Tulsa. Tulsa chamber officials also played a major role in the creation of the event, with H. O. McClure, president of the chamber, and William Holden, the chamber's general secretary, both serving on the exposition's board of directors. A general committee of more than seventy-five Tulsans was organized into twelve working committees: finance, attractions, exposition, convention, parades and pagents, transportation, scientific and technical exhibits, public safety, auditing, entertainment, decorations, and building and grounds. Among those serving on the various committees were L. D. Armstrong, A. L. Beekly, Carl Blackman, A. V. Bourque, O. V. Borden, A. F. Bourne, F. W. Bryant, D. E. Buchanan, John Campion [sic], Fred S. Clinton, M.D., O. L. Cordell, E. H. Cornelius, A. B. C. Dague, W. E. Espy, C. T. Everett, Charles F. Farren, E. R. Filley, T. M. Pariss [sic], N. R. Graham, R. D. Gwynne, J. Burr Gibbons, Frank Glasscock, E. Bee Guthrie, R. L. Ginter, H. H. Goddard, J. H. Gardner, J. M. Hayner, William Holden, Alf G. Heggem, T. J. Hartman, A. W. Hurley, Frank Hinderliter, Summers Hardy, Richard Hughes, A. C. Holmes, W. R. Hamilton, R. P. Humes, L. B. Jackson, Cornelius Kroll, L. E. Kennedy, I. G. Long, W. L. Lewis, C. M. Lemason, J. J. McGraw, H. O. McClure, W. A. Melton, P. M. Miskell, Charles Meyers, Everett Manning, J. H. McBirney, D. W. Moffitt, G. L. Matson, R. L. McFarland, T. F. Mayer, C. M. Murray, H. E. McElroy, J. S. McKelvey, Hollis P. Porter, Asa E. Ramsey, I. G. Rosser, E. A. Richards, Ralph C. Riley, Harold E. Roe, W. R. Ritchie, Harry Smith, J. A. Sartori, E. T. Tucker, J. A. Udden, W. A. Vanderver, M. M. Valerius, E. H. Wiet, Allan Whiteside, J. S. Warren, W. M. Welch, J. E. Minger, and John Zink. These members of the

William G. Skelly became first president of the International Petroleum Exposition in 1925 and oversaw its development and growth for more than a quarter of a century. *Courtesy of the Oklahoma Heritage Association*.

general committee were representatives of "all organizations connected with the oil industry, local business and civic organizations," and many of them were leading early-day Oklahoma oilmen. Skelly, destined to remain a staunch supporter of the exposition throughout his lifetime, served as the organization's president for more than thirty years.[2]

Officers of the initial International Petroleum Exposition were L. B. Jackson, who served as president; J. M. Hayner and W. A. Vanderver, who were elected vice-presidents; J. J. McGraw, treasurer; and W. A. Holden, who was selected as secretary. Other members of the board of directors included "H. O. McClure, A. V. Bourque, Alf G. Heggem, W. A. Melton, J. H. Gardner, and T. J. Hartman." Apparently a number of the early records of the organizational period have been lost and many details are, therefore, sketchy; however, Jackson did appoint Alva J. Niles, J. E. Crawford, and I. E. Cornelius to fomulate plans for the establishment of a permanent IPE organization. In an effort to create "general good feeling among the oilmen," within a short time had recommended the selection of E. R. Perry, vice-president of Cosden and Company, and W. H. Grey, president of the National Association of the Independent Oil Producers, to serve as chairmen, and A. V. Bourque, secretary of the Association of Natural Gasoline Manufacturers, to be chosen as permanent secretary of the organization.[3]

Plans for the extravaganza progressed rapidly, and the exposition was scheduled for October 8 through 14, 1923. It was very carefully pointed out that, although the IPE was "fathered by Tulsa, conceived in the minds of Tulsa oilmen and brought to birth by Tulsa energy," the undertaking was "the ward of the entire Oil World," not simply a Tulsa project. As the concept of the exposition took shape, its purposes were more clearly defined: it was to "provide an annual meeting place . . . [for] representatives of every branch of the industry in general and their own branch in particular"; it was designed to "promote a feeling of cooperation between large and small producers, refiners, manufacturers and marketers"; and it served to provide a place at which the manufactured articles, tools, supplies, machinery, and accessories used in the industry may be exhibited to the crowds attending the exposition and congress, their good points demonstrated and the service offered by each exhibitor explained to the prospective customers." The members of the organization hoped that the meeting would also "hasten the standardization of tools and equipment and thus eliminate one of the great sources of waste and inefficiency now hampering the industry."[4]

In an effort to "promote in the oil field workers a sense of pride in their work by encouraging rapid and efficient performances of their tasks," the exposition was to offer "contests of skill between crews and individuals." In addition the founders scheduled entertainment to attract "oil men in all walks of life from all parts of the world" and planned for "an annual reunion of old friends and a meeting place for new acquaintances." It also hoped to present the opportunity of educating the public, sought to

The beginning of a parade at one of the early International Petroleum Expositions in the 1920s. *Courtesy of Beryl D. Ford Collection.*

"eliminate fake promoters from the oil business" and present "the true facts concerning the industry." Other goals were the formation of "the nucleus of a Museum of Petrology . . . [and the assembly of] articles of historic value showing the development of the industry from its beginnings," as well as "recognize and reward distinguished service on the part of oil field employees by the awarding of a medal or medals annually for the performance of outstanding actions in saving life or property or such other acts." These lofty ideas resulted in the growth of the International Petroleum Exposition and Congress into a world-wide gathering of oilmen, in which the latest technological innovations of the industry were available for inspection.[5]

Surrounded by a carnival atmosphere, the First International Petroleum Exposition and Congress opened on October 7, 1923. In the days immediately preceding the extravaganza, visitors from throughout the oil world poured into Tulsa. Within a short time, all the hotels of the city were filled, and the housing committee was predicting that another three or four thousand rooms would be needed to handle the flood of visitors. It was indeed an international spectacle. Newspapers noted the arrival of Lucio Baldo and Manuel Gonafels of Venezuela and Paul Ehrhardt of France, and announced the imminent arrival of "R. Staechelin and Herr. Dr. Melamid of Basel, Switzerland," as well as "Senor Ricardo A. Deusta of Lima, Peru," expected to "arrive within a day or two." A special train was engaged to carry "the performers, scenery, animals and paraphernalia" for the pre-exposition showing of the "world's hippodrome" which kicked off the festivities on Sunday evening, October 6.[6]

The official opening was heralded by a parade in

The International Petroleum Exposition attracted huge crowds during early years. *Courtesy of the Beryl D. Ford Collection.*

which "King Petroleum" rode a gaily decorated float through downtown Tulsa while en route to the Convention Center where he was enthroned to oversee the festivities. In his role as "King," Judge S. H. Klinge formally opened the exposition shortly before 11:00 A.M., and newspapers reported that in excess of five thousand people crowded through the exhibits along Boulder Avenue and Brady Street during the first day. That evening, a formal reception was held for all the delegates, both foreign and American, at the Tulsa Country Club. Parades continued to attract thousands of visitors throughout the week of the congress, and "long lines of persons formed at the gates outside, waiting their turn to buy tickets." Miss Dorothy Vensel (the daughter of Verne Vensel, a Tulsa operator) made an appearance as "Queen Petrolia" on October 11, attended by Misses May Reisling, Rosalind Hollow, Katherine Gavin, Lillian Randal, Cordelia Ann Kennedy, and Nellie Cook, who acted as "Duchesses."[7]

Held on October 12, 1923, the General Assembly of the Congress was attended not only by oilmen, but business and community leaders from throughout the Southwest. Great predictions were made for the future of the region. "The next 10 years will witness in Oklahoma, Missouri, Kansas, Texas, and Arkansas, the exploitation of natural resources and the building of industries and institutions on a larger scale than ever before," declared the participants. This would be made possible, they proclaimed, because "The Southwest is an undeveloped industrial empire within itself possessing deposits of coal, oil, natural gas, zinc, lead, cement, gypsum, building materials, and other resources unequaled in most states." Although threatening clouds dampened the affair somewhat, "rain insurance" provided by the officials of the exposition covered any damage incurred, and by Sunday, October 14, 1923, the event came to a close.[8]

The initial exposition was hailed as a great success. Before it had ended it was announced that the IPE would become an annual event. More than $1,500,000 worth of exhibits had been displayed, and in excess of five thousand people from outside the

"The oil capital." The great success of the International Petroleum Exposition was one of the reasons why Tulsa was proclaimed the "oil capital of the world." *Courtesy of the Beryl D. Ford Collection.*

Tulsa area, including many from foreign countries, had attended the week of festivities. Although rain had forced the cancellation of the largest parade planned, as well as the "Queen's" coronation ceremony, and had undoubtedly reduced the number of people in attendance and the amount of revenue generated, little hesitance was encountered in giving the International Petroleum Exposition and Congress a permanent home. Tulsa had been called "the oil capital of the world" since 1905 and was strengthening its claim to the title.[9]

The Second Annual Petroleum Exposition and Congress promised to be twice as large as the first. Opened October 2, 1924, it featured more than ten million dollars' worth of displays supplied by 461 exhibitors, and thousands of visitors thronged through the gates on the first day. On October 6, ten cars of a special train carried oilmen from El Dorado, Arkansas, one of the nation's greatest producing fields of the 1920s, to Tulsa to participate in the festivities and campaign for their own special candidate for "Queen." Oklahoma's Governor M. E. Trapp addressed the oilmen during the exposition, and severely criticized "too much government in business [and] too much politics in government." During the week of activities, special cars were added to trains to handle those traveling from Chicago, Kansas City, and St. Louis. Many state delegations from oil-producing areas were in attendance, and oilmen from Kansas and Texas were especially prominent.[10]

The Second Exposition insured the continuation of

the event. Having grown in a single year to cover seven and one-half acres of ground, the 1924 show offered two and one-half miles of exhibitions. Most people expressed the same desire as that of Paul H. Ehrhardt, a notable French mining engineer, who declared "I hope . . . that the exposition will be continued in Tulsa, the world's oil capital, yearly so that the oilmen of the world can gather and exchange their ideas." Indeed, the exposition, as much as anything else, was marking Tulsa as the center of the world's petroleum industry in the minds of oilmen around the world.[11]

The third year, a disastrous fire destroyed all parade floats only eleven days prior to the opening of the event and nearly wrecked preparations for the pageant. Enough of the floats were hastily rebuilt or redesigned, however, to enable the opening parade to take place. Journalists once again proclaimed that "Petroleum rules Tulsa!" Special trains from throughout America headed toward the "oil capital of the world" carrying interested individuals to the Third Annual International Petroleum Exposition and Congress, by now world-famous. Once again, rain plagued the show, but sunshine returned the latter part of the week, and the "whole exposition territory . . . [took] on a . . . carnival feeling."[12]

One facet of the petroleum industry given great exposure at the Third IPE was the role of the inventor. "Men who have been unable to get into the offices of the larger oil companies for a chance to display their inventions are finding the exposition the best medium to bring their ideas before the oil world," reported one newspaper. Many oil firms took advantage of the situation, as "leaders in every branch of the petroleum industry had their experts on the grounds where the "inventions are conspicuously displayed." Amazed at the new innovations unveiled, the experts declared that "there are more new inventions being shown this year than in the previous two years combined."[13]

When the accompanying congress convened at the Mayo Hotel in downtown Tulsa on October 5, 1925, the representatives listened as Governor Trapp again condemned government regulation of the oil industry. Although there had been some talk of abandoning the congress the previous year, the 1925 session was the best attended in the history of the event. J. Edgar Pew, president of the American Petroleum Institute, presented a very "optimistic address on the natural supply of crude oil." In addition, a Petroleum Fair was inaugurated that year in conjunction with the exposition, with Phillips Petroleum Company winning the silver trophy. Again, the show was an unqualified success, with a "Million Dollars in Sales and Prospects" reported for the Third IPE.[14]

Having scheduled future expositions on a biennial basis in 1925, the industry did not hold an expo again until September 24, 1927. In an "Impressive ceremony," President Calivin Coolidge pressed a "golden key in the executive office in Washington promptly at 2 P.M. Tulsa time" to begin the Fourth International Petroleum Exposition. The key was linked to the speakers' platform by telegraph wires, and "when it clicked, a plug was loosened from a steam pipe . . . releasing . . . pressure on" a replica of E. L. Drake's first oil well in America. In turn, this triggered a pump which spouted oil "from a producing well on a lease adjacent to the Drake well . . . over the top of the derrick." "The well was 'shot' and the exposition was on!" One of the highlights of the Fourth Exposition was the selection of "the oldest man, in point of service, in the industry." There were to be two awards presented, both donated by John D. Rockefeller: a gold medal for the oldest oilman living, and a silver medal for the oldest oilman attending the affair. J. H. Wagoner, active in the petroleum industry since 1867, won the silver prize.[15]

After the highly successful exposition in 1927, the biennial schedule was dropped, and it was decided to return to an annual basis. Opened on October 20, 1928, by Charles M. Schwab, known as the "King of Steel," the Fifth IPE featured the organization of "The Pioneers of the Oil Industry," designed to recognize those individuals who "made the petroleum industry possible." Headed by James Amm, a New York millionaire selected as the "Grand Old Man of the Industry," the organization honored Andrew Jackson Sanders, who started in the "oil business as a driller 71 years ago." The forty-two original members of the "Pioneers" had a total of 2,208 years of experience in oil field work among them. Among the charter members were many early-day Oklahoma oil leaders including: "A. J. Sanders, J. H. 'Uncle Joe' Evans, Alex Stephenson, E. L. Fowler, James Briody, J. A. Connolly, D. M. Elliott, Hugh B. McGivern, I. W. Hunter, Martin C. O'Brien, W. G. McClune, J. F. Brown, George C. Partridge, H. E. Partridge, H. E. Osborne, W. E. Klumph, C. N. Lawson, W. P. Roach, L. D. Gwynne, C. M. Trowbridge, D. O. McCormick, J. A. Kennedy, Fred G. Viger, Barney Harrington, John H. Markham, and W. H. McFadden."[16]

An " 'On-to-Tulsa' air derby" featuring a "mammoth fleet of 21 planes" marked the beginning of the

Equipment displays. Perhaps the most useful aspect of the International Petroleum Exposition was the opportunity it provided for various petroleum companies to display their latest equipment. *Courtesy of the Beryl D. Ford Collection.*

Sixth Annual International Petroleum Exposition on October 5, 1929. Having been born in the 1920s, during the period of discovery of numerous prolific fields, the IPE was now beginning to reflect the growing concern of many oilmen over the questions of proration, regulation, and conservation. During this session of the IPE, the Independent Producers Association met to discuss the question of an oil tariff and the curtailment of crude production in an attempt to stop the downward spiral of petroleum prices. Although the depressed market was having an adverse effect on the industry, plans were made for the continuation of the IPE the following year.[17]

The Seventh Annual International Petroleum Exposition opened on October 4, 1930, featuring exhibit equipment valued at thirteen million dollars. Officially opened by United States Secretary of Commerce Robert P. Lamont, this session saw the

The International Petroleum Exposition Building in Tulsa, one of the modern structures constructed to ensure the show's continued success. *Courtesy of Leslie Brooks and Associates.*

inauguration of an "award for the most outstanding advancement made in the petroleum industry" to be presented by Henry L. Doherty at future IPE gatherings. Once again, the specter of the Great Depression had a dampening effect. A parade, one-half mile long, was held in protest of proration and its effect on oilmen, while abroad "production and prosperity" reigned because there was no American oil tariff. When the 1930 show closed on October 11, plans had been made for a 1932 gathering; however, the devastating effects of the Great Depression forced its postponement, and it was not until May of 1934 that the IPE met again.[18]

Although the Great Depression was still gripping the nation, there had been some improvement in the plight of the oil industry when the expo opened its gates on May 12, 1934, after a lapse of three and one-half years. Attempts were made to make the Eighth IPE as great an attraction as the previous expositions; however, the carnival atmosphere of the show began to be replaced more and more by demonstrations of equipment and illustrations of the latest techniques. Little, if any, prestige was lost, however, with more than "20 oil-producing nations . . . [and] more than 400 exhibitors" being represented. Much attention was focused on a speech delivered by United States Secretary of the Interior Harold L. Ickes, as well as the national oil bill designed to speed the economic recovery of the petroleum industry. Attendance at the 1934 IPE was pegged at 142,863, and William G. Skelly, president of the IPE, expressed hope for a revitalization of the show on a regular basis.[19]

There was some indecision as to whether the IPE would be resumed on an annual or biennial schedule; however, it was eventually decided that biennial meetings were more advantageous, and the next IPE convened on May 16, 1936. Although struck by the "worst storm that has ever lashed an International Petroleum Exposition," the Ninth IPE set a new attendance record of 180,235, with sales estimated at twenty million dollars. The future of the exposition now seemed secure, and elaborate plans were made for the 1938 edition.

Opening on May 14, 1938, the Tenth IPE was expected to bring between $1,000,000 and $1,500,000 into the Tulsa area and included "fifteen miles of exhibits to see." The show proved to be a "real gusher," as perhaps the most astonishing event of the show was the actual striking of oil at a depth of 540 feet by the Franks Manufacturing Corporation's portable rotary rig while crewmen drilled a shallow well bore on the exposition grounds for "demonstration purposes" only. "It's the damnedest thing I ever saw," declared Fred S. Wilber, the oil editor of the Houston, Texas, *Chronicle*, when viewing the Franks discovery. A record-breaking crowd of more than two hundred thousand people attended the Tenth Exposition, greatly encouraging the effort.[20]

Great plans were made for the Eleventh International Petroleum Exposition scheduled for May, 1940, with Skelly announcing proposals for a great Hall of Science and a permanent Oil Merchandise Mart for the IPE. However, by the time the expo opened on May 19, 1940, World War II was imminent, and much of the exposition's rhetoric focused on the role of petroleum in the growing conflict. Some questions as to the continuation of the show were raised, especially its conflict with the Houston Oil World Exposition, but leaders in Tulsa declared that such fears were groundless and plans for the 1942 edition were already well under way. Regardless of their optimism, however, world events forced the cancellation of the 1942 exposition shortly after the United States entered World War II.[21]

After a lapse of eight years, the silver anniversary gathering of the International Petroleum Exposition opened on May 15, 1948, with a theme portraying "the great strides [made] since the last" exposition. "The newest in equipment, processes, ideas and techniques will be presented," IPE officials declared in order that "Every person affiliated with the oil industry—drillers, tool pushers, engineers, company executives, investment bankers and the wives and children—will find a stimulating version of what the industry signifies in services and business." United States Secretary of the Interior J. A. Krug officially opened the festivities that had grown in twenty-five years to cover 15 acres of land, attracting two thousand exhibitors and more than a hundred million dollars' worth of equipment. The revival of the exposition was an unqualified success. When the gates were closed on May 24, approximately 301,307 people had passed through the turnstiles to view the "marvels of the latest technological advancements" in the industry.[22]

During the run of the Twelfth IPE in 1948, questions concerning the scheduling of future expositions were widely discussed. Many oilmen believed that the two-year schedule was not necessary to keep abreast of the latest technological developments. Inasmuch as a greater time interval generally passed between significant alterations in equipment and methods, one plan advanced at the 1948 exposition advocated adoption of a quadrennial schedule. It was, however, eventually decided that a five-year plan would be more practical and also would allow enough time for advancing technology to warrant another gathering.[23]

Opening on May 14, 1953, under the direction of William G. Skelly, president of the IPE since 1925, the Thirteenth International Petroleum Exposition was scheduled for a ten-day run. Much had transpired in the period of time between the twelfth and thirteenth expositions, and there was a great amount of attention directed toward the improvement in oil field equipment on display. Numerous sales were reported throughout the event. Described as having the "extravagant razzle-dazzle of a world's fair coupled with the dead seriousness of a museum of science and industry," the 1953 show closed on May 23, with a record 369,835 people attending. Although the popularity of the show could not be denied, the five-year program was abandoned and a period of six years passed before the next expo opened on May 14, 1959.[24]

Convening 100 years after E. L. Drake's first oil well was drilled in Pennsylvania, the Fourteenth IPE represented a total investment 250,000 times greater than the two thousand dollar cost of the first oil well.

W. K. Warren, who succeeded William G. Skelly in the leadership of the International Petroleum Exposition.
*Courtesy of the Oklahoma Heritage Association.*

Attendance reached a total of 547,208 at the 1959 Exposition, featuring a theme centered around "progress." Because the cost of production was beginning to mount, the leaders of the event decreed that the next exposition would be centered around "more economical methods of deep drilling." Their fears proved well-founded, because the industry was swept by a recession resulting from a curtailment of drilling activity, and it was not until 1966 that the exposition again opened its gates.[25]

When the "Greatest Oil Show on Earth" unveiled its fifteenth edition on May 12, 1966, it silenced many "faint-hearted" oilmen who, during the industry's downward spiral, had predicted that "there would never be another International Oil Exposition." Undoubtedly, the thousands of spectators from throughout the world who poured through the exhibits helped to dispel the fears, as did comments such as "real fine prospects," "delighted," and "good response." When the 1966 expo closed, W. K. Warren, president of the IPE, and Martin Dwyer, the IPE manager, declared that "The central purpose of holding the show is to help exhibitors introduce new and improved equipment faster and to more more oilmen than can be done in other ways, and also to enable them to make contacts that lead to sales during or after the show." In this regard, they both were "pleased," and there seemed little doubt that the show would continue on a regular basis. Some suggested a three-year interval between shows, but the five-year rule remained in effect.[26]

When the 1971 IPE opened on May 15, United States Secretary of the Interior Rogers C. B. Morton participated in the ceremonies and welcomed oilmen to the exposition. Included in the program was a "five-day symposium on oil industry problems" that in part examined the situations encountered in the search for petroleum in the far corners of the world. On May 20, only five days after the beginning of the Sixteenth International Petroleum Exposition, IPE President F. R. Yost announced that the seventeenth edition of the show would be held in Tulsa in five years. This "early announcement" caught many oilmen by surprise; however, Yost declared that "after assessing the reaction to this year's event," the executive directors of the IPE "have decided to begin immediate plans for the 17th International Petroleum Exposition." Undoubtedly, this was in reaction to the "Crowds that flocked" to the show, as well as a desire to have the Seventeenth IPE coincide with the nation's impending Bicentennial celebration.[27]

Two billion dollars was the price tag placed on the 1976 International Petroleum Exposition that opened on May 17. Because of the serious energy questions facing the United States at this time, as well as the controversial policies of the Organization of Petroleum Exporting Countries, Randolph Yost announced that the expo's schedule would return to a fall program, with only three and one-half years to intervene between the seventeenth and the eighteenth exposition, which was scheduled for 1979. W. K. Warren, in commenting on the alteration, declared that "Fast-moving technological advances and the United States' goal to become self sufficient in oil and gas is speeding up improvements in equipment and methods, warranting the change to a 3 year cycle." The new time table would "provide ample time for the new and improved equipment to be developed and yet not put a yearly exhibit burden on the exhibitors," Warren continued. Thus, it appeared that after more than half a century, the International Oil Exposition, conceived and developed in Oklahoma, was clearly an intangible part of the petroleum industry of not only the state and nation, but the world.[28]

Earl Sneed's desire to create an organization to distribute knowledge of the oil industry, provide a forum for discussion of new innovations, and offer a gathering place for oilmen had undoubtedly succeeded. The show witnessed the introduction of many forms of innovative technology to the industry, therefore hastening the scientific advancement of the industry. In surpassing Sneed's wildest dreams, the IPE became recognized as the foremost petroleum exhibition in the world.[29]

## Chapter 15.
# Martial Law and Proration

THE LEGAL MORASS resulting from a tangle of suits and conflicting opinions as to the best method of halting the downward spiral of crude prices had created an almost impossible situation by the summer of 1931. Although Oklahoma had once been in the forefront of oil conservation legislation, it was obvious by the 1930s that the acts were outdated and did not provide the corporation commission with enough authority. In June, 1931, the Champlin Refining Company secured an injunction in federal court preventing the OCC from enforcing the proration order. Although the Oklahoma attorney general requested an immediate hearing to clarify the issues, and the injunction was suspended on June 30, the issue still awaited a final decision by the federal judges. Oklahoma officials were somewhat distressed when, on July 28, a federal court overturned a proration order by the Texas Railroad Commission, and the Oklahomans feared the Texas decision would influence the action pending in their own state.[1]

Murray and other political leaders of Oklahoma were not the only individuals concerned with the necessity of a workable conservation program. Most oilmen realized that such an endeavor would be of great benefit to them and supported some sort of action. Charles P. Dimit, longtime vice-president of production for Phillips Petroleum Company, recalled that "everyone who could be used to any importance connected with Phillips Petroleum Company regardless of their departmental responsibilities was asked to aid in the matter of bringing about a permanent conservation plan of operations." The production engineers, research staff, and legal advisers "were working night and day figuring out what plan could be used that would be a practical application" of the technology and legal questions involved. The oilmen and state officials could not, however, arrive at a plan acceptable to all parties.[2]

On August 4, 1931, Governor Murray declared martial law in reaction to the situation, as well as an alleged claim that Harry F. Sinclair and his oil company planned with "secret intrigue to consider the possibility of bribing 40 members of the legislature and impeaching the governor." The National Guard was placed in control of an area covering "a distance of 50 feet around each of the 3,106 wells, in 27 Oklahoma oil fields where crude oil production . . . [was] under proration." In the executive order placing the oil fields under the control of the militia, Murray cited sections of the Oklahoma Constitution enjoining the governor to enforce the laws of the state, and a 1921 statute authorizing the chief executive to call out state troops to prevent the "forcible obstructing of the execution of the laws or reasonable apprehension thereof." As added leverage, the governor pointed out that the legislation also provided that "Any person who shall willfully and unnecessarily interfere with the militia or any part thereof . . . in the performance of any military duty shall be guilty of a misdemeanor."[3]

Oilmen contended that the governor's action would throw approximately 7,500 men out of work and force many of the companies to go on "half-time basis, in order to aid their employees and keep a semblance of an organization intact." According to several sources, nearly 3,600 men employed in the Greater Seminole field and approximately 3,000 workers in the Oklahoma City Pool would suffer under Murray's edict. Not only would the men in the fields be unemployed, but office clerks and "virtually all classes of workers, from superintendents and department heads down." Although the largest number of those affected would involve production units, as many as "1,000 employees in the natural gasoline industry. . . . Several hundred refinery workers and many pipeline workers also will go on the unemployed." In addition, with an average wage of $3.50 a day per worker lost, oilmen argued that the economy of the entire state would soon suffer.[4]

Murray steadfastly refused to budge, however. While many of the independent producers willingly closed their wells, he contended, others had refused, and this was not only to the "detriment of the adjoining lands operated by the independent producers," but also involved taking "property without 'due process of law.'" In justifying his actions, the governor also maintained that the independent producers had no other recourse but to "rely upon the chief magistrate of the state to exercise the supreme executive power, in order to secure that equity and justice to which they are entitled under the constitution and laws." Murray argued that by shutting down all the wells, neither side suffered undue damage or injury and did not "take from interstate commerce any oil." This was important, for it was obvious that the United States Constitution reserved the control of interstate commerce for the federal government, not state authorities. Undoubtedly hoping to avoid federal interference in the matter, the governor contended that his order "conserves . . . natural resources . . . in which the state is vitally interested, possessing a vested property right therein, for the reason that when these natural resources, oil and gas, are exhausted, it can no longer collect revenues therefrom; and at present market, the percent of the tax collected is too insignificant for the wholesome public benefit to which the state is entitled." In addition, Murray maintained that without such an action, "the future children of the state" would not receive "that amount of royalty to which they are entitled, by reason of the discovery of oil and gas under that portion of the domain of the state owned by the state for common schools."[5]

It was not the contention of Oklahoma's governor that a corporation was "an artificial person" with "no powers outside of the expressed authority of the law, while a sovereign state, or the people thereof, possess all powers not by them delegated away." However, he charged that "some of the inferior federal courts have for decades assumed for corporations a footing above the sovereign states," and that this legal standing has been greatly abused "doubtless because of the well-known fact that many such judges owe their possessions to the influence of corporations." When such actions occur "in a contest between them [corporations] and the welfare of a sovereign state and its people," Murray argued, "there is but one course for an officer sworn to obey the constitution of his state

Oklahoma Governor William H. ("Alfalfa Bill") Murray ordered the Oklahoma National Guard into the oil fields after declaring martial law "for a distance of 50 feet around each of the 3,106 wells, in 27 . . . oil fields." *Courtesy of the Oklahoma Publishing Company*.

and of the United States—usurpation by judges, nullifying legislative acts—may take from the citizen for a time, his constitutional guarantees, but does not change the constitution or the law." The governor maintained that "every well-regulated government possesses the power to meet and correct every evil that may confront such government and its people; otherwise," he reasoned, "it is not a government."[6]

Specifically, Murray accused the "Sinclair Oil and Gas Co., the Champlin Refining Co., and their associates, and the pipeline companies" as acting "under the assumption that they were superior to and possessed power beyond the sovereign power of the state of Oklahoma [and] its 3,500,000 people." Because of their actions, it was "necessary for the governor of the state . . . to give this order, closing down all oil wells except the stripper wells, in order to protect the school children of the state, conserve its natural resources, and prevent their exhaustion and waste." In addition, Murray contended that his action was designed to "protect the rights of the tax-

Enforced proration. To ensure that his edict was enforced, Governor Murray appointed his cousin, Cicero Murray *(shown here in the foreground)*, a lieutenant colonel and placed him in charge of the enforced proration. *Courtesy of the Oklahoma Publishing Company*.

payers . . . to uphold the guarantees of the Fourteenth Amendment to the Constitution of the United States to the independent producers against taking their property, without 'due process of law,' and to establish justice and equity for them," as well as "to protect the citizens of the state from the general monopolistic control of their natural resources by corrupt combinations of capital, who produce, transport, refine, and sell oil." Murray declared that this constituted "a power of control monopolistic in its nature, destructive of the independent producer, whose ingenuity and enterprise, with the use of all the funds at his command, have made possible all the development of the natural resources of this state." Claiming that such independent producers pioneered the oil business, the governor maintained that their rights were being taken from them "by 'judicial juggling,' through injunction," which have "left the umpire and proration officers powerless."[7]

Murray's action had been urged for some time by a number of the state's independent oilmen, and from every indication, he believed that "there was a movement in the state, led by Sinclair Co. . . . to destroy the proration law . . . and . . . the independent operators." As a result, his proclamation of martial law specifically explained the plight of the independents who, according to the governor, "have promised to keep their men employed, while the monopolistic companies have cut down wages and reduced the number of employed, some with 10 to 15 years' service to their companies." Because of this, there had been "threats of dynamiting their pipelines and other destruction of their properties," which he had learned about through his "secret scouts." Of course, this presented a direct threat to the larger companies and their holdings which, the governor reasoned, "it is my duty to protect." In addition, Murray concluded that such policies could lead only to "a continuation of present conditions," which were "likely to produce mob violence unless all these wells are at once closed down."[8]

Sinclair wasted little time in renouncing the governor's charges and, in a statement issued by his company, declared that he "Bitterly Resents Murray [sic] Accusations of Sinister Tactics." As to the use of the militia to raise oil prices, Sinclair commented that "all the proclamations and troops in the world will not add one cent to the price of oil"; however, he did state that "his companies would obey, 'under duress' the shutdown order." The attitude of many independent operators was in direct opposition to Sinclair's reaction, and they praised the governor for closing down the wells in an attempt to boost the price of crude. As time passed, however, some of the early optimism began to lag.[9]

Murray readily acknowledged that one of the motivations behind his use of the National Guard was what he considered the interference of federal judges in the realm of state power. However, on the day after he ordered the closing of Oklahoma's oil wells, the federal court in Guthrie ruled in favor of his actions by a two-to-one margin. Judges Orie L. Phillips and John H. Cotteral prepared the majority decision in the case, while the dissenting statement was written by Judge Franklin E. Kennamer. Kennamer's argument was based on what he considered a violation of the Fourteenth Amendment of the United States Constitution, but Phillips and Cotteral viewed the question in a different light. They declared that "Oil and gas are natural resources which cannot be replaced and the power of the state to impose reasonable regulations to prevent waste in the production, handling, and marketing thereof is undoubted." Basically, the federal judges ruled that the "corporation commission has power to make rules and regulations to prevent waste of oil and gas"; however, the "commission has no power to levy penalties for violations of its rules." "Penalties," the judges contended, had to "be assessed by mandamus or injunction procedure in state courts."[10]

Not only did the federal judges rule that the corporation commission did not have the power to assess penalties for violations of its regulations, they also condemned some of the commission's methods. They

The National Guard in the Oklahoma oil fields. *Courtesy of the Oklahoma Publishing Company.*

"rapped sharply the practice of the state and corporation commission in employing umpires who are paid by the oil companies interested in proration." Nevertheless, Murray believed that he had won an important victory in his effort to drive the price of crude above the one dollar per barrel mark, as well as the conservation of Oklahoma's oil and gas reserves. But opposition to his tactics began to mount, challenges of his use of martial law in federal court were being muttered by both the large companies and a number of independents. Production did decline to some extent after seizure of the fields, but the re-

National Guard camp in the Oklahoma oil fields during the governor's declaration of martial law. *Courtesy of the Petroleum Publishing Company.*

fineries wasted little time in finding new sources to replace the Oklahoma crude that was no longer available.[11]

As time passed, it became apparent to many that Sinclair's prediction was correct, and a number of Murray's early supporters began to question the wisdom of closing the state's oil fields. Even though several refineries offered seventy-five cents a barrel for crude on August 19, 1931, the governor remained unwavering in his goal of one dollar per barrel, despite mounting criticism. By September there was talk of eighty-five cents a barrel, but Murray still refused to compromise. Many of those who had predicted "that the crude purchasers would be down on their knees and begging for oil" shortly after the closing of the wells, now began to have second thoughts. The Oklahoma supply was soon being replaced by oil from Texas, and although Murray relaxed the ban after a period of seventeen days, oil production was still significantly curtailed. Seeing that no good was being accomplished, oilmen sought redress in the courts against military control, and Murray allowed the wells to resume production on October 10, despite the fact that the price of crude ranged from only sixty-six cents to seventy cents per barrel.[12]

With the fields reopened, the majority of the troops were sent home; nevertheless, the state's crude production remained regulated, with Oklahoma's oilmen allowed 546,000 barrels daily. Shortly afterward, the OCC issued two new orders affecting the state's production: one prevented oilmen from recuperating by additional production the losses incurred during the period they shut down, and the other created special allowances for certain regions of the Oklahoma City field. In May, 1932, the United States Supreme Court upheld the state's proration regulations, and to many this seemed to end the question of who controlled Oklahoma's petroleum production. Dissatisfaction among oilmen continued

Governor E. W. Marland sent the National Guard back into the Oklahoma City field in a dispute with municipal officials over his authority to allow wells to be drilled on public land near the state capitol. *Courtesy of the Oklahoma Publishing Company.*

to grow, however. Allegations were made that some companies were illegally producing more crude than they were entitled to, and as the rumors spread so did displeasure with the corporation commission. When it appeared to Governor Murray that the situation had deteriorated to the point that the commission could no longer enforce its regulations, he again resorted to martial law, and on June 21, 1932, dispatched troops into the oil fields.[13]

Murray's proclamation stated that the action was taken "not only for the purpose of watching the enforcement of orders of proration, but for the purpose of actually prorating the oil output in the Oklahoma City field and controlling the transport of such oil through the pipelines and its storage tanks." At the same time, the governor created a proration board to consist of Cicero I. Murray, the proration umpire; a member selected by those producers not having pipelines or refineries; one member from those producers controlling refineries; and a member chosen by pipeline owners. The proration board was charged with "investigating orders, for equitable proration, and was to recommend to the governor . . . the allowable for each of said wells in said field." The umpire was empowered to make excavations around oil wells where he suspected that there were secret pipes allowing for illegal production. In addition, oil pipelines were required to install "meters sufficient to determine as accurately as possible the amount and flow of oil through . . . [the] pipes." Basically, Murray's proclamation removed the power of enforcement of proration orders from the corporation commission and gave them to the newly created proration board. Also, it provided for "a special levy of one-fourth of a cent a barrel" on "all oil produced" to defray the expenses of this new regulation.[14]

Despite Murray's actions, however, the situation in the Oklahoma oil fields continued to deteriorate. While some oilmen turned unsuccessfully to the courts for relief from military control, others simply continued to produce outside the limitations. Several companies were called before the corporation commission to answer charges of illegal production, but the situation remained chaotic.[15]

By this time, there were few individuals connected with the state's petroleum industry who did not believe that Oklahoma's oil code was seriously outdated, and that new legislation was needed to bring about stability in the oil fields. Thus, when the Fourteenth Oklahoma Legislature convened in January of 1933, much hope was placed in its ability to provide the industry with a new set of rules. However, as it became more apparent that an updated series of regulations would be enacted, the tempo of proration violations increased. Rumors were rampant in the oil fields, and a number of oilmen were determined to pump as much crude from the earth as possible prior to passage of the new oil code.[16]

To cope with the unstable conditions, the governor ordered the militia back into the Oklahoma City field on March 4. Pressure had been placed on Murray to take some sort of action, since it was obvious that the pool was in the midst of a period of "wild production." So frantic was the activity that "the roar of escaping gas could be heard for five or six miles." In justifying his decision, Murray declared that "certain persons are not disposed to cooperate with their fellow-producers, but seek every excuse or opportunity to take more than a reasonable share" of crude, and they "are producing in excess of their allowable under the most recent order, and giving excuse . . . to others to exceed their prorata allowables." If "permitted to continue," he contended, "it would destroy the structure price and break down entirely all enforcement of proration." Therefore, the governor ordered Oklahoma's adjutant general "to close down completely all petroleum and oil wells in the Oklahoma City field, excepting only water wells . . . and to prevent the production or flow of oil from . . . the said field, until the fourteenth legislature . . . shall have enacted into law a complete statute governing proration." Murray pointed out that "the law of the state, in effect, prohibits the production of oil when it falls below cost of production," and that the action was necessary in order to eliminate the "tremendous waste in natural gas by reason of the violation of [the] proration order." Once the militia reentered the field, it was only a matter of days before the legislature produced the long-awaited oil code, which was approved on April 10, 1933.[17]

Born of the desperate situation in the Oklahoma oil fields, "House Bill No. 481" was passed by decisive margins in both the House of Representatives and Senate. The legislation gave the corporation commission the needed authority to enforce its rulings and, at the same time, took away much of the power of the governor to regulate the oil industry. Providing the OCC with a method of punishing those in violation of conservation laws, the legislation specifically outlawed the production of "hot oil" or crude that was pumped in excess of proration allotments. Many of the outmoded provisions of the 1915 legislation were updated, and because of the confusion reigning in the oil fields, the act was declared an emergency measure, and therefore it became effective immediately.[18]

Many Oklahoma oilmen disagreed with Governor Murray's stand on proration. Friends since the early days of statehood, Herbert H. Champlin, founder of Champlin Refining Company, and Murray found themselves on opposite sides of the proration issue. Because of the importance of the prospect to the petroleum industry, the question strained their relationship to a point that, when Murray signed the proration statute into law, he reportedly turned to those in the room and said, "Take this pen up to Enid and give it to Herb Champlin."[19]

Entitled "An Act Defining and Prohibiting Waste of Crude Petroleum and Natural Gas, and providing for the Ratable and Equitable taking thereof from Common Sources of Supply . . .," the legislation

With the protection of the National Guard, several oil wells were drilled on the grounds of the state capitol in Oklahoma City. *Courtesy of the Oklahoma Publishing Company.*

defined "waste" as including "economic waste, underground waste, . . . surface waste, and waste incident to the production of oil in excess of transportation or marketing facilities, or reasonable market demands." For gas, it was specified as "the unreasonable production and/or the inefficient or wasteful utilization of gas in the operation of oil wells . . . in excess of the amount necessary in the efficient drilling." The OCC was directed to "limit the production of gas from wells producing gas only or gas and gasoline only to a percentage of the daily open flow capacity of such wells."[20]

The act granted to the Oklahoma Corporation Commission the "Power to Regulate Taking of Oil . . . so as to prevent the inequitable or unfair taking from any common source of supply, and to prevent unreasonable discrimination in favor of any such common source of supply against another." In making rules and regulations, the commission was prohibited from permitting "unreasonable discrimination." To insure that the oil removed from prorated fields was done so in an equitable manner, wells were permitted to produce "only that proportion of the allowable as their total production relates to total allowable." The "capacity of each well in any prorated common source of supply . . . shall be as certained and determined at regular intervals"; however, not more than six months were to elapse between the determinations of a well's potential. In addition, the commission was prohibited from restricting a well's production to fewer than "twenty-five . . . barrels of oil per day or its full production if it is incapable of producing twenty-five . . . barrels per day." To provide an adequate means of measuring the amount of crude shipped to refineries, loading racks, and storage facilities, the OCC was empowered to require the installation of meters on pipelines in prorated fields.[21]

"House Bill No. 481" allowed state officials to require periodical "reports of all oil purchased and/or transported . . . from prorated fields." All operators in a prorated field were obligated to maintain books which accurately recorded "the amount of oil produced daily . . . and . . . the amount of oil sold or otherwise disposed of each day." The locations of "all pipelines, connections, pumps and tanks," as well as their "size and capacity," were to be reported to the commission, and all details of production and sales, including the name of the carrier who purchased the petroleum and its destination. Failure to comply could result, in addition to other penalties, in an order to "discontinue to produce any oil from any leasehold, property or well with respect to which such operator has failed or refused to make and file such report, until he or it shall have filed same." All such site maps or reports had to be verified by a member, agent, or employee of the firm making the report or by a responsible person approved by the commission.[22]

The corporation commission was given the "jurisidiction to make any and all such orders, rules and regulations authorized and/or provided for" in the act, but only "after a hearing before the commission." In addition to being posted ten days before a hearing, a notice had to be signed by a majority of the commission, had to designate the time and place for the hearing, and contain a brief statement of the general nature of the regulations sought, as well as the names of the common sources of supply that might be affected. To aid in its work, the corporation commission was granted the following powers:

(1) Of visitation and of a court of record; (2) To administer oaths; (3) To compel attendance of witnesses; (4) To compel the production of books and records; (5) To punish for contempt any person guilty of any disrespectful or disorderly conduct in the presence of the commission while in session; (6) To punish as for contempt any disobedience or violation of the provisions of this Act and of any of its orders, rules, regulations and judgements made or rendered by it under and in pursuance of the provisions of this Act; (7) To enforce the provisions of this act, and in compliance with any of its orders, rules, regulations or judgement by appropriate process, and by shutdown orders, ordering and directing the shutting down or discontinuance of production of oil from any well or wells . . .; (8) To appoint or designate one of its agents or employees or one of the deputies to the proration umpire to act as marshal of the commission, . . . and when directed by it shall serve and execute orders, subpoenas, commitments and other process issuing from it.

The taking of depositions from witnesses was also provided for, as were specific penalties for contempt.[23]

The proration umpire, as well as his assistant and deputies, were authorized access to all properties producing oil or gas, pipelines, tank farms, and pump stations for the purpose of ascertaining rule violations. Provisions were also made for the closing of wells judged to be overproducing, as well as the means to enforce such a ruling. Anyone sworn to oath before the commission and who "willfully and contrary to such oath state any material matter which he knows to be false" was to be guilty of perjury and subject, upon conviction, to "imprisonment in the State penitentiary of not more than five years." Likewise, anyone presenting false information under oath on "any report, map or drawing or other statement or document authorized . . . by this Act" was, if found guilty, to "be punished by imprisonment in the State penitentiary for not less than two . . . years, nor more than ten . . . years." The only appeal from "orders, rules and regulations of the commission" was to the Oklahoma Supreme Court; however, appeals to "repeal, amend, modify, or supplement" legislative or administrative orders of the corporation commission were to be made to the commission itself.[24]

Not only were stringent provisions made for perjury, but also for a myriad of situations which the commission might face in the discharge of its duties. "Any person who knowingly and willfully delays or obstructs" the proration umpire, his assistant, or deputy, or any agent or employee of the commission or any public officer in performing their duty could be fined up to five hundred dollars, sent to the county jail for six months, or both. Should firearms be used in the attempt, the violator could be sent to prison for a year. For conspiring "to violate any provision of this

Act, or any lawful order, rule or regulation of the Commission," the penalty was a fine of not more than five thousand dollars or imprisonment in the penitentiary for a period not to exceed five years, or both. The commissioners were empowered to seek injunctions for continued violations of regulations; any bribery in an effort to influence the commission was punishable by a fine not to exceed five thousand dollars and five years in prison. To accept such a bribe invited a fine of up to ten thousand dollars and 10 years in the penitentiary. Clearly, the new oil code had "teeth."[25]

Requiring a bond of fifty thousand dollars, the office of proration umpire was established. Charged not only with the specific duties spelled out in "House Bill No. 481," he was also required to investigate "all charges and complaints of violations of this Act, and any orders, rules and regulations of the commission." An assistant proration umpire was created to aid the proration umpire in the fulfillment of his duties, and a proration attorney was provided to represent the commission in legal dealings. All were to be appointed by the corporation commission. Whenever the members of the commission believed it necessary, they were authorized to employ up to thirty deputies, selected by the proration umpire, to assist him in his work. Stenographic and clerical help was allotted to the OCC, and the proration umpire, as well as his assistant and deputies, was granted "all of the powers and authorities of a peace officer, with the right to bear firearms." In addition, the commission was given the "right from time to time to designate one or more of its agents or employees to act as inspector for the commission . . . for the purpose of ascertaining whether the provisions of this act, and the orders, rules, regulations and judgements of the Commission . . . are being complied with."[26]

The necessary funding was provided by companion legislation, "House Bill No. 483." Basically an excise tax on petroleum, it levied a cost of "one-eighth of one cent per barrel on each and every barrel of petroleum oil produced in the State of Oklahoma," for the period of time between the passage of the act and June 30, 1935. Collected by the Oklahoma Tax Commission, the money was to be "paid by the purchaser." Provisions were made, however, to allow the tax to be deducted "in making settlement with the producer and/or royalty owner." If the oil was not sold, the producer became obligated to pay the tax. All funds raised by the tax were to be "deposited in the State Treasury to the credit of a . . . 'Proration Fund,'" which was to be used in paying the "salaries and expenses . . . office expenses and office supplies . . . and all other items of expense" according to the provision of "House Bill No. 481." Payable the first of each month, a lien could be placed on the property of those tardy in paying the tax. Such a delinquency was also subjected to "interest at the rate of eighteen per centum per annum." Anyone making a "false oath to any report required by the provisions of . . . [the excise tax legislation] shall be deemed guilty of perjury." To insure that the tax was paid, the Oklahoma Tax Commission was authorized to examine all books and records necessary to determine a violation. A person or company refusing to file "any sworn statement or report required by the provisions" of "House Bill No. 483" could be fined up to $500 per day for the period of failure to comply. Like its companion, the excise tax legislation was approved on April 10, 1933, and declared an emergency.[27]

The Oklahoma Corporation Commission wasted little time in adjusting its procedures to conform with the new legislation; however, reaction among oilmen was diverse. Many believed that the inequities of the 1915 act had finally been abolished and demonstrated a satisfaction with the new oil code, while others continued to resist the exertion of state authority over the petroleum industry and took legal action to prevent the implementation of the act. In general their legal moves were rebuffed by the courts, and the commission was allowed a free hand. Most oilmen realized that the "new proration bill . . . passed by the Oklahoma Legislature . . . contains sharp teeth," and one writer in the *Oil and Gas Journal* went so far as to state that "Enactment of a new proration law in Oklahoma which provides severe penalties for violations of its provisions has cleared up the production in that state to a very great extent."[28]

The 1933 oil conservation code was to be, with minimal additions, the basis of the state's petroleum policy for many years to come. There were some alterations, however. In 1935, the Fifteenth Legislature enacted six pieces of legislation concerning oil and gas. "House Bill No. 187" dealt with the spacing of wells to prevent waste; "House Bill No. 188" redistributed the proceeds of the excise tax on petroleum; "House Bill No. 274" changed the title of the proration umpire to conservation officer and made other alterations in the bureau for the conservation of oil and gas; and "House Bill No. 2," "Senate Bill No. 346," and "Senate Bill No. 208" dealt with the establishment of the Interstate Oil Compact. Six years later, in 1941, the Eighteenth Oklahoma Legislature approved "Senate Bill No. 10," which authorized the governor to extend the Interstate Oil Compact agreement, and "Senate Bill No. 124," which reduced the number of officials involved in the regulation of the oil industry. But it was the action of the Fourteenth Legislature in 1933 that laid the groundwork for an acceptable method of regulation.[29]

In reworking the 1915 oil code, the 1933 action established the Oklahoma Corporation Commission as the major regulatory agency of the petroleum industry within the state. Although not originally conceived as such, the OCC had slowly evolved into its regulatory status and, with the passage of the 1915 legislation, assumed the role which it maintained until Murray resorted to martial law. In the intervening years the 1915 laws had become outdated, and the commission was somewhat ineffective in maintaining stability in the oil fields. As the situation deteriorated, many oilmen appealed to state officials to end the chaos. This and other factors moved Murray to take drastic action, and the National Guard was ordered into the oil fields to enforce proration and raise the price of crude. It soon became obvious that the militia could not arbitrarily solve the question, and the state legislature enacted "House Bill No. 481," which gave the OCC the "sharp teeth" needed to restore order to the state's petroleum industry. From 1933 forward, the corporation commission was able to establish a firm and equitable policy in dealing with Oklahoma's oilmen, and the future statutes, rules, and regulations were, in general, simply embellishments of the 1933 legislation.[30]

## Chapter 16.
# Conservation and Cooperation

IT HAD BEEN SEVERAL YEARS since the federal government played an active role in the regulation of the state's petroleum industry. After Oklahoma Territory and Indian Territory were joined to form the State of Oklahoma in 1907, the task was undertaken by local officials. With the election of Franklin D. Roosevelt as president during the 1930s, however, federal officials resumed their supervisory role of American oil production. The reentrance of federal officials into oil regulation was prompted to some extent by the desire of a number of oilmen to see a comprehensive petroleum program developed along the guidelines of Roosevelt's "New Deal" program.[1]

Actually, the basis for renewed federal intervention was established as early as 1924, when President Calvin Coolidge created the Federal Oil Conservation Board (FOCB), in order to bring industry leaders together with federal officials to discuss problems peculiar to the petroleum industry. Nonetheless, the majority of oilmen feared undue regulations and were reluctant to concede much of a role to the federal officials. The boom period of the 1920s, coupled with advancing technology and the partial success of voluntary prorationing, had filled the petroleum industry with optimism. But when the stock market crash of 1929 threw the American economy into the Great Depression, it dealt the oil business a crippling blow, and many oilmen were convinced that only federally enforced regulation would save the industry from total collapse.[2]

Implementing a policy in which the federal government would take the lead in conservation of the nation's petroleum resources in March, 1929, President Herbert Hoover closed the public domain to further leasing. In addition, many oil and gas permits that had been previously issued were canceled. Hoover also called for an Interstate Oil Conservation Conference to convene at Colorado Springs, Colorado, on June 10, 1929. Although the announcement of the meeting was greeted with enthusiasm, it quickly turned into one of the most regrettable events in the history of the oil industry. The gathering of between 300 and 400 oilmen, politicians, and federal officials soon divided into diverse segments demanding different actions, and the resulting factionalism doomed the meeting to "an endless wrangle," and ultimate failure. Hoover's ban on exploration and drilling on the public domain would prove slightly more effective, but it was removed in April, 1931.[3]

In the spring of 1930, the Federal Oil Conservation Board produced a plan for interstate cooperation among the nation's oil producers. The three major provisions of the proposal were: (1) the conservation and equitable apportionment of an individual pool's resources, preferably through unitization; (2) protection for pools within the same jurisdiction against unfair competition created by a wastefully produced flush pool, through uniform well spacing rules; and (3) the coordination of sources of supply not within the same jurisdiction to insure the protection of conservatively operated pools against destructive competition. Shortly afterward, Secretary of the Interior Ray L. Wilbur called for renewed cooperation among the oil-producing states, and Oklahoma Governor Murray urged the officials of the Mid-Continent region to consider a united attempt to end overproduction.[4]

As a result, delegates from Oklahoma, Texas, New Mexico, and Kansas convened at Fort Worth, Texas, in February, 1931, to consider proposals for the establishment of an Oil States Advisory Committee (OSAC). Murray was present, as were Texas Governor Ross Sterling and official representatives of the governors of New Mexico and Kansas, and they requested that President Hoover ask all major American oil companies to limit their import of foreign petroleum. Four days later Hoover replied that United States Secretary of Commerce Robert P. Lamont was currently negotiating with the leading oil producers for just such a limit. Murray attended a similar conference held at Texarkana, Texas, the following month, but it was not until April 9, 1931, that the Oil States Advisory Committee finally met with federal officials in the national capital to present a series of suggestions for stabilizing the production of crude oil.[5]

In order to achieve a balance in the public supply of petroleum products and eliminate the waste of a valuable and irreplaceable natural resource, Murray and the others recommended that a stabilization of the production of crude must be achieved. The members of the Oil States Advisory Committee also pointed out that no state could adequately protect the national interest by its individual action and that it was therefore necessary that a coordinated effort be undertaken by all the large oil-producing states. Otherwise, they argued, any proration action taken by one state could "be quickly nullified by flush production in another state." Moreover, they contended that increased importation of foreign crude made individual state action inadequate and inflicted a great amount of damage to the economy of oil-producing regions of the United States. The committee members were quick to point out, however, that the question of regulation and stabilization of the crude production with a state was "purely a problem for solution by that state and the industry therein, with such cooperation and advice as may be rendered by interstate advisory committees, and with such aid and assistance as the Federal Government may be able to give." Basically, the OSAC recommended a limitation of unnecessary drilling, elimination of the waste of natural gas, unitization of oil fields, a prorationing system for oil pools, and an equal opportunity to ship crude to market. In addition, the committee recommended that the legislatures of oil-producing states allow negotiation between the various states for a coordinated conservation effort, greater uniformity of laws, and stricter enforcement. Great pains were also taken to ensure against misinterpretation of the proposals, and the committee was "opposed to anything akin to monopoly."[6]

The suggestions of Murray and the others were met with mixed emotions by Secretary of the Interior Wilbur, who maintained that the Federal Oil Conservation Board did not have the power to regulate the oil industry; nevertheless, he did agree to the recommendation for cooperation among the oil-producing states and endorsed the proposal for uniformity of state laws. The ideals of the Oil States Advisory Committee were, however, soon dashed by the huge production of Texas and California, as well as the economic realities of the Great Depression. Although the OSAC continued to function for more than a year, the election of Franklin D. Roosevelt as president and his implementation of the "New Deal" spelled its downfall. Nonetheless, the Oil States Advisory Committee marked the beginning of state cooperation on the questions of oil production and conservation.[7]

Roosevelt took office on March 4, 1933, and three days later, America's leading oilmen appealed to Secretary of the Interior Harold L. Ickes for a program of stringent regulation to promote stability in the oil fields. Ickes reacted by summoning a conference in Washington, D.C., of the heads of large petroleum companies and political leaders of the major oil-producing states in late March, 1933. At the same time Wirt Franklin, a prominent Oklahoma oilman and president of the Independent Petroleum Association of America (IPAA), also scheduled a meeting of his organization in the nation's capital. It appeared that some action would finally take place. But when Ickes's conference adopted a statement recommending the shutting down of all new and highly productive oil fields in Oklahoma and four other states, the action was not supported by all delegates, and a group of dissidents left the conference in disgust. Likewise, when Franklin spoke in favor of federal regulation to members of the IPAA, he was shouted down. It was clear that the nation's oilmen were not unanimous in their position regarding federal intervention in the petroleum industry.[8]

In May, 1933, Ernest W. Marland, a pioneer Oklahoma oilman and congressman, introduced a resolution in the United States House of Representatives which would empower the secretary of the interior to set prices, wages, and hours of laborers in the oil fields; limit the production of crude; and control the importation of petroleum. Backed by both Roosevelt and Ickes, the measure was introduced into the United States Senate; however, the National Industrial Recovery Act (NIRA) was already before Congress, and Marland's measure was ultimately incorporated as Section 9c into the NIRA legislation which was signed into law on June 16, 1933.[9]

Section 9c provided for the regulation of pipelines by the federal government and the prohibition of the shipment of hot oil in interstate commerce. It stated that "The President is authorized to prohibit the transportation in interstate and foreign commerce of petroleum and products thereof . . . in excess of the amount permitted to be produced or withdrawn from storage by any state law . . . or other duly authorized agency of a State." Furthermore, the president could separate any pipeline company from its holding company in the event that a pipeline firm charged high rates or monopolized operations in an area. The passage of the NIRA's Section 9c by Congress was to have a lasting impression on the oil industry, not only because it introduced the federal government directly into the question of regulation, but because it also provided officials with a means of doing so, a means previously unavailable to individual states.[10]

Shortly after passage of the NIRA, Roosevelt issued an executive order affirming the right of the federal government to prohibit excess oil production and control the interstate and foreign commerce of petroleum. Supervision and enforcement of Section 9c was initially given to the Bureau of Mines and later to the Petroleum Administration Board (PAB). Moreover, Ickes hoped to establish a "Code of Fair Competition" as a companion measure to the NIRA legislation. Previously, Axtell J. Byles, president of the American Petroleum Institute, had undertaken the job of creating such a code and invited representatives of several major producers to meet in Chicago, Illinois, in June, 1933, but the many diverse opinions within the oil business made it a formidable task. Ickes held a series of hearings among the various sections of the industry during the month following passage of Section 9c; however, controversy abounded. A number of different codes were proposed by various groups before the final "President's Code" was signed by Roosevelt in late August, 1933. Eventually, the majority of those participating in the long and arduous assignment of preparing the code accepted the president's recommendations, and by August 20, 1933, most oil companies had also signed the document.[11]

The Oil Code allowed the president to prescribe a base price for gasoline during a ninety-day test period, and to sell at a lower cost was declared "an unfair practice." A formula was developed to determine the cost of a barrel of crude oil: eighteen and one-half times the cost of a gallon of gasoline at the refinery. Minimum wages of between forty-five and fifty cents per hour were set for workers in the production and refining branches of the industry, depending on the location. Clerical employees were limited to forty hours per week, and all other workers, with the exception of executives, supervisors, and stripper well pumpers, were limited to not more than seventy-two work hours in fourteen consecutive days and no more than sixteen hours in two days. A minimum wage of between forty and forty-seven cents an hour was established for the employees of the marketing branch of the industry, and with some exceptions, work was limited to forty hours per week. Service station workers were limited to a forty-eight hour week and a minimum wage of between twelve and fifteen dollars per week. In order to control the industry, the Oil Code created an oil administrator and a fifteen-member planning and coordination committee. In addition, a federal agency was to be established by the president to work with the planning and coordination committee; however, the majority of its duties were to be involved with production control. With the approval of the president, the agency could establish production quotas for states, and the chief executive could control the shipment of petroleum or oil products to enforce the Oil Code.[12]

In general, it appeared that most Oklahoma oilmen realized the necessity of Roosevelt's extension of federal regulation over the state's oil industry. Claude Barrow declared that the actions of the federal government had put a stop to the illegal production and sale of hot oil from the Oklahoma City field. In addition, between May and October of 1933, the price of a barrel of crude oil rose from 25 cents to $1.08, and most of Oklahoma's oilmen credited the enforcement of federal regulations for the increase in price. The Oklahoma Corporation Commission wasted little time in complying with the new federal guidelines, and during September, 1933, three separate orders were issued to bring the state into compliance. Opposition to federal intervention in the oil industry remained, however, and Section 9c was ultimately challenged in the courts and ruled void by the United States Supreme Court in May, 1935.[13]

For a period of time, oilmen were unsure of just what role the federal government would play in the regulatory process, and several legislative bills were introduced into Congress to continue prohibition of the transportation of hot oil. Oklahoma's representatives and senators played a role in several pieces of such legislation, and because the state was a leading oil producer, the matter occupied a considerable anount of their time. United States Senator Elmer Thomas of Oklahoma introduced a bill in April, 1934, that called for unrestricted and permanent control over the oil industry by the federal government. In addition, Oklahoma Congressman Wesley

E. Disney proposed legislation to reestablish federal regulations over the production of crude. The actions of Thomas and Disney eventually gave rise to a series of congressional hearings that were the most comprehensive discussions up to that date regarding the question of a federally-regulated petroleum industry. Oklahoma's other United States Senator, Thomas P. Gore, continued to urge a resumption of federal regulations in the Senate, and although he eventually withdrew his proposed legislation, it was later resubmitted and passed with the added support of Senator Thomas T. Connally of Texas. Essentially a redraft of Section 9c, it authorized the Department of the Interior to prepare measures that would restrict the interstate shipment of hot oil.[14]

In the midst of the confusion resulting from the attempted extension of federal control over the oil fields, the voiding of the legislation by the Supreme Court, and subsequent attempts to reestablish federal control via other legislation, another movement began to take shape in an attempt to bring stability to the petroleum industry through the combined efforts of the major oil-producing states. In November, 1934, the American Petroleum Institute (API) held its annual meeting in Dallas, Texas, and after hearing an address by Secretary Ickes, members of the conference adopted a report recommending continued federal regulation over hot oil and foreign importation of crude. Furthermore, the report suggested that the United States Congress be petitioned to allow the establishment of a compact among oil-producing states to aid in the regulation and conservation of petroleum and petroleum products. At the same time, Oklahoma Governor Ernest W. Marland scheduled a meeting of the governors of these states, or their representatives, at his home in Ponca City to discuss the oil question.[15]

Convening in Ponca City on December 3, 1934, those attending the meeting included Marland; the attorney general of Oklahoma, J. Berry King; James V. Allred, the governor-elect of Texas; Kansas Governor Alf M. Landon; and representatives from California, New Mexico, Louisiana, Arkansas, and Wyoming. The majority of those present opposed extended federal control of the oil fields; however, recommendations were made to limit imported crude. Marland played a strong role of leadership in the conference and introduced a resolution calling for the creation of an agency composed of federal officials and representatives of the oil-producing states to estimate the petroleum demand of the United States and allocate to the various states a fair proportion of the demand. He also pressed for restrictions on the importation of foreign crude; limitation on the interstate shipment of petroleum not produced under the allotment system; and authorization for the petroleum-producing states to participate in such an agency. Marland declared that "In my opinion, our oil-producing states should perform their duties of sovereignty not only to their local individual interest, but in co-operation with other states for the common welfare of the nation." Continuing, he concluded that "This end can be achieved not by the creation of new and untried powers, but by the exercise of powers already vested in the states." He hoped that "as a result of this meeting and conference, we will see fit by resolution to request Congress to grant permission for the states to enter into an interstate pact."[16]

At Marland's invitation, the conference reconvened at Ponca City for a second meeting on January 3, 1935. Those attending the second session included Marland and representatives from Texas, New Mexico, Kansas, Louisiana, Arkansas, California, and Illinois. A resolution was unanimously adopted "requesting Congress to pass legislation permitting the states to enter into an interstate compact for the conservation of petroleum and the prevention of waste of petroleum." In addition, the various legislators were asked to authorize their representatives to meet and to adopt a compact of a similar nature. There was some disagreement among the members; however, this was expected because of the wide-ranging and diverse nature of the petroleum industry. Compromises were worked out on the most pressing issues, and Marland summoned the representatives to a third session in Dallas, Texas, on February 15 and 16, 1935.[17]

At the Dallas meeting, officials of Oklahoma, Texas, California, New Mexico, Arkansas, Colorado, Illinois, Michigan, and Kansas signed the Interstate Oil Compact (IOC), and agreed to begin working immediately with their individual state legislatures to secure approval of the agreement. In addition, work was begun in the United States Congress to secure acceptance of the agreement by the federal government. The compact was to take effect upon approval of any three of the states of Oklahoma, Texas, California, Kansas, or New Mexico and when approval was received from Congress. Specifically, the IOC stated that the "purpose of this compact is to conserve oil and gas by the prevention of physical waste thereof from any cause," and to do so, "Each state bound hereby agrees that within a reasonable time it will enact laws, or if laws have been enacted,

The January, 1935, oil parley held at Ponca City, Oklahoma, which helped to complete the plans for the Interstate Oil Compact Commission. *Standing, left to right:* Northcutt Ely, Oklahoma; Jeff Davis, Arkansas; John W. Olvey, Arkansas; John S. Farrell, Louisiana; Ralph H. Cummins, Louisiana; Hiram M. Dow, New Mexico; William Bell, Illinois; E. B. Shawver, Kansas; Patrick J. Hurley, Oklahoma; and T. C. Johnson, Kansas. *Seated, left to right:* Jack Blalock, Texas; E. W. Marland, Oklahoma; and Robert L. Patterson, California. *Courtesy of the Interstate Oil Compact Commission.*

then it agrees to continue the same in force . . . to accomplish the prevention" of wasteful practices. In detailing the legislation to be enacted by the signatory states, the compact required the prohibition of the "operation of any oil well with an inefficient gas-oil ratio," the "drowning with water of any stratum capable of producing oil or gas, or both oil and gas in paying quantities," the "creation of unnecessary fire hazards," the "drilling, equipping, locating, spacing or operating of a well or wells so as to bring about physical waste of oil or gas or the loss in the ultimate recovery thereof," and the "inefficient, excessive or improper use of the reservoir energy in producing any well." If it had not already done so, each state signing the compact agreed to enact legislation within a reasonable time, "providing in effect that oil produced in violation of its valid oil and/or gas conservation statutes or any valid rule, order or regulation promulgated thereunder, shall be denied access to commerce; and providing stringent penalties for the waste of either oil or gas."[18]

To alleviate criticism that the compact was a form of monopoly, the agreement declared that "It is not the purpose of this compact to authorize the states joining herein to limit the production of oil or gas for the purpose of stabilizing or fixing the price thereof, or to create or perpetuate monopoly." Specifically, it stated that it "is limited to the purpose of conserving oil and gas and preventing the avoidable waste thereof within reasonable limits." The organization

The group of men who presented the Interstate Oil Compact to President Franklin D. Roosevelt at the White House on July 31, 1941. *Front row, left to right:* Oren Harris, Arkansas; Joseph Guffey, Pennsylvania; Edward Kelly, Illinois; Jed Johnson, Oklahoma; Charles L. Orr, Oklahoma; W. J. Holloway, Oklahoma; Carl A. Hatch, New Mexico; and J. C. Hunter, Texas. *Rear row, left to right:* Russell B. Brown, Independent Petroleum Association of America; Sam M. Russell, Texas; William P. Cole, Jr., Maryland; Jared Y. Sanders, Jr., Louisiana; Elmer Thomas, Oklahoma; Josh Lee, Oklahoma; Tom Connally, Texas; and Jack Houston, Kansas. *Courtesy of the Interstate Oil Compact Commission.*

created by the compact was designated the Interstate Oil Compact Commission (IOCC), and each state joining was entitled to appoint one representative. Charged with making inquiries and to "ascertain from time to time such methods, practices, circumstances and conditions as may be disclosed for bringing about conservation and the prevention of physical waste of oil and gas," the IOCC was to report "at such intervals as . . . [it] deems beneficial . . . the findings and recommendations to the several states for adoption or rejection." Granted the power to "recommend the coordination of the exercise of the police powers of the . . . states . . . to promote the maximum ultimate recovery from the petroleum re-

The 1946 Executive Committee of the Interstate Oil Compact Commission. *Left to right:* Hiram Dow, Sidney Latham, Robert S. Kerr, Sam Jones, and Andrew Schoeppel. *Courtesy of the Interstate Oil Compact Commission.*

serves," the commission was also to "recommend measures for the maximum ultimate recovery of oil and gas." To do so, it was to "organize and adopt suitable rules and regulations for the conduct of its business"; however, no action could be taken by the IOCC except "by the affirmative votes of the majority . . . of the compacting states represented at any meeting and . . . by a concurring vote of a majority in interest of the compacting states at said meeting." In order to determine the interest of each state in the voting, it was decreed that "such vote of each state shall be in the decimal proportion fixed by the ratio of its daily average production during the preceding calendar half-year to the daily average production of the compacting states during said period."[19]

No state that signed the compact was to "become financially obligated to any other state." Likewise, "the breach of the terms" of the agreement in no way placed a state in "financial responsibility to the other states." Born at Dallas in February, 1935, the compact was to expire on September 1, 1937; however, any signatory state could withdraw "upon sixty . . . days' notice." The original document was deposited in the National Archives in Washington, D.C., and copies were given to the various state representatives. Membership to the Interstate Oil Compact Commis-

Eleven governors who attended the 1946 Interstate Oil Compact Commission meeting in Oklahoma City, with Commission Secretary Earl Foster. *Left to right:* Chauncey Sparks, Alabama; John C. Vivian, Colorado, Andrew F. Schoeppel, Kansas; Sam C. Ford, Montana; Harry F. Kelly, Michigan; Robert S. Kerr, Oklahoma; Ben Laney, Arkansas; Frank J. Lauche, Ohio; Edward Martin, Pennsylvania; Millard F. Caldwell, Florida; Clarence W. Meadows, West Virginia; and Earl Foster. *Courtesy of the Interstate Oil Compact Commission.*

sion was opened to "Any oil-producing state," that signed the compact.[20]

Oklahoma's governor, one of the leading forces behind the organization of the commission, was the first to sign the accord on February 16, 1935. He was followed by representatives of Texas, California, New Mexico, Arkansas, Colorado, Illinois, Michigan, and Kansas. New Mexico was the first state to approve the compact on February 25, 1935; however, Colorado, Illinois, Kansas, Oklahoma, and Texas were quick to follow. Michigan, Arkansas, and California withheld approval until 1939, 1941, and 1975 respectively, although California generally sent an observer to meetings prior to becoming a full member. The United States Congress gave its consent on August 27, 1935, and an organizational meeting was held in Oklahoma City on September 12 of that year to make the compact a reality.[21]

The Oklahoma Legislature gave its approval to Marland's efforts in a series of actions between January and April in 1935. "House Bill No. 2," passed with an emergency clause on January 30, 1935, authorized the governor or his representative "to meet with representatives of the governors of other petroleum-producing states, and of the United States, for the purpose of agreeing upon a compact." The consultation was to have as its objective the establishment of a joint federal/state agency to "make periodic findings . . . of the demand for petroleum to be produced with the United States," and for voluntary reduction of production and the formulation of uniform conservation measures and tax laws. However, "House Bill No. 2" specifically declared that the IOCC would not be binding upon Oklahoma unless "Said compact shall be ratified by the Legislature of this State, and Congress." In addition, the Oklahoma legislators stated that Congress was to make provisions for the "control and limitation of importations" of oil, and "provide for the control of interstate movements of petroleum" produced under the regulations or in excess of the rules.²²

"Senate Bill No. 208," also containing an emergency clause, was approved on March 6, 1935, ratifying the "Interstate Compact to Conserve Oil and Gas." However, it was "Senate Bill No. 346," approved on April 19, 1935, that amended "Senate Bill No. 208" to read the "Interstate Oil Compact Commission," and designated the governor "as the official representative of the State of Oklahoma." The governor was to "exercise and perform . . . all the powers and duties imposed by the . . . compact," and was granted the authority to "appoint an assistant representative who shall act as the official representative of the State of Oklahoma . . . when the authority to act is delegated to him by the Governor." "House Bill No. 346" also authorized the assistant to employ "such technical and legal assistants as may be necessary," and established in the State Treasury " a special fund to be known as 'The Interstate Oil Compact Fund of Oklahoma' " to pay the expenses of Oklahoma's participation in the program.²³

Established as an administrative agency to carry out the functions of the compact, the IOCC was originally created for a period of two years; however, it was consistently renewed for two-year periods until 1943, when it was given a four-year term. It has been renewed at the end of each four-year cycle since that time. Oklahoma City serves as headquarters for the organization. The states that originally formed the commission were soon joined by Alabama, Alaska, Arizona, Florida, Indiana, Kentucky, Louisiana, Maryland, Mississippi, Montana, Nebraska, Nevada, New York, North Dakota, Ohio, Pennsylvania, South Dakota, Tennessee, Utah, West Virginia, and Wyoming. In addition, Georgia, Idaho, North Carolina, Oregon, South Carolina, and Washington are associate members.²⁴

In following the precedent set by the passage of "House Bill No. 481" in 1933, which defined and prohibited the wasting of crude oil or natural gas and provided for the ratable and equitable production of petroleum, the Oklahoma Legislature passed a series of acts during the following years to strengthen state control of the oil industry. "House Bill No. 187," adopted on April 22, 1935, established regulations dealing with the spacing of wells. Drilling units were limited to ten acres in size, unless eight percent of those holding leases "embraced within the probable producing area of the common source of supply agree

Oklahoma Governor Johnston Murray, *left*, and Kansas Governor Edward F. Arn, *right*, at the dedication of the Interstate Oil Compact Commission headquarters in Oklahoma City, 1954. *Courtesy of the Oil Compact Commission*.

to a larger unit"; however, in no event could "a drilling unit exceed forty . . . acres." Six years later, on June 4, 1941, "Senate Bill No. 124" was approved with provisions that reorganized the enforcement personnel of the 1933 legislation and created a state fuel inspector "for the purpose of enforcing the fuel inspection laws of the State of Oklahoma." That same year, legislators approved "House Bill No. 390," which brought regulation to the liquefied petroleum gas industry. In 1945, the Oklahoma Legislature updated many of the earlier conservation laws and authorized the corporation commission to expand its staff with more professional personnel. Other acts of the Twentieth Legislature increased the regulatory power of the corporation commission and strengthened the fuel inspection laws.²⁵

Many of these legislative actions were recommended to other IOCC states as examples of regulatory practices consistent with the times and did much to keep Oklahoma in the forefront of petroleum con-

Interstate Oil Compact Commission headquarters in Oklahoma City. *Courtesy of the Interstate Oil Compact Commission.*

servation. The large numbers of conservation statutes enacted by the legislature and regulatory orders issued by the corporation commission were climaxed on May 21, 1947, when "House Bill No. 172" was approved. This act charged the OCC "to prepare an annotated compilation of the oil and gas laws of the State of Oklahoma and the rules and regulations of the Corporation Commission," and provided for their publication. This furnished state oilmen with a comprehensive listing of the rules and regulations governing Oklahoma's oil fields. Throughout the attempt to bring stabilization to the petroleum industry, Oklahoma oilmen were among the leaders of the movement and, as a result of their efforts, were able to glean "vast amounts of oil and gas left . . . by the . . . early unrestricted days of the industry." Because of their actions, "astounding progress was made toward solution of the many questions . . . that bedeviled the course of regulation."[26]

## Chapter 17.
# The War Years

OVERPRODUCTION, COUPLED WITH the economic effect of the Great Depression, had a dramatic impact on Oklahoma's petroleum production. Nonetheless, exploration remained at a fairly high level during the first few years of the era, and between 1930 and 1934 new oil discoveries were made in twenty-one of the state's seventy-seven counties. Strikes were made at the Edmond field in 1930, the Lucien field in Noble County in 1932, and at the Crescent field in Logan County in 1933. One of the largest strikes made during this period was located in east-central Oklahoma, where an Ada geologist, John Fitts, convinced B. C. King and W. A. Delaney that "oil could be found along the Arbuckle fault" in southern Pontotoc County. The money needed to finance an exploratory well in this untested area was difficult to obtain, however. While searching for participants to help finance the wildcat well, King and Delaney approached an executive with a large Tulsa-based oil firm who, after listening to their proposal, declared, "I can drink all the oil that will ever be produced south of the South Canadian River." Nevertheless, the necessary funds were raised, and on September 1, 1933, the No. 1 Wirick was brought in as Fitts had predicted. Eventually expanded to cover more than 5,000 acres, the Fitts field was a prolific producer, much to the embarrassment of at least one oil executive. Even so, the legal questions of prorationing, the general economic uncertainty, and the shift in political philosophy resulting from a change in presidential administrations made the plight of the Oklahoma oilman uncertain.[1]

During the 1930s, the number of barrels of oil produced in Oklahoma ranged from a low of 153,244,000 in 1932 to a high of 288,839,000 in 1937, with a corresponding fluctuation in value of between $119,200,000 in 1931 and $283,500,000 in 1937. Despite the fact that the decade's total production of 1,928,251,000 barrels of oil exceeded the 1,843,477,000 barrels produced during the 1920s, the cash value dropped significantly from $3,221,504,000 to $1,921,270,000. At the same time, the average price of a barrel of Oklahoma oil plummeted from $1.43 in 1929 to a low of 66 cents in 1931 and 1933 before rising to $1.24 in 1937; in 1939, it was quoted as $1.04. It was not until the end of the 1930s that more stability was achieved and "crude output... equated to market demand."[2]

Although the era of the Great Depression had a traumatic effect on the state's oil business, it was also a period of great technological innovation in the production and marketing aspects of the industry. This was primarily a result of the Sun Oil Company's development of Houdry's catalytic cracking system in March, 1937. Eugene Houdry, a Frenchman, had originally developed his idea for producing high octane gasoline with the help of the French government until when confronted "with high operating costs" it withdrew support in 1929. Turning to American companies for backing, Houdry had considerable difficulty in generating financial support until the Vacuum Oil Company established a plant and laboratory for him at Paulsboro, New Jersey, in 1930. When the country was struck by the depression, however, this support ended, and Houdry "again faced the task of obtaining sufficient funds to continue" his research.[3]

It was not until 1933 that Houdry convinced the Sun Oil Company (one of the pioneering petroleum concerns with ties to the early-day Oklahoma oil companies of Cosden, Barnsdall, Sunray, and D-X) to back his experiments. But Houdry's theory soon proved its worth. With an input of fifty to one hundred barrels of petroleum stock daily, Houdry's pilot plant in New Jersey yielded twenty-three to twenty-four percent of ninety-one octane gasoline.

Eugene Houdry's process for producing high-octane gasoline eventually led to such facilities as Cities Service's huge catalytic cracker and auxiliary units at East Chicago, Indiana. *Courtesy of Cities Service Oil Company.*

Although this output was lower than that produced by the thermal process, the gasoline's anti-knock rating was higher, and Sun Oil's engineers were convinced that it could be developed into a commercially profitable process. Within a relatively short time, the output was increased to forty-three percent, nearly equal to the yield produced by the thermal process. By 1938, the Sun Oil Company announced that three catalytic cracking units were in operation producing quality high octane gasoline and, along with Socony-Vacuum Oil Company, planned to invest $35,000,000 in the construction of ten additional units. Eventually, Sun Oil was able to crack all but the final ten percent of crude fractions.[4]

Not only a boon to Oklahoma's oil industry but nationally as well, this development allowed the production of high quality products. Gasoline made with the Houdry process had "an octane number . . . of 88 compared with 72 for thermally-cracked gasoline." In addition, the "heavier-than-gasoline

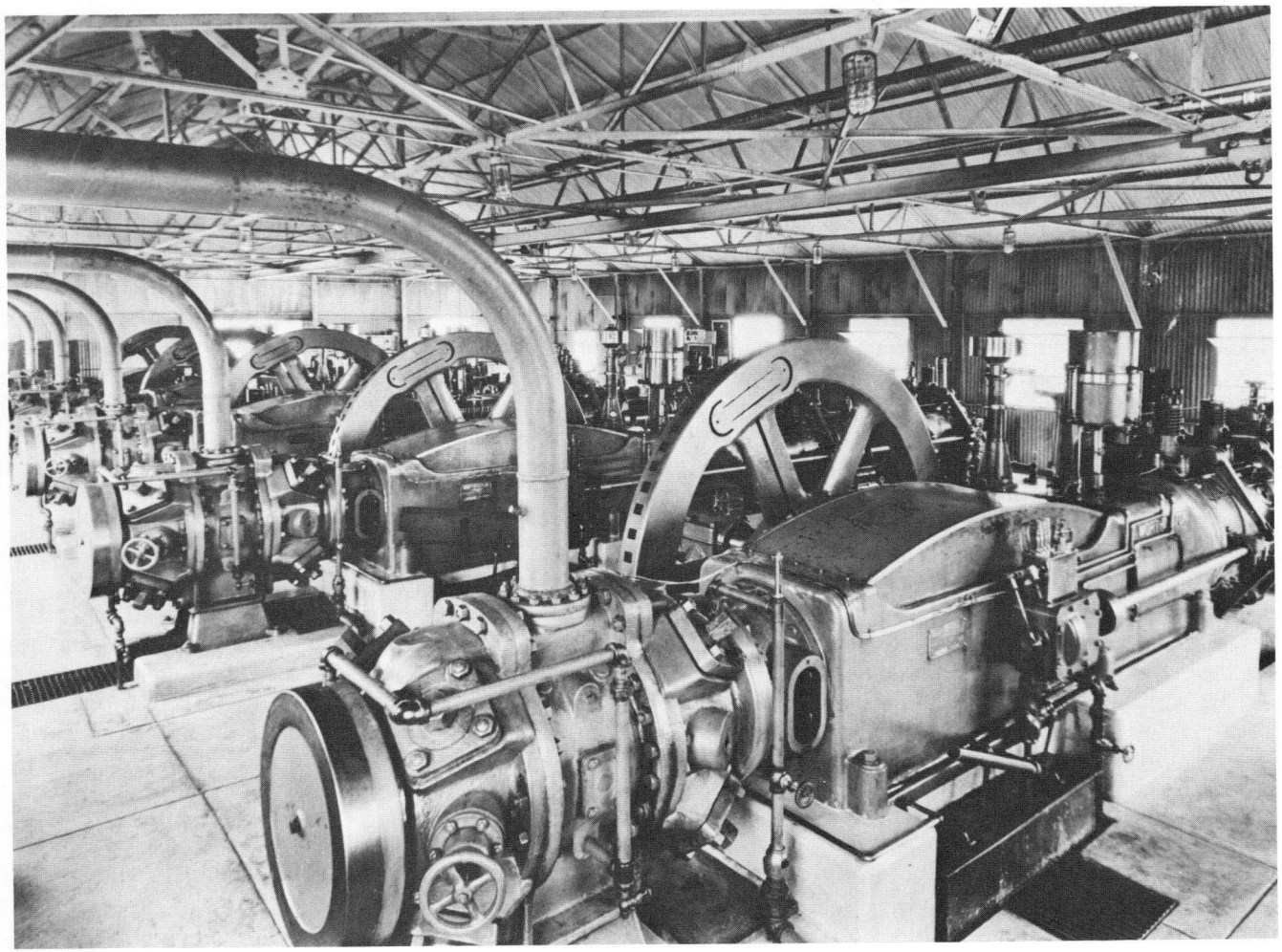

Gas booster station. With the outbreak of World War II, Oklahoma's huge natural-gas production and operations such as this Consolidated Gas booster station at Carpenter became more and more important to the war effort. *Courtesy of the Western Oklahoma Historical Society.*

material" produced by the Houdry process could be sold as "No. 2 furnace oil, whereas the heavy material from the thermal cracking process could be sold only as heavy fuel oil." Moreover the high octane quality of the catalytic stock allowed Sun Oil to "improve the quality of its motor fuel without adding tetraethyl lead"; thus the company "was able to sell an unleaded premium quality gasoline in competition with regular leaded grade."[5]

Despite the rapid technological advancement made in the 1930s, there remained a number of pressing unresolved issues: allocation complexities, questions concerning well spacing, and foreign imports. These were, however, quickly cast aside at the onset of the conflict in Europe. With the involvement of the United States in World War II, Oklahoma's petroleum industry once again entered a period of prosperity and expansion. The contribution of the state to the war effort was as diverse as it was important. Not only was Oklahoma's production of petroleum utilized for a modern mechanized conflict, the technology of the state's oilmen was also used to increase production throughout the Allied World.[6]

The science of warfare had made huge strides since the close of World War I, and the mechanization of modern armies and conversion of navies to oil-burning ships were among the most important. As a result, many oilmen anticipated an increase in the world's consumption of American-exported crude; however, just the opposite occurred. Exports actually declined and markets dropped until America began to rearm in early 1940. This increase in the domestic consumption of crude offset the loss in exports. As the international situation continued to deteriorate during 1940 and the early months of 1941, it became apparent that the United States was being drawn closer to the conflict. On May 28, 1941, President Roosevelt appointed former Secretary of the Interior Harold L. Ickes as petroleum coordinator for national defense and charged him with undertaking the preparation of the petroleum industry for a wartime situation. Later, on November 28, 1941, the Office of Petroleum Coordinator (OPC) appointed the Petroleum Industry Council for National Defense—which later became the Petroleum Industry War Council (PIWC)—and for easier control, the United States was divided into five districts under the Office of Petroleum Coordinator for National Defense. Tulsa was selected as the sub-office of District Two.[7]

The military market for Oklahoma crude began with the entrance of the United States into World War II, and the conflict dealt a crippling blow to the domestic marketing of petroleum goods. The rationing of gasoline and other petroleum products began in some parts of America soon after Pearl Harbor. On December 1, 1942, gasoline rationing was extended to the entire domestic market and remained in effect until August, 1945. The problem of increasing the output of the nation's oil fields became critical in order to supply the necessary petroleum products for twentieth-century warfare. Throughout the first six years of the decade, encompassing the war years, Oklahoma's oil fields produced approximately 838,623,000 barrels of crude oil; 1,993,845,000 gallons of natural gasoline; and over 1,714,847,000,000 cubic feet of natural gas. Furthermore, the state furnished thousands of skilled oil field workers, as well as the expertise and technological knowledge of the geological and petroleum engineering staff of the University of Oklahoma.[8]

Although oil and petroleum products were essential to the war effort, it became increasingly difficult for Oklahoma oilmen to get badly needed equipment and parts because of federal regulations and military priority. Melvin B. Heine, manager of the purchasing department of Phillips Petroleum Company for many years, recalled some of the problems connected with maintaining the increased output demanded by the war. Salvage operations designed to reclaim previously used equipment became an important part of the state's petroleum industry. According to Heine, "many, many, many tons of various types of oil field equipment, machinery, and so forth" were refurbished and used again. The effort was so successful that Phillips Petroleum Company built one natural gasoline plant utilizing nothing but salvaged material.[9]

Perhaps one of the most interesting and secretive projects undertaken by Oklahoma oilmen during World War II was the petroleum production carried out in Great Britain. When it entered the war against Germany, England faced a critical oil shortage, and every effort was made to tap all potential sources of petroleum within the British Isles. The attempt was, however, limited by the inadequate and outdated equipment available. As a result, British officials turned to American oilmen, and C. A. P. Southwell was sent to the United States in September, 1942, to seek aid. During meetings with American officials, he was introduced to Lloyd Noble, president of the Noble Drilling Corporation, C. C. Forbes, vice-president of the Noble Drilling Corporation, and Frank Porter, president of the Fain-Porter Drilling Company, all Oklahoma oilmen.[10]

Apparently Southwell had hoped to secure aid from the Oklahomans in upgrading the British petroleum effort, but because of prior commitments, Noble, Forbes, and Porter were limited in what they could offer. Even so, Southwell was apparently convinced that Noble's organization was just what the embattled British needed to get their petroleum production moving. Accordingly, he visited Noble in Oklahoma in another attempt to solicit the oilman's aid. Flying to Dallas, Southwell drove to Noble's home in Ardmore, Oklahoma, in an attempt to convince Noble of the urgency of the situation. Undoubtedly surprised when the unexpected guest appeared at his door, Noble invited the Britisher in and, while Noble showered and shaved, Southwell presented his arguments again. Convinced, Noble gave his consent to the proposal with the stipulation that Porter could be persuaded to participate as well. In addition, he surprised the Britisher by declining all profit in the venture, and instead, declared it to be the contribution of the Noble Drilling Corporation and the Fain-Porter Company to the winning of the war.[11]

Noble then contacted Porter and obtained his agreement to meet with Southwell in Washington, D.C., and discuss the project. Noble also pointed out to Southwell the necessity of acquiring the cooperation of the United States government in securing the necessary equipment and men for the task. Encouraged, the Britisher returned to the East Coast to await developments and clear the plan with American officials. A few days later, Ed Hold and P. M. Jones with the Noble Corporation, George Otey, an Ardmore attorney who represented the Oklahoma oilmen, and Porter met with Southwell and other government officials in the nation's capital. After Southwell presented the British plan that called for the completion of at least 100 wells within the next 12 months, a satisfactory agreement was quickly reached.[12]

Noble, in partnership with the Fain-Porter Drilling Company, agreed to undertake the exploitation of the British fields and soon began recruiting men for the task. Eugene P. Rosser, an early-day Oklahoma oilman who had been a longtime employee of the Noble Drilling Corporation, was placed in charge of the rigs in Great Britain, and another Oklahoman, Don Walker, was made his assistant. Working under wartime conditions in the British Isles, they received deferments from the military draft. All involved agreed that only skilled oil field workers would be recruited for the task and put on the payrolls of the Noble and Fain-Porter enterprises. Each employee was hired on a work contract and required to obtain a passport and a British visa. Their salaries were set and approved by the Office of Price Administration. Organized into four-man crews: a driller, a derrick man, a floor man, and an engine man, the men were scheduled to work twelve-hour tours, seven days a week. In addition, the recruits were warned that their employment constituted "top secret" war work, and they were forbidden to discuss their duties with anyone outside of their immediate families. Several Noble workers were employed, and others were hired in the oil fields of Oklahoma, Texas, Louisiana, Mississippi, and Arkansas.[13]

Although the British were anxious to begin work as quickly as possible, the Americans refused to ship anyone overseas until the details of the work contract were completed. By the end of January, 1943, most of those selected had received the needed deferments from the draft board and permission to leave the United States. Because of wartime shortages, however, problems were encountered in acquiring the necessary equipment. Both Rosser and Walker had hoped to secure airplane priority for the trip to England, but this was not possible, and although the activity of German submarines made the prospect of a sea crossing unattractive, the men reconciled themselves to the voyage. On February 24, 1943, the first group set sail from New York City, and the forty-four American men were soon reunited in Great Britain.[14]

Lloyd Noble, the president of the Ardmore-based Noble Drilling Corporation, who worked closely with C. A. P. Southwell to send an American drilling crew to England during World War II to boost British petroleum production. *Courtesy of the William A. ("Mac") McGalliard Collection.*

The English did not waste time in getting on with the task of drilling the badly needed wells. The Americans were to work and live in Nottinghamshire in the midst of the legendary Sherwood Forest to the north of London in the English Midlands. Here among the remaining massive oaks, geologists had discovered good prospects for petroleum deposits. The Anglican monastery in Kelham was selected as the site to house and feed the oilmen. Originally, the structure had been the "mother house" of the Society of the Sacred Mission and had later served as a theological seminary for ministerial candidates; however, several alterations made for the military units previously stationed there made it especially attractive to the Americans. The recent installation of two large bathrooms, each containing four hot showers, offered added comfort to the crews working on the drilling rigs. While offering adequate housing space, the site afforded comparative security in the form of isolation as well. A nearby pub, the Fox Inn, offered additional recreation. Although it appeared that the monks welcomed the oilmen, Walker, a member of the Anglican Church, was concerned about their reaction to the "rough-and-ready" oil field workers from America.[15]

Rosser and Walker wasted little time in making preparations to begin drilling operations. The men were issued identity cards by the Sheriff of Nottinghamshire as soon as they arrived, and on March 9, 1943, they were notified that part of their equipment had landed at Liverpool. While crewmen waited for the remainder of the machinery to arrive, the well sites were leveled, timber was gathered, and the mud pit was dug for the first American drilling location: Eakring 98. Once the preliminary preparations were completed, however, the crews settled down to await the arrival of the remainder of the equipment. Although the citizens of the community were courteous to the oilmen, and every effort was made to provide them with adequate facilities at the monastery in the interim, the period of inactivity began to affect the crew's morale. Their boredom was alleviated to some extent when the men began work on the outdated English rigs, found to be equipped with "loud-sounding electric gongs" that were used to signal air raids.[16]

Work on the rigs was not pleasant. The Midlands were swept by rain and harsh winds from the North Sea in March, which turned the roads into "knee-deep quagmires." Proper lighting, under the provisions of a wartime blackout, was another problem for the work area. Only two small shaded lights, each limited to "one candlepower for each foot above the floor on which it was placed," were permitted on opposite corners of the drilling floor, and two equally dim sources of light were allowed in the doghouse and near the mud pumps. In addition, the continuing rains made the laying of fuel and water lines a particularly nasty chore. Nevertheless the crews were able to relax at the Fox Inn during off hours where, although beer was rationed and ale was in short supply, "bitters" was generally available.[17]

As work progressed, the war slowly turned in favor of the Allies, and Rosser and Walker began to receive more and better equipment, despite military red tape which seemed to strangle all efforts to speed the project along. A portion of the new equipment shipped to the Oklahoma operation was lost in a submarine attack, compounding the problems faced by the drillers and causing endless delays. In addition, the shortage of food was a constant problem. A few of the men were farsighted enough to bring along seeds and, within a short time, had cultivated a small vegetable garden to supplement their diet; however, fruits, meats, and sweets were all but impossible to obtain. To offset an accompanying shortage of water, the Oklahomans drilled two nearby water wells. By late April, 1943, Rosser had three sites in operation, and drilling activities were moving along at an acceptable pace. Nonetheless, the meager wartime diet and the twelve-hour shifts were beginning to take their toll on the crew. Rosser dropped twenty-five pounds the first month on the job, and one member of the crew, Bob Cristie, lost thirty-two pounds in six weeks.[18]

A shortage of replacement equipment also plagued the operation, and Rosser was continually searching for materials such as mud and cement. Yet, despite the handicaps, the two Oklahoma-based companies were completing an average of one well per week. By June, 1943, almost one million barrels of crude had been shipped to refineries, and on July 5, Rosser and Walker had completed a total of five wells. But the serious problem of obtaining an adequate food supply reached crisis proportions that month and threatened the entire project. The cook announced that there was no food left. In desperation, Walker turned to the military which was persuaded to furnish the men with army rations, and the situation immediately improved.[19]

Noble had made plans to visit his company's British project on several occasions, but the press of wartime activities prevented the trip. Nonetheless, he did go to considerable lengths to keep abreast of the latest happenings of Rosser, Walker, and the others, and made it a point to keep them informed on events transpiring in Oklahoma. By late summer and early fall in 1943, the task of drilling wells in Sherwood Forest had evolved into a common routine, even though excitement was undoubtedly provided by a German bombing raid over the Nottinghamshire area on October 31 of that year. Only the discovery of a new source of crude would prevent the project of drilling 100 or more wells from being finished on schedule, and most of the wells previously drilled outside of the Eakring field had proved disappointing.[20]

Just as it seemed the two Oklahoma companies had

finally resolved a majority of the problems, tragedy struck. Herman Douthit fell from the mast of a drilling rig in Duke's Woods on November 13, 1943, and was killed. It was a bitter blow for Rosser and the Americans, especially when the task was nearing completion; yet, some of the gloom was dispelled by activities associated with the approach of the Christmas season. With the close of 1943, pressure on Rosser and Walker for the production of more oil began to slacken as the petroleum crisis passed.[21]

Although the Nottinghamshire region was "not much of an oil field" by American standards, the two Oklahoma companies, Noble and Fain-Porter, had completed ninety-four wells between April 12, 1943, and January 16, 1944. Seventy-six of these wells were producers. It was obvious that the operation would be completed around March 1, and most of the men desired to return home when the undertaking was completed; only thirteen men seemed willing to remain in England on a temporary basis. As the contract neared fulfillment, the men were allowed to tour the countryside, with the exception of Rosser, who was overwhelmed with paper work involved in closing out the project. When their efforts were eventually totaled, the Oklahomans were credited with 106 wells, one well for each 2,500 feet drilled. Because of their contribution to the British venture, the peak daily production for the English Midlands surpassed 3,000 barrels of oil in 1943, and by the end of the year, the Eakring-Duke's Woods and Formby fields had produced 2,289,207 barrels of high-grade paraffin base oil. By the end of 1945, the wells would produce another 1,231,346 barrels of oil. Their job completed, the American crew of Noble and Fain-Porter sailed for home on March 3, 1944. Rosser remained behind a few weeks longer to close out the operation.[22]

Back in Oklahoma, petroleum production continued its downward spiral during the first few years of the 1940s, despite the greater demand made by the outbreak of war. There was an increased effort to locate new pools to compensate for the declining output of the older fields, however. In 1940, there were 1,011 wells completed in the state, with an average initial daily production of 204 barrels of oil. The following year, the number of new wells had increased to 1,099, with an average initial production of 226 barrels of oil per day. One of the most lucrative discoveries occurred in June, 1941, in Caddo County, when the Texas Company drilled a well west of Apache in the SW¼ NW¼ of Section 2-T5N-R12W, near another highly-productive well brought in by the same company in October, 1938.[23]

The Texas Company's well, completed in the Simpson Bromide Sand at 3,360 feet, flowed between 200 and 300 barrels of oil a day; however, it later dropped to an estimated 125 barrels per day. Shortly thereafter, the well "blew out" and flowed approximately two million cubic feet of gas into the air daily from a zone located at a depth of 2,200 to 2,400 feet. The flow gradually increased to an estimated seventy-five million cubic feet of gas per day, and the well ran wild for nearly a week before a special blow-out preventer brought it under control. The find generated considerable excitement among the state's oilmen, and additional drilling was undertaken soon after the initial discovery. Within a short time, seven wells were producing in the region, with an average daily production of 4,012 barrels of oil.[24]

Other new locations in Pottawatomie, Pawnee, Carter, and Garvin counties allowed the state to maintain a fairly consistent rate of production during the early years of World War II. In addition, new efforts were made in previously discovered producing fields. Nonetheless, the long decline of Oklahoma's oil ouput continued and, as the outbreak of war had created new outlets for production, a major discovery was needed to reverse the downward trend.[25]

It came in 1943 with the completion of the discovery well in the West Edmond field. Convinced that there was oil beneath the surface in an area to the west of Edmond, Oklahoma, Ace Gutowsky claimed to have gained knowledge of the oil deposit by "doodlebug" exploration and had secured numerous leases in the region; however, he found it difficult to persuade oilmen that the prospect contained a favorable oil structure.[26]

Lacking the necessary funds to finance the drilling of a test well, Gutowsky tried unsuccessfully to raise the necessary backing among various oil company landmen and independent oil producers. He then attempted to sell a portion of his lease holdings to finance the oil well, but because of the lack of adequate geological or geophysical data, oilmen were apparently reluctant to place their faith and funds in the hands of Gutowsky's "doodlebug" exploration. Undeterred, he continued his search for backers outside of the state and eventually induced B. D. Bourland of San Antonio, Texas, to provide financial aid. Bourland succeeded in gaining the support of a number of Chicago backers and raised enough capital to drill a well to a depth of 7,350 feet, where Gutowsky was convinced that an oil-producing zone was located.[27]

West Edmond field. One of the most important petroleum discoveries of the 1940s was the West Edmond field located by Ace Gutowsky in 1943. *Courtesy of the Interstate Oil Compact Commission.*

Gutowsky's discovery well, the No. 1 Wagner, was spudded in on January 2, 1943, in the NW¼ NW¼ SW¼ of Section 22-T14N-R4W, and by April 5, 1943, Gutowsky, Bourland, and the other backers were rewarded. At the spot where most geologists had predicted there was no oil, Gutowsky brought in the No. 1 Wagner with an initial estimated flow of 50 to 70 barrels of crude a day and 5,000,000 cubic feet of gas per day. Although Gutowsky was known throughout the industry as a "doodlebug" and had drilled a number of dry holes on the basis of his "bugdope," he had the final laugh on "a lot of reputable geologists [who had] said there wasn't a chance" of finding oil in the West Edmond area.[28]

The No. 1 Wagner flowed 522 barrels of 41° gravity oil during the first twenty-four hour period. It was the most significant oil discovery in Oklahoma during World War II. The Champlin Refining Company, bought the field and initially used trucks to transport the oil to refineries, but little time was wasted in tapping the field to a pipeline. It was not long before the entire region was the site of a major development effort, and eleven more wells were soon producing from the Hunton Limestone in the West Edmond field.[29]

The downward trend of Oklahoma petroleum production in the years immediately prior to the outbreak of World War II, had not encouraged new oil companies to invest within the state; however, an exception was the Sohio Oil Company, a subsidiary of

Standard Oil Company of Ohio. A peculiar set of circumstances brought the firm to Oklahoma. When the United States Supreme Court ordered the dissolution of Standard Oil Company in 1911, Sohio was left with only a marketing outlet in Ohio and no crude production. By the mid-1930s the firm was actively searching for areas of potential production in which to enter the exploration and production phases of the petroleum industry. This search eventually brought Sohio to Oklahoma. Active in the initial development of the West Edmond field, Sohio owned an extensive holding of leases in the area, and under unitization Sohio later became the pool's operator.[30]

The output of the West Edmond field helped reverse the state's decline in petroleum production and marked a major contribution to Oklahoma's war effort. By the end of 1944 the state's output of crude reached 124,616,000 barrels, an increase of approximately 1,500,000 barrels over the preceding year, and much of this increase could be attributed to the West Edmond field. Before long, "the greatest concentration of rotary rigs in the world were boring down into West Edmond's red earth." The Cimarron Valley Pipe Line Company quickly ran an outlet to the field and was soon joined by another pipeline constructed jointly by Phillips and Sohio. By 1945, Oklahoma had become the third-largest producer of petroleum in the United States, largely as a result of the West Edmond discovery.[31]

Not only were Oklahoma oilmen and petroleum companies vital to the war effort, they were also a source of patriotic fervor. Many of the state's citizens were exempted from service in the military and found highly productive and important positions in the oil industry. These men and women contributed greatly to the war effort, not only by their work in the transportation, production, refining, and supply of petroleum products, but also in their voluntary contributions. An example of this was the drive to raise funds through the sale of war bonds to finance the construction of the light cruiser U.S.S. *Oklahoma City*. The Phillips Petroleum Company purchased $1,000,000 in bonds toward the goal and urged all employees of the firm to participate in the effort. Frank Phillips, chairman of the board, declared that "It makes me and all of the thousands of employees in Phillips Petroleum Co. in Oklahoma happy that our company can be a part of the campaign to pay for the cruiser *Oklahoma City* to carry our capital city's name on the high seas."[32]

When the war ended in 1945, Oklahoma's oil industry had made an impressive contribution to-

Oil journalist Claude V. Barrow, *left*, covered the Oklahoma oil scene for the *Daily Oklahoman* for thirty-five years; *on the right*, Oklahoma Governor Roy J. Turner. *Courtesy of Mrs. Claude V. Barrow.*

wards the victory over the Axis powers. Millions of barrels of crude had been taken from beneath the state to power the nation's war machinery, and thousands of Oklahomans had contributed endless hours in the oil fields, refineries, and laboratories. In addition, at a meeting in Tulsa in March, 1942, W. Alton Jones, president of Cities Service, formed a committee of national oil executives to develop a comprehensive pipeline project for the war effort. With the close of the conflict, however, most of the state's oilmen undoubtedly welcomed the end of federal wartime regulations and welcomed the rise in crude prices. In January, 1947, oil sold for $1.64 per barrel, and by the end of that year, it had climbed to an impressive $2.44 per barrel.[33]

Oklahoma's petroleum production climbed steadily after peace was restored, increasing in output nearly every year for the remainder of the decade. This was, of course, partially the result of the tremendous technological advances made during the war, as well as the desire of many Americans to secure the luxuries that had been denied them by wartime shortages and rationing. In 1945, the demand for petroleum in the United States had increased by twenty-two percent when compared to the pre-war demand peak in 1941, and every indication was that it would rise even higher. Oklahoma's output of 134,789,000 barrels of crude in 1946, valued at

Offshore drilling in Lake Texoma. After the close of World War II the search for crude in Oklahoma was renewed. One aspect of the effort was the beginning of offshore drilling operations in Lake Texoma on the state's southern border. *Courtesy of Halliburton Services.*

$194,100,000, had grown to 151,902,000 barrels in 1949, with a price tag totaling $388,870,000. By the end of 1948, oil was selling at $2.59 per barrel; however, with the removal of wartime price restrictions, the fast increase in the price of oil led to a demand for an investigation into the pricing practices of the petroleum industry by Congress and turned much of the attention of national leaders from the issue of conservation of petroleum reserves to the realm of a more strict control of marketing and pricing aspects of the oil industry. Needless to say, lower prices were not welcomed by an industry that had suffered the devastating economic blow of the Great Depression, as well as the subsequent controlled rationing and price regulation during the war years.[34]

World War II had brought new life to Oklahoma's petroleum industry, and state oilmen struggled to supply the needs of a nation at war. In 1947, two years after the close of fighting, there were 110 major oil fields in the United States with an output in excess of one hundred million barrels of oil, and fifteen of these pools were located in Oklahoma. The renewed drilling activity continued into the postwar period, and more wells were drilled in 1949 than in any other year in the state's history; however, the heyday of the oil boom was over. True, there would be additional discoveries made within the state in subsequent years, but nothing on the scale of Red Fork and Glenn Pool during the bustling era at the turn of the century or of Cushing and Healdton during the roaring twenties.[35]

Although the spectacular boom period of the Oklahoma petroleumm industry ended in the post-World War II era, men such as Sam K. Viersen, Jr., were still willing to invest in the oil business and involve themselves in the American free enterprise system. With headquarters in Okmulgee, Oklahoma, the Viersen and Cochran Drilling Company was originally established by A. A. Viersen, who, after arriving from Iowa, became involved in the Okmulgee boom. Viersen, together with his son, Sam K. Viersen, Sr., and Ad D. Cochran, expanded the firm's interests throughout the Mid-Continent and Rocky Mountain regions. In 1948, Sam K. Viersen, Jr., became the third generation of the family to be involved in the business.[36]

In 1948, the Continental Oil Company opened the West Short Junction field, approximately six miles west of Moore, Oklahoma, near the junction of State Highway 37 and United States Highways 62 and 277. Although production was found at a depth of 8,200 feet, the well was not considered good enough at that time to justify offsetting. Five years later, however, interest was resurrected in the region, and drilling activity was renewed. In February, 1955, Continental completed its No. 1 Castlebury "in the Hunton at 8,106 feet," establishing production in the field and indicating the location of a large pool. During the following months, nearly forty wells were drilled in the area without encountering a single dry hole. During this period, the production of the West Short Junction field grew from a modest daily output of 35 barrels of oil per day to more than 2,000 barrels daily, and by October, 1955, the cumulative production of the pool was approximately 270,000 barrels. There would be other locations throughout the state in the years to follow, but with advancing technology, much of the efforts of Oklahoma oilmen were turning to increasing the output of already existing fields or reopening old locations with the aid of modern drilling technology that allowed them to reach petroleum deposits that had been too deep in the past.[37]

# Chapter 18.
# Science and Technology

As the oil industry of America neared the end of its first century, it was obvious that the tremendous technological advances in petroleum production had made the business one of the most specialized in the world. When Edwin L. Drake drilled the first commercial oil well at Titusville, Pennsylvania, in 1859, he dug the first sixteen feet of the well by hand before employing a crude cable tool drilling rig to complete the hole. The cable connected the engine to the drilling tools, and a twenty-five foot length of stem was fastened to a steel-lined bit joint to drill the hole. A simple sand pump was employed to clear sediment from the well. After the drilling equipment was pulled from the well and unfastened from the cable, a thread on the upper end of the pump was attached to the cable socket and lowered into the hole, a time-consuming operation. Modern drilling methods bear little resemblance to such a crude beginning, and many of the technological advances made in the industry were the direct result of experiences of Oklahoma oilmen and unique situations encountered in the state's oil fields.[1]

Among the outstanding contributions the state has made to the petroleum industry was its pioneering work in geology and geological survey. Geological surveys have a long history in Oklahoma, where much of the early work was conducted by the United States government, as well as individuals. The expedition of First Lieutenant James B. Wilkinson, a member of First Lieutenant Zebulon Montgomery Pike's party, was the first. Wilkinson entered Oklahoma from present-day Kansas and followed the Arkansas River through the northeastern portion of the state in 1806. Later, in 1811, George Champlin Sibley made the first of two visits to the region and explored the saline deposits in northwestern Oklahoma. Again, in 1825–1826, Sibley entered the area of the Oklahoma Panhandle to survey the Santa Fe Trail. In 1817, Major Stephen Harriman Long followed the Kiamichi and Poteau rivers in southeastern Oklahoma to near present-day Fort Smith, Arkansas.[2]

However, the first published reference on the geology of Oklahoma was made by a botanist and ornithologist, Thomas Nuttall, who toured the state in 1819. Visiting the region in eastern Oklahoma, he first traveled southwestward from Fort Smith with a party of soldiers, then northwestward along the Arkansas, Grand, Verdigris, Canadian, North Canadian, Deep Fork, and Cimarron rivers. In 1820, he presented his findings to the Academy of Sciences in Philadelphia, Pennsylvania, and fourteen years later, in 1834, in a paper entitled "Collections toward a Flora of the Territory of Arkansas" at a meeting of the American Philosophical Society. Nuttall's work was supplemented somewhat by Edwin James who traveled down the Arkansas River through Oklahoma with Captain John R. Bell. These expeditions were followed by others: Thomas James, Hugh Glenn, and Jacob Fowler between 1821 and 1823; Washington Irving, Henry Ellsworth, Albert-Alexandre de Pourtales, and Charles Latrobe in 1832; Brigadier General Henry Leavenworth and Colonel Henry Dodge from 1834 to 1835; Josiah Gregg between 1839 and 1840; and Captain Nathan Boone in 1843. The observations of all these men added to the geological knowledge of the state.[3]

Among the early expeditions, the most comprehensive geological account of Oklahoma was provided by Captain Randolph B. Marcy. Marcy first traveled through the area in 1849, while escorting a group of emigrants to California; however, in 1852, he made an extensive examination of southwestern Oklahoma, visiting the North Fork of the Red River, the Wichita Mountains, the Red River, and south-central Oklahoma. Dr. George C. Shumard, the expedition's surgeon, made an extensive collection of fossils, rocks, and minerals during the trip, as well as "copious geological notes of the region." His findings were eventually analyzed by Edward Hitchcock of Amherst College and Dr. Benjamin F. Shumard. Marcy's report was later printed by the United States Congress and made available to all interested persons.[4]

After Marcy's expedition, Oklahoma's geology did not receive much attention for several years. However, prior to the Civil War, an additional effort was made by First Lieutenant Amiel Weeks Whipple's 1853 expedition to the region to survey a feasible railroad route across the state. H. B. Molhousen, a well-known artist, accompanied Whipple and left a series of sketches and watercolors depicting portions of the region's geological features. Other than the work of the Pacific Railway Survey, very little was done in the way of a systematic survey of Oklahoma's geology until the 1890s, with the exception of the attempts of several Texas geologists. Most of the surrounding states had established geological surveys by that time, and this stimulated a renewed interest in the area. In 1890, the United States Geological Survey sent several parties into the state under the initial direction of Robert T. Hill, and later Joseph A. Taff. This was in conjunction with the effort of the federal government to allot the lands of the Five Civilized Tribes, and an accurate knowledge of the region's mineral deposits was needed to ensure an equitable division of land. As a result, between 1890 and 1905 a series of topographic maps was prepared for the eastern part of the state. A doctoral dissertation by N. F. Drake for the Leland Stanford University in Stanford, California, was another individual attempt. Entitled "A Geological Reconnaissance of the Coal Fields of Indian Territory," it was published by the American Philosophical Society in 1897 and in a monograph series on biology by Stanford University. This was one of the earliest scientific descriptions of the geological formations of the eastern and northeastern portion of present-day Oklahoma.[5]

The renewed interest in the geology of Oklahoma, together with a concern over petroleum deposits, led to the creation of a Department of Geology and Natural History by the Oklahoma Territorial Legislature in 1899. Specifically, the position was "established for the purpose of beginning and continuing the geological and scientific survey of this Territory, . . . of discovering and developing its natural resources, and disseminating information in regard to its agricultural, mining and manufacturing advantages." The Professor of Biological and Kindred Sciences of the Territorial University was named Territorial Geologist and made chief of the department by the legislation; however, it specified that he "shall perform the services . . . without compensation." It was the duty of the territorial geologist to make an extensive survey of the territory and act as "curator of the geological cabinet, museum, apparatus, and library," which were to be located at the university. In addition, an appropriation of $200 was made "to be expended entirely in making such surveys and collections of specimens of minerals, organic remains and other objects of natural history peculiar to this Territory." Although it was a modest beginning, the Department of Geology and Natural History was the forerunner of the Oklahoma Geological Survey, which was to play a significant role in the development of Oklahoma's petroleum industry.[6]

Appointed Territorial Geologist, A. H. Van Vleet had a staff including Charles N. Gould, geologist; Edwin DeBarr, chemist; R. S. Sherwin, assistant in geology and chemistry; Paul J. White, collector in botany; and C. D. Bunker as collector in zoology. All were employed by the Territorial University at Norman. As time passed, interest in geology grew, and more and more geological studies were published as the petroleum industry developed, including works by Joseph A. Taff, Charles N. Gould, and H. Foster Bain on the Arbuckle and Wichita mountains.[7]

The Department of Geology and Natural History continued to function until statehood when the First Oklahoma Legislature replaced it with the Oklahoma Geological Survey. Under the direction of the State Geological Commission, the survey was charged with:

> First: A study of the geological formations of the State with special reference to its mineral deposits, including coal, oil, gas, asphalt, gypsum, salt, cement, stone, clay, lead, zinc, iron, sand, road building material, water resources, and all other mineral resources.
>
> Second: The preparation and publication of bulletins and reports, accompanied with necessary illustrations and maps, including both general and detailed descriptions of the geological structure and mineral resources of the State.
>
> Third: The consideration of such other scientific and economic questions [that] in the judgment of the commission, shall be deemed of value to the people.

In order to carry out its duties, all employees of the survey were authorized to "enter and cross all lands within the State; Provided, that in so doing no damage is done to private property."[8]

Frank Buttram, a pioneer Oklahoma oilman, who started his petroleum career with the Oklahoma Geological Survey. *Courtesy of the Oklahoma Heritage Association.*

Governor Charles N. Haskell, State Superintendent of Public Instruction E. D. Cameron, and President of the University of Oklahoma Dr. A. Grant Evans (members of the first State Geological Commission) selected Dr. Gould as Director of the Geological Survey. Gould was also authorized "to arrange with the Board of Regents of the State University for such rooms, laboratories, libraries, and testing apparatus as may be necessary for the work of the survey." Although the survey was allowed to languish to some extent (there was no funding provided for the survey in 1923 and 1931) it provided a great service to the petroleum industry when the oil boom swept Oklahoma in the twentieth century. Not only did it conduct field work to locate potential petroleum producing areas, but the publication of the *Oklahoma Geological Survey Bulletin* made the information available to Oklahoma oilmen and their geological staffs.[9]

In addition, many positions in the survey were filled by Oklahoma's early-day oilmen. Both D. W. O'Hern and Frank Buttram worked for the survey before entering the independent oil business. Buttram prepared a plane-table survey of Cushing for the survey in 1913, and in 1916, he joined with O'Hern to conduct extensive work near Cement in Caddo County. In 1913, the Gypsy Oil Company created a separate geological department in the firm under M. J. Munn and opened an entirely new segment of the petroleum industry within the state. Gypsy was soon followed by the Empire Gas and Fuel Company, and within a short time a geological department was an important aspect in all petroleum-producing firms. In 1923, the Oklahoma Geological Survey listed 243 oil and gas-producing areas across the state and stimulated additional development by making such information available to interested parties. The survey, as well as its counterpart in private industry, provided an important contribution to the Oklahoma oil industry. The ranks of these pioneering geologists included such men as Carl D. Smith, Robert H. Wood, Everett Carpenter, Pierce Larkin, L. L. Hutchison, Irving Perrine, Charles H. Taylor, L. C. Snyder, Charles E. Decker, and Robert H. Dott.[10]

Another important scientific and technological contribution of the state to the petroleum industry was the adoption of geology and other oil-related topics as part of the educational program at both the University of Oklahoma and the University of Tulsa. Such courses at two of the State's major institutions of higher learning produced the vast number of trained specialists needed by the petroleum industry. The University of Oklahoma, a pioneer in the study of petroleum geology, began offering courses in geology in 1901 and produced its first petroleum graduate, Charles T. Kirk, in 1904. Located in the "oil capital of the world," the University of Tulsa's School of Petroleum Engineering, which started in the laboratory of the Carter Oil Company, was among the

---

*Facing page:* The indoor drilling rig used for training petroleum students at the University of Tulsa. *Courtesy of Guy Logsdon.*

A diamond core drill used to provide geologists with underground-formation samples. The concept was first suggested by George E. Burton in 1919. *Courtesy of the Interstate Oil Compact Commission.*

nation's leaders in training petroleum engineers. The courses offered by the schools covered a wide spectrum of the petroleum business, including petroleum engineering, geological engineering, petroleum geophysics, petroleum land management, and petroleum sciences. The application and distribution of useful information by these two universities greatly stimulated the oil industry of the state and thrust Oklahoma to the forefront of the petroleum industry in terms of science and technology.[11]

In 1912, Charles N. Gould was retained as a geologist by Cities Service. He was joined the following year by Everett Carpenter, and the two scientists were assigned to undertake the first wide-area geological exploration and survey for oil and gas. Beginning in Oklahoma, the survey expanded into Kansas where

The International Petroleum Exposition, held in Tulsa since the 1920s, offered Oklahoma-based oil companies the opportunity to display their innovations and distribute their newly developed technology to oilmen throughout the world. *Courtesy of the Beryl D. Ford Collection.*

Gould and Carpenter discovered the rich Augusta and El Dorado fields. The success of their work did much to raise the stature of geologist as oil finders, and Cities Service, impressed by the utilization of "scientific" exploration and the accuracy demonstrated by Gould and Carpenter in locating the El Dorado anticline, established the "first in-house geological training school." Within a short time the company had developed the largest scientific geological staff in the industry at that time.[12]

The great emphasis placed on geology by the oil industry in Oklahoma was to pay huge dividends. In 1916–1917, the Sinclair Oil and Gas Company undertook preliminary exploratory work in the Garber area, about twenty-five miles southwest of the Tonkawa oil field. The company's geologists were convinced that the structure of the Garber area held a source of petroleum. As a result of the geologists' study, the firm began drilling operations and, in the fall of 1917, brought in a producer, from what was to be known as the Hoy Sand, with an initial flow of 200 barrels of oil per day. This was the first important oil discovery to be located "purely on the basis of geology," and ushered in a new chapter in the history of the petroleum industry.[13]

After the discovery of the El Dorado field, Cities Service continued to develop staffs of specialists, including petroleum engineers and reservoir engineers. The company also pioneered the development of the air oxidation approach to petrochemical processing and established the earliest petrochemical plant within the state at Tallant, Oklahoma. This plant was described as the "petrochemical patriarch of the Southwest." Cities Service revolutionized the petroleum industry with the development of a down-the-hole well treatment that was the forerunner of current wash detergents.[14]

Henry L. Doherty, the founder of Cities Service and an early-day proponent of conservation with the petroleum industry, instituted scientific research at the firm's Bartlesville facilities, proving that natural gas increased the oil's flowing qualities when dissolved in a reservoir of crude. As a result, he fought vigorously for compulsory unitization of oil fields and conservation of natural gas. To Doherty, it was astounding that petroleum producers would allow such a waste of natural gas, and by doing so, greatly diminish the amount of crude flowing from wells. When he presented his views in a paper read before the Federal Oil Conservation Board in 1926, however, his ideas met scant approval from producers. Nonetheless, later developments dramatically proved the accuracy of his arguments.[15]

Doherty's revolutionary ideas and farsighted planning fostered metallurgy research that allowed the development of large volume pipelines for petroleum and natural gas. Aware that the small diameter and low pressure of early-day pipelines restricted the amount of crude or gas that was transported to market, he suggested the development of large, high tensile pipelines which altered the petroleum industry not only in Oklahoma, but nationwide. His theory ultimately created an efficient method for transporting petroleum and natural gas to markets and refining centers. Although much of the technological research was conducted by another company, Cities Service was joint owner in the first cross-country pipeline in the Mid-Continent region constructed according to Doherty's ideas.[16]

However, not all early-day Oklahoma oilmen were impressed with the pioneering efforts of geologists in the state and they viewed the application of scientific principle to petroleum exploration with suspicion. After all, they argued, the oil industry had successfully discovered field after field by relying on instinct and intuition. For years, early-day oilmen had adhered to such theories as "creekology," the belief that streams changed course to flow around an underground formation; therefore one of the best places to sink a hole was in the bend of a creek—and they did not surrender their long-used methods easily. Perhaps one of the best examples of this lack of understanding was displayed by the attitude of Tom Slick, who had played a prominent role in the developing of several paying pools within the state, including the prolific Cushing field. On one occasion, Slick and a young geologist were in the field searching for a promising site on which to drill a well. The scientist pointed out a number of the geological features of the terrain in the area, indicating that it would be a good spot to drill. Slick's disdain for the science of geology was well-known, and after listening to the geologist's recommendation Slick pointed to a white horse and suggested that they follow the animal to the next pile of manure which would mark the site of the new well. Although history does not record whether or not the well was a producer, the story graphically illustrates the problems faced by many early-day geologists in gaining recognition among the oilmen.[17]

Early in 1915, Everette L. DeGolyer, who had graduated from the University of Oklahoma in 1911 and then taken a position as chief geologist of the Mexican Eagle Oil Co., Ltd., at Tampico, Mexico, returned to the university for a visit with Professor Charles H. Taylor. During their conversations, DeGolyer advanced the idea of organizing the geologists of the Southwest for the purpose of holding a meeting to discuss problems of mutual interest. Taylor agreed and began making plans. At the same time J. Elmer Thomas, of Tulsa, fostered the same idea and invited thirty friends, all of whom were active petroleum geologists, to dinner. During the course of the evening, James H. Gardner also suggested the formation of a geological society.[18]

Both DeGolyer's and Gardner's proposals generated interest, and shortly afterward, on October 15, 1915, Taylor invited all geologists working in the Southwest to convene in Norman in early January of the following year. In response, about sixty geologists met with Taylor on January 7–8, 1916, and appointed a committee to organize an even greater gathering at Tulsa the next year. Meeting in the "oil capital of the world" on February 10, 1917, the geologists organized the Southwestern Association of Petroleum Geologists. The following year the group changed its name to the American Association of Petroleum Geologists (AAPG), and agreed to establish headquarters in Tulsa. From the time of its conception, the AAPG "foster[ed] the spirit of scientific research throughout its membership," and made invaluable contributions to the advancement of the petroleum industry. Because of AAPG activities in disseminating new information and methods regard-

Cities Service's plant at Tallant was the "petrochemical patriarch of the Southwest" and the first such facility in the state. *Courtesy of Cities Service Oil Company*.

ing geology, the organization of the group must rank high on any list of Oklahoma's contributions to the oil industry.[19]

Another area in which Oklahoma led the petroleum industry for many years was the application of advancing scientific and technological methods to the conservation of both petroleum and natural gas. In 1913 the Oklahoma Legislature enacted the first state law providing for ratable production from common reservoirs of natural gas. This milestone was followed in 1914 by the Oklahoma Corporation Commission's "Order No. 813," the first proration order issued by a state regulatory body in America. The comprehensive gas conservation act passed by the Oklahoma Legislature in 1915 to replace the defective 1913 gas act was the first elaborate gas conservation measure adopted in the nation. That same year on June 5 the OCC issued "Order No. 920," the first state regulatory proration order that applied directly or indirectly "to the production of oil and not to the marketing of the same." In 1928 the corporation commission issued "Order No. 4430," which set a statewide demand of 700,000 barrels of crude oil, and divided the need between established fields (each of which was allocated 275,000 barrels of oil daily) and flush pools, allocated 425,000 barrels per day. This allocation was the first statewide oil proration order issued in the United States.[20]

Another facet of the petroleum business in which Oklahoma led the nation was that of well spacing.

The rapid decline of oil prices during the Great Depression made the need for such regulation apparent. Previously, the wasteful practice of rival producers in drilling wells as close together as possible had resulted in much unnecessary drilling and increased costs. Oklahoma became the first state to enact a well spacing law when in 1935 it gave the corporation commission the power to regulate the size and shape of drilling units, to locate individual wells within a unit, and to control the royalty interest within a producing unit. The 1935 legislation was amended in 1945 to authorize the forced pooling of the working interest (whether divided or undivided) of a field. If they could not reach an agreement among themselves, the OCC was empowered to fix the equities of the various interests and a limit of one well per drill unit was established. This was the first time a state had assumed a "Pool-wide compulsory pooling" arrangement.[21]

Although Oklahoma entered the petroleum picture several years after other producing states in the East, little time was wasted before Oklahoma companies were introducing innovations to the industry. In addition, the state gained an advantage over other producing areas of the nation in its role as host of the International Petroleum Exposition, held more or less regularly in Tulsa after 1923. Designed to distribute knowledge of new technology and equipment to oilmen from throughout the world, the IPE featured innovations in petroleum production and "had its place in accentuating the technology of the Industry and spreading it on a more widely-used basis and faster than would have been otherwise" possible. The IPE introduced to the oil industry "rotary drilling; the steel drilling derrick . . .; the steel well-servicing derrick; the truck-mounted, over-the-road well-servicing unit . . .; the gas/gasoline engine . . .; portable drilling rigs . . .; the cantilever-type derrick . . .; equipment for water-flood . . .; console-type and remote drilling, and many other innovations which have come to the industry through the years."[22]

Natural (or casinghead) gasoline was produced from the natural gas that accompanied the flow of crude from a well. Of higher volatility than ordinary gasoline, it was removed either by compression or absorption and was useful as a blending material in the manufacturing of motor fuel. By the removal of lighter hydrocarbons such as propane, isobutane, and butane, it was used as stock for the manufacturing of city gas or the hydrocarbon process. As Oklahoma was proved a source of natural gasoline, D. W. Franchot and a number of other investors built the first "casinghead" gasoline plant in the Southwest at Kiefer, Oklahoma, in 1909. Utilizing the output of Glenn Pool, Franchot was able to produce a "few hundred gallons weekly." Natural gasoline continued to be big business in Oklahoma, and by 1944, there were 112 natural gasoline processing plants within the state. In the decade that followed, Sinclair Oil and Gas Company, having originated in Oklahoma, pioneered the effort for the underground storage of propane. By 1955, the firm had approximately 110,000 barrels of propane stored in a huge underground facility between Seminole and Wewoka. Here, beneath the Bell City Lime, a huge cavern was mined out of the Nellie Bly Shale at a depth of 380 feet. Interconnecting rooms, 8 feet wide and 26 feet high, branched out from the vertical entrance shaft where liquefied propane, under pressure of 150 pounds per square inch, was stored.[23]

Oklahoma companies were in the forefront of innovative technology during the early decades of the twentieth century. In 1910, the Oklahoma Natural Gas Company built the first compressor station in the Southwest, and on September 9, 1913, the first recorded dual completion on a well was reported in Tulsa County's Wicey Pool. In 1915 air repressuring in Oklahoma was first attempted at an old stripper well in Rogers County, and the recycling of natural gas was an innovation of the Osage oil fields in 1924. One of the most useful developments to evolve from Oklahoma was the diamond coring rig suggested by George E. Burton in 1919. A well near Sayre, Oklahoma, previously drilled to a depth of 2,200 feet, was deepened to 4,920 feet by the adaptation of Burton's diamond bit on a cable tool rig which provided the operator with precise and definite detail of the formations encountered. Burton's innovation revolutionized the petroleum industry and greatly expanded the drilling capabilities of twentieth century oilmen.[24]

The birth of the reflection seismograph, one of the state's most important contributions to the petroleum industry occurred in 1917 when Dr. W. P. Haseman, the Head of the Department of Physics at the University of Oklahoma, met with one of his former students, J. C. Karcher, at the National Bureau of Standards in Washington, D.C., to discuss the possible use of reflection sound waves (seismic waves) to locate underground deposits of petroleum. As Haseman's and Karcher's concept was developed, it became known as the seismograph and it was utilized in the oil fields to gauge the depth of oil-bearing reflective structures and to map subsurface formations and structures. A charge was exploded in shot holes which had been drilled at selected locations in a prospective oil-producing region. The detonation caused seismic waves to travel through the earth until they struck the underlying hard formations. Reflected by these structures, the waves bounced back toward the surface where they were received by special equipment capable of accurately measuring to one-thousandth of a second the time interval between the explosion and the appearance of the reflected waves. The information, when properly interpreted by trained specialists, provided a useful underground map and became the most extensively used method of geophysical prospecting. Field tests which confirmed the validity of the reflection seismograph method of prospecting for oil were conducted near Oklahoma City on June 4, 1921, and were virtually echoed around the world. In 1925, the Geophysical Research Corporation surveyed the Earlsboro structure, making it the first oil field located by the reflection seismograph technique in the Southwest.[25]

Another area in which Oklahoma led the petroleum industry was in the utilization of cement in drilling operations. The cementing of wells was necessary for a variety of reasons: to prevent the drilling fluid from circulating outside of the casing and causing surface erosion; to seal off and protect fresh water formations; to provide an anchor for blowout preventer equipment; to give surface support for deep strings of casing; to seal off abnormal pressure formations; to isolate incomplete formations; to seal off zones of lost circulations; and countless other operations. Erle P. Halliburton, an early-day Oklahoma oil man, introduced the two-plug method of cementing to the Mid-Continent region in 1919, and in 1922, he patented the jet-cement mixer. Studies made by the United States Bureau of Mines in the Cushing Pool in 1919 led to the use of cementing to exclude water from producing wells, and by 1930, the use of cement was increasing significantly in petroleum production. In 1939, Halliburton's high pressure cementing pump was first used in the field by the Phillips Petroleum Company. Later, Halliburton's contributions led to the establishment of the Halliburton Oil Well Cementing Company, which first introduced the petroleum industry to gypsum cement in 1940.[26]

Halliburton and his firm made several other contributions to the oil business. In 1926, he invented a "well-sounding device," and in 1949, the Halliburton Oil Well Cementing Company first commercialized the hydraulic fracturing process, using napalm soap. In such a process, the object was to rupture the rock formation surrounding the bore hole and, in doing so, break down the oil-bearing formation and increase production. In 1953, Halliburton's firm introduced Pozmix, a revolutionary new form of cement which was a "mixture of possolan and portland cement."[27]

The state has produced several notable firsts in the pipeline aspect of the industry. In 1920, the first acetylene welding on oil field pipelines was performed on an eight-inch line that ran between Bartlesville and Ramona, Oklahoma, by Noah E. Wagner with the Prairie Pipe Line Company. Because of the state's central location, it became a hub for the development of a pipeline network in the Southwest. In 1933, the first electric welded pipeline, without the use of a backup ring inside the pipe, was constructed from Oklahoma City to Thrall, Kansas, for the Phillips Petroleum Company by the H. C. Price Company.[28]

In 1917, secondary recovery operations were inaugurated in the Southwest in Nowata County, Oklahoma, by gas repressuring. Seven years later in 1924, compressed air was introduced in the Bartlesville oil field and used extensively to stimulate crude production. The Carter Oil Company, on April 29, 1937, ran the first flow string of all-welded casing in the Mid-Continent region, and in 1946, the firm devised the "oil pool analyzer," designed to predict the productivity of certain oil fields as much as a quarter of a century into the future. During the late 1940s, the search for petroleum deposits on the continental shelf in the Gulf of Mexico was intensified, and on December 13, 1947, the world's first offshore producing well out of the sight of land was completed by Kerr-McGee, an Oklahoma-based oil company. Dean A. McGee recalled that the firm was a small concern during the late 1940s, and as such, was searching for drilling sites not already covered by lease holdings of larger companies. Most geologists realized that oil-bearing formations did not end abruptly at the water's edge, McGee declared, and after some seismographic exploration, Kerr-McGee drilled a well approximately 12 miles off Terrebonne Parish, Louisiana, in water less than 20 feet deep. The No. 1 State Lease 764, in Block 32 of the Ship Shoal area, came in flowing 922 barrels of oil daily, through an 11/16 inch choke, from production that topped at 1,732 feet. The successful completion of this well heralded an entirely new phase of oil production and opened a vast new area to exploration.[29]

The myriad of oil-well service operations offered by Halliburton Services, such as this acid-fracturing job near Sayre, Oklahoma, can trace their origins to the genius of Erle P. Halliburton, a pioneer oilman. *Courtesy of Halliburton Services.*

As for scientific contributions to the petroleum business, in 1919, C. E. Van Orstrand, working in cooperation with Dr. G. C. Matson, the Chief Geologist of the Gypsy Oil Company, made his first attempt in Oklahoma to correlate temperature with structure. Eleven years later, in 1930, K. C. Head made a series of temperature measurements in the Tonkawa and Burbank oil fields; however, he failed to show any irregularities that would correspond with the localization of heat near oil-bearing sands and concluded that "abnormal amounts of heat are not associated with deposits of petroleum." In 1937, F. B. Nichols, employed as a development geologist in the Seminole region by the Indian Territory Illuminating Oil Company, invented the Geolograph; however, it did not become generally available until

The expansion of offshore drilling operations led to the development by Halliburton Services of offshore oil-well service units. The first such operation was accomplished simply by lashing a land-based vehicle to a barge and towing it to the well site. *Courtesy of Halliburton Services*.

1943. Well Surveys ran the first commercial radioactivity well log at the Polo field in Oklahoma on April 1, 1940. Although the HF alkylation process was patented by Universal Oil Products, it was Oklahoma's Phillips Petroleum Company that developed, engineered and constructed the first commercial alkylation process unit at Borger, Texas. Instead of sulphuric acid, the HF process used hydrogen fluoride as a catalyst to join isobutane and olefin in the production of a base-blending stock for aviation gasoline. Phillips' plant facilities at Borger also produced and shipped the first furnace carbon black for use in tire-type rubber on December 25, 1943, and in September of 1946, the firm established the first American plant equipped with refrigerated reactors capable of producing a cold rubber used in making commercial tire-type synthetic rubber from butane.[30]

Oklahoma oilmen also led the way in the development of drilling equipment. D. D. Wertzberger of Tulsa was one of the earliest proponents of the steel drilling derrick, and another Tulsa firm, the Franks Manufacturing Company, produced what may have been the first portable well-servicing unit transported on a truck; another firm in Enid (Failing) produced a smaller version at approximately the same time. In addition, the Franks Company led the way in designing the original cantilever derrick capable of being raised from the ground; however, because of its small size (it had only an eight-foot working base) it was not very successful. But Lee C. Moore, also a resident of Tulsa, developed a larger and more useful version that was quickly adopted by most oilmen. Later, in 1954, the Franks Manufacturing Division of Cabot Shops constructed the first telescoping aluminum mast for Ray's Well Servicing of Maysville, Oklahoma. Another great contribution to the petroleum industry was the development of the "Reda" down-the-hole pumping unit by Armais Arutunoff, which made it possible to produce economically large volumes of liquids to recover a relatively small volume of oil. Oklahoma oilmen also led in the development of pneumatic rotary drilling, or pressure drilling with gasified water. Although

Members of a Cities Service geological survey crew that worked in Oklahoma and Kansas. *Courtesy of Cities Service Oil Company.*

the process had previously been tried in the Big Lake, Texas, oil field, it was successfully used as early as 1934 in the Fitts field of Oklahoma.[31]

Oklahoma was also the home of several important petroleum-related organizations that made great contributions to the development and expansion of the industry, such as the American Association of Petroleum Geologists, the Gas Processors Association, the Petroleum Equipment Institute, the Society of Exploration Geophysicists, and the Interstate Oil Compact Commission. The National Stripper Well Association is also based in the Sooner State. Organized on May 16, 1934, at a meeting held in Tulsa, it was designed as an umbrella organization for various oil and gas associations in stripper-well producing areas, to "promote and protect the interests of stripper well owners and operators, their employees and the owners of royalties under such wells." Other pioneering efforts in related fields, such as the firm of Leslie Brooks and Associates, one of the earliest concerns to specialize in oil and gas marketing research should not go unnoticed.[32]

One of the main reasons that Oklahoma's scientific and technological contributions to the petroleum industry rank among the most important in the world is the existence of such facilities as the Bartlesville Energy Technology Center. Established by the United States government on March 28, 1918, the Petroleum Experiment Station provided pioneering scientific and engineering research for the industry. Focusing on oil and gas field problems, the center developed specialists in petroleum engineering and technology for use throughout the oil business. Later, as the industry's needs expanded, the center contributed greatly to orderly oil- and gas-field development, secondary and tertiary methods of recovery, and more efficient methods of use.[33]

Playing an important role in the entire spectrum of the petroleum industry, the innovations contributed by Oklahomans did much to keep the state in the forefront of energy production. In addition, their efforts allowed the state to maintain a continuous position of leadership among the world's energy producers. Because of this leadership, the value of petroleum extracted by Oklahoma oilmen during the first fifty years of the state's history was ten times the value of Alaska's gold production and three times the total worth of all of the California gold mined between 1848 and 1957.[34]

## Chapter 19.
# Oklahoma Petroleum-Related Industries

ALTHOUGH THE PRODUCTION, processing, and marketing phases of the Oklahoma petroleum industry have poured hundreds of millions of dollars into the state's economy and provided tens of thousands of jobs for its citizens, even more money and opportunities for employment have been provided by companies allied with the oil industry. Perhaps one of the best examples of this economic coalition is the Halliburton Services, an oil field service firm with headquarters in Duncan, Oklahoma. When they founded the business during the oil boom in the early twentieth century, Erle P. Halliburton and his wife, Vida, started with the help of a single employee, hired on a daily basis. Ultimately, the firm was so successful that, in 1977, it employed more than 13,000 people, and "led the world as an oil field services organization."[1]

Halliburton received his initial training in the petroleum industry in 1916, as an employee with Perkins Cementing, Inc., of California. In 1919, the lure of the promising Mid-Continent region brought him to Wichita Falls, Texas, where he worked as a well-cementer and a consulting engineer on drilling problems in the Burkburnett field. However, the process of cementing a well, a procedure that protects oil and gas reservoirs by forcing cement down the hole to prevent the migration of water from one subterranean level to another, was still new, and many of the early-day oilmen were not interested in investing money in the procedure.[2]

Undaunted, the Halliburtons (Erle as the "salesman, manager, laborer, and capitalist" and Vida as the "silent partner") embarked on what would become in less than fifty years a multi-billion dollar enterprise. With a working capital of $1,500 and a borrowed pump, Erle and Vida opened the Better Method Oil Well Cementing Company. In addition to the pump, Halliburton obtained "an oil tank, a clothesline to measure depth and some wooden plugs" and litterly "begged permission to demonstrate his well cementing technique to operators." Although offering "to give 100 percent satisfaction or make no charge for his work," he still had to work at odd jobs to make ends meet. In January, 1920, the opportunity to prove the value of his services came when W. G. Skelly, the president of Skelly Oil Company, was faced with a wild well on one of the company's leases in the Hewitt-Wilson field. Contracting with Skelly, Halliburton was able to control the well successfully, and the future of his business was assured. Several months later, on May 7, 1920, the Halliburton Oil Well Cementing Company was formed and quickly became well-known in the state.[3]

Always seeking ways to improve his services, Halliburton was an innovator as well. The time-consuming task of manually mixing the cement often resulted in a product of inconsistent quality which sometimes hardened before it could be forced into the wells, and Halliburton soon developed a Jet-Mixer which increased the speed and quality of the mixing process. In addition, he developed a Sack Cutter in order that the cement might be introduced into the well at a faster rate. Halliburton's clothesline also evolved into an accurate depth-measuring device. Although he began the business with wagons and teams, Halliburton quickly adapted his equipment to fit four-wheel-drive trucks.[4]

The early years of the organization offer a classic example of the American free enterprise system and the accomplishments of individuals willing to endure hard work and long hours to achieve success. On one occasion, Vida Halliburton persuaded her husband to bring home empty cement sacks in order that she might wash and fold them for reuse. Because time was extremely important in the process of cementing a well, it was necessary for Halliburton to guarantee

Erle P. Halliburton, whose creative genius helped Halliburton Services become one of the world's leading oil-well service firms. *Courtesy of Halliburton Services.*

delivery at a specified time, and during the early part of the company's history, this was not a simple task. The condition of the existing roads in the state often made a number of the oil fields virtually inaccessible. Trucks were often stuck in mud and teams of mules were required to transport the equipment to the drilling site.[5]

As the firm expanded, Halliburton's crew grew quickly. Several units often operated at the same time in widely separated areas of the state. Generally, when a crew arrived in a town located near an oil field, a telephone was installed in a rented house which served as temporary field headquarters. A superintendent was placed in charge of the crew and, in addition to his supervisory role, was required to be a skilled mechanic, and to have a basic knowledge of geology as well. Independent of the home office, with the exception of general policies, replacements, and salaries, the superintendent was also required to be an accountant, crew chief, and company representative.[6]

When Halliburton hired his first employee, Frank Kempf, on February 1, 1920, his initial bank statement reflected a total deposit of $1,100, out of which more than $900.00 was spent for a truck, over $180.00 in expenses and $5.00 for labor, leaving a balance of $50.27. The company established headquarters in Duncan in 1921, and by the end of 1923, twenty trucks were at work in the oil fields. By the time the Halliburton Oil Well Cementing Company was incorporated on July 1, 1924, its employees numbered fifty-six. A primary factor in the phenomenal growth of the company was Halliburton's decision not to compete with the oil companies on which he depended for business. Instead, efforts were concentrated on expanding the services his facility offered the industry. When Halliburton died thirty-seven years later, the original bank balance of $50.27 had grown to more than $156,000,000 in assets, $194,000,000 worth of business, and a net profit of more than $19,000,000.[7]

When the company was converted from a husband-wife partnership in 1924, the oil companies of Magnolia, Texas, Gulf, Humble, Sun, Pure, and Atlantic indicated a willingness to purchase a portion of the stock. Halliburton was insistent that the business remain under his principal ownership, however, and a number of the large petroleum companies were hesitant to leave fifty-two percent of the stock in the "hands of this aggressive young man with the fires of ambition in his eyes and caked cement under his fingernails." As a result, Halliburton agreed that he and his wife would hold only equal voting rights with the companies. Thus, each of the seven oil companies purchased approximately seven percent of the stock of the Halliburton Oil Well Cementing Company, Erle and Vida Halliburton retained approximately forty-nine percent, and the balance was placed in a voting trust.[8]

By 1932, Halliburton had seventy-five cementing units operating in seven states. In fulfilling his slogan of "We will get there, somehow," he continued to utilize only the most modern equipment in his operations and, by the early 1930s, four airplanes were in the service of the company, by now comprising five divisions: East Texas-North Louisiana, Oklahoma-North Texas, Kansas, Gulf Coast, and West Texas-

New Mexico. In 1926, Halliburton Oil Well Cementing Company, Ltd., was established in Canada, and in 1940, the firm expanded into South America. During World War II, the company produced material for the Allies, developed new pumps and well servicing techniques, and expanded to countries new in oil exploration and production. By its twentieth anniversary, it was selling products to thirty-six different countries. After the war, Halliburton continued to expand at a rapid pace, and by 1978 had performed services or sold products in eighty-four countries.[9]

The pioneering innovations of Erle P. Halliburton were carried on by the Halliburton Services Company in its effort to provide the best technology available to the petroleum producers it served. In 1938, when production was established in the offshore region of the Gulf of Mexico, Halliburton was quick to provide the same services as those offered on land. Although the first offshore service units consisted of only "lashing a pump unit on a seagoing barge and towing it to the well site," the firm quickly acquired a fleet of eight oceangoing ships and two barges designed specifically for servicing offshore drilling units. In 1978, Halliburton's seaborne units offered acidization, cement fracturization, sand consolidation, gravel packing, nitrogen service, and well stimulation services, not only in the Gulf of Mexico, but in the Middle East and the North Sea as well.[10]

In addition to offshore drilling facilities, Halliburton Services Company contributed greatly to the Oklahoma petroleum industry through its ongoing technological research. Its series of high pressure, high volume pumps refined the elementary cementing process into a highly technical procedure. In 1978, Halliburton was the only service company with an extensive network of field laboratories, and as part of a long-range program began construction on the Halliburton Energy Institute at Duncan, designed to make the most up-to-date servicing technology available to interested personnel in the petroleum industry. Halliburton can count among its contributions to the oil business the most powerful portable pumps available; the Densometer weighing system; a laboratory program that was the first to recreate actual downhole conditions; the initial X-ray diffraction for core analysis; the original utilization of surfactants for cleaning formations; and breaking water and emulsion blocks. The accomplishments of the company are unparalleled in the field of cement development, and include: the use of Pozzolans in oil well cement; gypsum cements; multiple-state cementer; jet-mixer; water shut-off materials; retarders and accelerators; bulk cementing; skid cementing units; and at-the-well storage bins. In addition to developing more than thirty-five additives for oil field cement, Halliburton Services Company also produced many "firsts" in tools and floating equipment, such as the hydraulic jet gun, high-pressure squeeze tools, Roto Wall Cleaners, and many others. In the field of testing, the company was the first to perform "commercial oil well testing in 1926," and has never relinquished its lead in this area. The initial company to perform commercial fracturing, Halliburton also pioneered the continuous sand proportioner, radioactive sand, and water fracturing of oil and gas wells. The initial emphasis of Halliburton in regard to research and development, later continued by the Oklahoma-based firm, contributed greatly to the petroleum industry, not only in the state and nation, but worldwide.[11]

Among the most overlooked aspects of the petroleum industry in landlocked Oklahoma are the offshore drilling operations based in the state. A graphic example of this is the Reading & Bates Offshore Drilling Company, with headquarters in Tulsa. Originally a land-based firm, the enterprise was organized by J. W. Bates, Sr., and George M. Reading in 1937 with three rigs purchased on credit. "You learned fast in those days," Bates reminisced, "because you worked twelve hours a day, seven days a week."[12]

Both Bates and Reading had previous experience in the oil fields. After graduation from Dartmouth College in Hanover, New Hampshire, in 1910, Bates entered the oil industry as a teamster in California, driving a team of mules in the oil fields. Working his way up as a roustabout, tool dresser, and drilling superintendent, he became in 1914 a general superintendent for Roxanna Oil Company, predecessor to Shell Oil Company in the Mid-Continent region. At the same time, Reading worked his way through the University of California before entering the oil business as a mining superintendent in Arizona, and at the age of forty-two, he worked in a trucking operation in Texas. Later, while Reading was working as a drilling contractor for Shell, the two men met and began a lifelong friendship and business partnership.[13]

Upon returning from the Navy following World War II, J. W. Bates, Jr., joined Clyde Baker in organizing the B & B Drilling Company, which was absorbed into the firm of Reading & Bates in 1949. C. E. Thornton joined the company in 1951 and

Dowell Well Incorporated, another internationally known service company that traces its beginnings to the Oklahoma oil-boom era. *Courtesy of the William A. ("Mac") McGalliard Collection.*

expanded the company's operations into Canada, where it enjoyed much success. Reading & Bates continued to grow, and in 1955, Thornton and the younger Bates persuaded the others to enter the offshore drilling phase of the petroleum industry. "Jack and Charlie Thornton came up with the idea of going offshore," the elder Bates recalled; "I can't claim any clairvoyance. . . . I hadn't foreseen it would be such a booming thing, but Jack never wanted to do anything else." The younger Bates and Thornton traveled to the Gulf Coast of Louisiana to explore the offshore facet of the oil business and returned convinced of its potential. In order to finance the venture, the Reading & Bates Offshore Drilling Company was organized and stock sold to the public. When the necessary funds were raised, the firm's first offshore operation was started.[14]

Two Levinston-type drilling tenders, the *George M. Reading* and the *J. W. Bates*, were constructed for the firm. With a length of 260 feet, a beam of 54 feet, a depth of 16 feet and 3 inches, and a loaded draft of 12 feet, the two non-self-propelled, flume-stabilized drilling tenders launched the Reading & Bates Offshore Drilling Company's operations in the Gulf of Mexico. Provided with mooring equipment in both the bow and stern, the ships accommodated fifty-two men and had storage facilities for water, mud, cement and fuel. Drilling operations were carried out by a multiple well derrick 140 feet high and a rotary drilling rig with a 465-ton capacity. At that time, these two ships represented perhaps the most advanced technology in offshore drilling operations.[15]

From this modest beginning, the firm underwent tremendous growth until, in 1978, twenty-eight offshore drilling units were operating in the Gulf of Mexico, the North Sea, and off the shores of Indonesia, Iran, Brazil, Brunei, Abu Dhabi, Italy, the Ivory Coast, Tunisia, Egypt, Japan, and also Singapore. Entering into offshore activities during its early phase of development, Reading & Bates endeavored to keep in the forefront of technological advancement. In recalling the rapid expansion of offshore operations, the elder Bates declared, "The biggest changes I've seen in the industry are in communications and transportation. The boys from Tulsa can get to one of our rigs in the North Sea as fast as I could get from Tulsa to . . . Perry, Oklahoma, in 1922." Another aspect of the advancing technology was the increased cost: "The expense of offshore drilling is, to a veteran of simpler times, staggering," the senior Bates declared, "You have eight to ten million dollars invested before you even get started."[16]

Since the initial effort was made by Reading & Bates, offshore drilling operations have dictated a great variety of equipment. As a result, the modern fleet maintained by the firm has evolved into a diversified collection adaptable "to types of equipment and geographical market representation around the world." The rig types utilized by Reading & Bates include tenders, platforms, tender jack-ups, and drill ships. Tenders such as the company's *W. D. Kent* contain air-conditioned accommodations for 103 men, and have an overall length of three hundred feet and a beam of seventy feet. Equipped with two cranes and six different types of mooring equipment, they have a depth rating of twenty thousand feet and are equipped with an emergency generator, water distillation units, a cementing unit, mud conditioning equipment, and drilling apparatus. For use in up to 600 feet of water, self-propelled drill ships, such as the *J. W. Bates*, are utilized. With a cruising speed of

Founded by J. W. Bates, Sr., and George M. Reading, the Reading & Bates Offshore Drilling Company operates units, such as this OS-7 platform, throughout the world. *Courtesy of Reading & Bates Offshore Drilling Company.*

approximately nine knots and a crew of 100, the *J. W. Bates* was capable of accommodating a Sikorsky S-61 helicopter and had a depth-rating of twenty thousand feet.[17]

Primarily "engaged in international contract drilling of oil and gas wells," the Reading & Bates Offshore Drilling Company conducts business from its corporate headquarters in Tulsa. Drilling in the Gulf of Mexico, the North Sea, the Mediterranean,

Reading & Bates Offshore Drilling Company's drill ship *J. W. Bates. Courtesy of Reading & Bates Offshore Drilling Company.*

and the South China Sea is far removed from the prairies and plains of Oklahoma; nevertheless, the Reading & Bates Offshore Drilling Company is but one example of a unique aspect of the Oklahoma petroleum industry which, in the second half-century of its existence, has expanded throughout the world.[18]

With Oklahoma being one of the major centers of the international petroleum industry, it is easy to understand why the state would also be a focal point for the publication of oil information. This provided Oklahoma with another oil-related business which had a great impact on the industry as a whole. With headquarters in Tulsa, one of the most impressive examples of this facet of the industry was the Petroleum Publishing Company and its widely-known publication the *Oil & Gas Journal*, which has been issued regularly since 1902.[19]

First printed in Beaumont, Texas, on May 24, 1902, as *The Oil Investors Journal*, the publication was founded by newspaperman Holland S. Reavis, a resident of St. Louis, Missouri, who was drawn into the petroleum industry by the excitement of Spindletop. As a publisher, Reavis quickly recognized the need to provide petroleum information to investors in Texas and neighboring states in addition to data appearing in eastern newspapers. Promising "The truth and nothing but the truth concerning the Beaumont field and other southwestern oil fields," Reavis launched the publication. Although much of its early emphasis was on regional investment potential, it soon began to broaden "its scope beyond the Southwest . . . [and] also switched from its financial orientation to focus on the information needs of operating oil men." During this stage of development, the *Journal* "stuck to" its publication promise and carried "reports from the field, economic analysis, and statistics on drilling, stocks, shipments, and prices of crude oil and refined products." In addition, operators were allowed to present their views "on the best methods of finding, developing, and producing oil and gas." This later became "the kind of editorial fare that was . . . a *Journal* staple over the years." In January, 1905, *The Oil Investors Journal* issued the "first of its annual review and forecast issues . . . beginning the longest-running and most-successful series in the magazine's history."[20]

When activity in the Glenn Pool area focused the attention of the petroleum industry on Oklahoma and touched off the Mid-Continent boom, *The Oil Investors Journal* opened a branch office in Tulsa. The publication was soon occupying more time than Reavis could contribute, and on April 20, 1910, he announced the sale of the publication to the Petro-

Reading & Bates Offshore Drilling Company's tender *George M. Reading*. Courtesy of Reading & Bates Offshore Drilling Company.

leum Publishing Company. Owned by Patrick C. Boyle, a Pennsylvania newspaperman who had built a national reputation as publisher of the *Oil City Derrick*, the Petroleum Publishing Company paid twenty thousand dollars for the publication, changed the name to the *Oil & Gas Journal*, and moved the entire operation to Tulsa. In addition, Boyle changed the magazine's publication schedule from bimonthly to weekly and tied it even more closely to the oil and gas industry. A goal of the new owner was the conversion of the periodical from a "personal organ of the editor into an institution serving the industry, with its own permanent character, independent of the personality of any individual," and this goal has been continued by four generations of proprietors.[21]

Boyle died in 1920, and his son-in-law, Frank T. Lauinger, succeeded him as head of the enterprise. Lauinger, a businessman in Pittsburgh, Pennsylvania, introduced "profit-oriented business practices to balance the atmosphere of frontier journalism which surrounded the company's beginnings," and greatly expanded the magazine's geographical coverage. In addition, he "opened the *Journal*'s pages more widely to the rapid advances in petroleum technology" during the era of tremendous growth. During Lauinger's eleven-year tenure as head of the organization, the oil business grew, not only in the United

# Oil Investors' Journal

THE TRUTH AND NOTHING BUT THE TRUTH CONCERNING THE BEAUMONT FIELD AND OTHER SOUTHWESTERN OIL FIELDS.

VOL. 1. SEMI-MONTHLY     BEAUMONT, TEXAS, U. S. A., MAY 24, 1902.     NO. 1. ONE DOLLAR A YEAR.

### GET THE FACTS.

Never before has there been such widespread demand for absolutely accurate and trustworthy information concerning the Beaumont oil field and the various companies operating in it as exists today.

Reports published in Northern and Eastern newspapers early in the present month to the effect that the gushers, as the result of a South American earthquake, had ceased gushing as suddenly as the Lucas well started spouting, have accentuated the already keen desire for a true statement of the condition of Spindle Top and the facts relative to the gas pressure.

Oil men have in no wise been disturbed by these reports—they realize that the profits of the oil business in a pumping field are far more sure than profits in a gusher field where two hundred or more companies have holdings.

It is the man that has been led to purchase stock on the single inducement that the corporation whose shares he buys "owns a great gusher" who is anxious to know. And his name is legion.

With conditions as they are—such a general craving for the facts—it is not too much to assume that THE OIL INVESTORS' JOURNAL begins its career at an opportune moment. It comes to supply a want, to occupy its place in the field of legitimate journalism which has been created by the natural sequence of events following the discovery of a marvelously productive oil field. Adhering to a strict policy of honesty and accuracy in its news reports, publishing the truth and nothing but the truth without fear or favor to any individual, firm or corporation, there is no reason why it should not receive the support of the people and become a permanent institution, worthy of the confidence of all good men and the enmity of all fakirs.

It will be an inviolable rule of the management of the JOURNAL in every instance to prevent the use of its news columns for advertising purposes. This paper can thrive only on truth and square dealing. It cannot hope to deceive people and still retain their confidence. It must seek to command the respect and confidence of all its readers above all other things. The public is easily fooled, but the man who does the fooling usually gets what is coming to him in the end.

To avoid any possibility of deceiving its readers the JOURNAL will designate all advertising matter as such. Advertisements will be accepted on the condition that they appear under proper classification and entirely distinct and separate from the news columns.

To its subscribers the JOURNAL will render reports on companies, giving the information requested in as complete and detailed form as it is possible to obtain it. This feature of its work will, no doubt, arouse some antagonism, but the JOURNAL has the satisfaction of knowing that this antagonism will come only from those who are afraid of the truth. Legitimate enterprises, conducted with the interest of the stockholder at heart, will fear nothing that may be written or said of them, and they will be glad to assist the JOURNAL in giving accurate information to those who ask it.

### CONDITION OF THE SPINDLE TOP FIELD.

Reports, worded in a manner calculated to cause apprehension on the part of those who have invested in the Beaumont field, have been scattered broadcast over the land. The most sensational story was, in effect, that the Spindle Top gushers had ceased to flow, all in a night, as the result of an earthquake in South America—the exact location of the disturbance in South America being omitted.

Telegrams by the score and letters by the hundred have poured into Beaumont since the publication of these reports, indicating the widespread uneasiness which has been caused. Almost without exception these inquiries have come from holders of "gusher" stock or from promoters of companies intent upon selling their shares.

To the best of its ability the JOURNAL will endeavor to describe the true situation at Spindle Top. Only information based upon what the eye has seen—not what the ear has heard—will be given.

On May 7th, the day after the New York Times and other leading newspapers published the dispatch stating that the gushers had quit flowing, a party of business men, correspondents and visitors went to Spindle Top for the express purpose of testing the wells to ascertain the actual conditions existing. Inspector George A. Hill superintended the tests.

First a six-inch well near the west line of Block 38, Hogg-Swayne tract, and adjoining the Heywood property, was opened. No artificial means were resorted to to agitate the well. A minute elapsed before the oil appeared. When it came it flowed in a heavy stream through the six-inch elbow, which had been tilted at an angle of 45 degrees. A discarded boiler, not less than 100 feet distant from the well, and not more than 150 feet away—no one measured the distance in exact figures—was drenched by the oil from the well. The flow was continuous and seemed to grow stronger as the well cleaned itself out. When all present had been satisfied that this well was still a gusher the oil was checked and the party proceeded across the

*Oil Investors' Journal*, which was the forerunner of the *Oil and Gas Journal*. *Courtesy of the Petroleum Publishing Company.*

States but worldwide, and he enlarged the publication's staff, added full-time editors based in Washington, D.C. and New York City, acquired additional refining and technical editors, and started a program of answering technological questions. He also rearranged the magazine to "de-emphasize field reports on drilling in order to feature other news and advances in technology . . . [and] devoted more effort to reporting technical developments as news."[22]

Because there was an absence of any industry information program, Frank Lauinger mounted a campaign "to expose false charges which were the basis [of federal regulation and] . . . takeover proposals." On March 20, 1924, the *Oil & Gas Journal* released a sixty-nine page special issue to "set the record straight on the facts of the business in language that anyone could understand." Sending copies of the material to all members of Congress, Lauinger produced a "best seller" that "led to the industry's first public relations program." Moreover, while he was serving as head of the organization, the *Oil & Gas Journal* issued its first international number in December of 1923, and "in the next two years, the magazine adopted a program of special issues devoted to different phases of the industry and regions of new activity."[23]

Upon the death of Frank Lauinger in 1931, his son, P. C. Lauinger, who had worked as an aide to his father for nine years, became president of the firm. At the same time, the Great Depression was affecting all aspects of the petroleum industry, and although he had an opportunity to sell the company, P. C. Lauinger stated that he would "try to make a go of it." He was the first of the chief executives to devote full time to the *Oil & Gas Journal*, and he moved from Pennsylvania to Tulsa to "take charge of the operation on the spot." Because of his competent leadership, the Petroleum Publishing Company was able to overcome the effects of the downward business spiral of the 1930s and become "a truly international publication."[24]

Of particular interest to P. C. Lauinger during the 1930s was the desire to provide a format for advancing technology, particularly reservoir engineering and conservation, a major theme of the *Oil & Gas Journal* during this period. Through his magazine, Lauinger argued that "Market-demand proration . . . should be based on engineering principles for the prevention of waste," and to a great extent, this "saved the industry from bankruptcy through overproduction." In addition, a number of geologists, petroleum engineers, chemical engineers, and economists were hired "to complement the staff of specialized petroleum writers." Assigned to the division of their specialization, these professional staff members added a great deal to the magazine's evaluation of the multifaceted aspects of the oil industry, including "drilling, production, exploration, pipelining, refining, gas processing, and . . . petrochemicals." The specialized editors were dispatched throughout the world to report on "the new scenes of oil and gas activity," and "Their reports brought readers around the globe first-hand information from the new centers of oil activity." As a result, the Petroleum Publishing Company underwent a sustained period of growth.[25]

Philip C. Lauinger, Jr., who succeeded his father, P. C. Lauinger, as president of the Petroleum Publishing Company. *Courtesy of the Petroleum Publishing Company.*

John H. Williams, who built The Williams Companies into a multinational operation. *Courtesy of the Oklahoma Heritage Association.*

Having previously worked in all aspects of the firm's publishing enterprises, Phillip Lauinger, Jr. assumed his father's role in the business when the elder Lauinger became chairman of the board in 1969. Focusing upon "diversification and restructuring of the company along modern, progressive lines to assure effective management of a growing enterprise," P. C. Lauinger, Jr., created an "integrated, diversified publishing and printing business" which, three-quarters of a century after its founding, produced four petroleum magazines, two newsletters, an encyclopedia, and several directories. Moreover, the firm carried on an extensive book publishing program that included such works as: *Production Operations*, *Seismic Exploration Fundamentals*, *Dynamic Positioning of Offshore Vessels*, *Underwater Engineering*, *Modern Petroleum*, *Trace Elements in Petroleum*, *The Technology of Artificial Lift Methods*, *Drilling Practices Manual*, *Guide to Refinery Operating Costs*, *The Properties of Petroleum Fluids*, *Enhanced Oil & Gas Recovery*, and many other titles in use throughout the petroleum-producing world. Two of its publications, *Petroleo Interamericano*, a magazine serving Latin America, and *Oil & Gas International*, directed to the Eastern Hemisphere, were designed for specific use outside the United States. However, the *Oil & Gas Journal* remained the mainstay of the firm, with an "editorial blend of news, statistics, and technology, prepared by a staff that is balanced between trained journalists and writers with technical degrees." Oilmen in 122 countries subscribed to the publication.[26]

The tremendous expansion of the science and technology of the petroleum industry has created a wide spectrum of separate businesses and pursuits. The Williams Companies of Tulsa, having focused its efforts on fertilizers, energy, and metals, is indicative of the diversification of many modern petroleum enterprises. Actively conducting drilling operations in Texas, Louisiana, the Gulf of Mexico, offshore California, Nevada, Arizona, Utah, Idaho, Wyoming, Montana, North Dakota, South Dakota, Arkansas, Mississippi, Alabama, Florida, and Oklahoma, the Williams Exploration Company includes both land-based and offshore facilities. Another aspect of its energy-related production is the Peabody Coal Company, the nation's largest coal-producing firm, of which The Williams Companies owned 27.5 percent. Its 48 active mines, when combined with the firm's reserve holdings, gave The Williams Companies operations in Montana, Wyoming, North Dakota, Utah, Colorado, New Mexico, Arizona, Arkansas, Missouri, Illinois, Indiana, Ohio, Kentucky, Alabama, and Oklahoma, with coal sales in 1977 of 64,700,000 tons.[27]

The fertilizer interest of The Williams Companies was vested in the Agrico Chemical Company, with major marketing channels in domestic and international agricultural sales. Several overseas operations maintained by the firm include a Korean fertilizer complex, twenty-five percent of which is owned by Agrico Chemical, and phosphate rock mining operations in Brazil. In addition, Agrico Chemical Company has maintained a marketing area over the eastern one-half of the United States, with major manufacturing facilities located in Florida, Louisiana, Arkansas, and Oklahoma, and a m．．ng operation in Florida. The Williams Companies' metal interests are operated by Edgcomb Metals Company, encompassing twenty-three metals service centers in thirty states east of the Mississippi River. Also included under the board management of The Williams Companies is the Williams Energy Company, serving 215,000 customers in twenty-seven states through 291 bulk distribution plants and 370 independent dealers. In addition, the Williams Pipe Line Company, involved in the transportation of petroleum products, crude oil, liquid petroleum gas, and fertilizer, has maintained a pipeline system stretching from Oklahoma to the Great Lakes, and eastward to Ohio.[28]

As demonstrated by the far-flung interests of The Williams Companies, the tremendous expansion of the petroleum industry into related fields is almost unlimited. That the modern petroleum industry of Oklahoma has far surpassed the early goals set by the production companies of the twentieth-century oil boom era is graphically illustrated by the development of the Halliburton Services, the worldwide operations of Reading & Bates Offshore Drilling Company, and the diversified interests of the Petroleum Publishing Company and The Williams Companies. In considering all the related aspects of the state's oil businesses (there are many more besides those discussed here) the sheer magnitude and complexity of the industry can be appreciated, and the huge investment in Oklahoma and its people by the petroleum industry can be more easily visualized. The Oklahoma oil industry and its allied enterprises are indeed invaluable resources of great benefit to the state.[29]

## Chapter 20.
# Oklahoma's Petroleum Legacy

WHAT HAS BEEN THE LEGACY of the oil and gas drawn from beneath Oklahoma's rich earth? One benefit lies in the amount of taxes paid by the oil companies to the state treasury. During the fiscal year 1976 a record $151,797,147 was paid in gross production taxes on oil and gas alone, and during the same period $270,636,612, amounting to 27.3 percent of the total state tax revenues, was paid by Oklahoma's petroleum industry in the form of taxes. While much of this money went into Oklahoma's common-school fund, a large part of the gasoline tax was used to upgrade the state's road system. Another facet of the state's oil legacy lies in the area of employment. In January, 1977, more than 85,000 citizens of the state were employed in the oil business, and of this number, 63,200 were directly involved in production, refining, transportation, and marketing operations. The salaries paid these workers comprised an important segment of the total salary income within the state and had a tremendous impact on the economy. In addition, according to a recent survey, 36,300 Oklahoma citizens are shareholders in various oil firms and a total of 11,750,000 Americans are investors. Furthermore, 91 colleges and universities, some 1,000 charitable and educational foundations, and 200 mutual insurance companies owned portions of oil companies throughout the country. During the fiscal year 1977, a sixth (16.9 percent) of all Oklahoma tax dollars came from gross production taxes on oil and natural gas, as well as petroleum excise taxes. When gasoline and other fuel-related taxes are taken into consideration, the Oklahoma petroleum industry contributed 27.8 percent of all taxes collected by the state that year.[1]

During the mid-1970s, the petroleum industry in Oklahoma employed thousands of citizens and pumped hundreds of millions of dollars into its economy. By this time, oil and gas accounted for 91.1 percent of the value of all mineral production in the state, and Oklahoma ranked fourth among the states in both the production of oil and in the number of refineries in operation. However, it was second in the actual number of producing oil wells in the state. By 1974, oil had been found in seventy-two of the state's seventy-seven counties, and out of a land area of 44,748,160 acres, 18.8 percent, or 2,824,420 acres, produced either oil or natural gas, or both.[2]

From this bountiful reserve of oil and natural gas, by 1973, there was a total of 72,880 producing oil wells averaging 7.2 barrels of crude per day. As of January 1, 1974, twelve refineries operated within the state at a total operating capacity of 481,000 barrels daily. In addition, there were eighty-six natural gas processing plants capable of processing 4,540,700,000 cubic feet of gas per day. Another aspect in which Oklahoma has a long history of research and development was the establishment of petrochemical plants, nine of which were operating within the state in 1973. The impact of the petroleum industry on the economy of the state was even more astounding. In 1976 a grand total of $1,226,040,000 was pumped into the main stream of the state's economy from bulk petroleum plants, terminals, liquified petroleum gas establishments, both wholesale and retail, gasoline service stations, and retail fuel oil outlets. The payrolls of these and other petroleum marketing businesses amounted to $130,746,000, an impressive figure. This was, however, only a portion of the total petroleum legacy inherited by the people of Oklahoma and does not include the educational, technological, medical, religious, and cultural gains reaped from the many philanthropic and research facilities financed by Oklahoma's oil.[3]

One of the most significant philanthropic contributions in the state was the establishment by Lloyd

Noble of the Samuel Roberts Noble Foundation as an Oklahoma common-law trust on September 19, 1945. A well-known state petroleum pioneer, Lloyd Noble first entered the oil industry on June 1, 1921, when he formed a partnership with A. O. Olson and established the Noble Drilling Company. The enterprise lasted until December, 1949, when the partnership was dissolved and Noble Drilling Company was incorporated solely under Noble's direction. A resounding success, the firm operated in twenty-five states, offshore in Texas and Louisiana, in Canada, Europe, and South America. In addition to his drilling operations, Noble organized several related petroleum enterprises such as Samedan Oil Corporation (named for his children, Sam, Ed, and Ann), an exploration and production company formed in 1931, and B. F. Walker, Inc., a trucking firm with headquarters in Denver, Colorado. All of these enterprises proved very successful and played an important role in the development of the Noble Foundation.[4]

After careful consideration of what he could do most for Oklahoma and the people of the state, Noble created the foundation in honor of his father, to "give recognition to the most charitable individual that I ever knew." After almost two years of careful research by Noble, the institution became an Oklahoma corporation on November 1, 1952. Noble's children—Sam, Ed, and Ann Noble Brown—took an active interest in the development of the foundation and continued to serve the organization.[5]

Noble's underlying philosophy in establishing the foundation can best be explained in his own words: "We have had the privilege of living in a marvelous age. We have seen many tools developed for the use of man. Any thinking individual must recognize, however, that the development of the human race itself in its moral concept is far behind the accomplishment of its scientists . . . who can say, if we keep on . . . along the line of attempting to do four our fellow man in a manner that will be truly helpful to him, how far-reaching may be our results." As the financial base for the organization, Noble gave to the foundation the Noble Drilling Company, Samedan Oil Corporation, and B. F. Walker, Inc.[6]

Many of the foundation's early efforts were in the field of agricultural research. A technical staff was employed in 1946, and a series of contests initiated to stimulate more interest in improving the area's agricultural economy. A chemical laboratory was established to analyze soil in order that maximum fertilizer use could be obtained, and that soil treatment on plant composition could be measured. The contests

Samuel Roberts Noble, the son of pioneer oilman Lloyd Noble. *Courtesy of the Oklahoma Heritage Association.*

were held over a three-year period, and prizes were awarded annually to individuals who made the greatest progress in cropland or pasture improvement, as well as the production of home gardens.[7]

Originally, the Noble Foundation was headquartered on the second floor of the McGhee Building in Ardmore, Oklahoma; however, in the summer of 1950, approximately two hundred acres of land two miles east of the town was acquired, and facilities were constructed to house the organization at this site. In compliance with the wishes of its board of directors to develop a more comprehensive program, the Noble Foundation signed a cooperative agreement in May, 1951, with the Oklahoma A&M College Board of Regents to organize a program designed to serve both rural and urban people in south-central Oklahoma. Concerned with erosion control, soil improvement, and water conservation, the foundation worked on the premise that "in a region of erratic rainfall conservation of water and the selection of cultivated crops which would produce a maximum yield of high quality feed during periods when rainfall is normally abundant would have an important influence on successful farm operations during many years." In fact, this was the "answer to the development of a more permanent agriculture in this region."[8]

Concentrating on consultation and making its facilities available to the citizens of south-central Oklahoma, the Noble Foundation did much in the field of developing pasture grasses, especially Elbon and Bonel ryes. By 1974, the foundation was working with approximately three hundred cooperators in an eight-county area of southern Oklahoma and north Texas. Consultation and research was provided free, with only the stipulation that adequate records be maintained and the information shared with others. Through these efforts, the foundation has raised the living standards of countless Oklahoma families and contributed greatly to the conservation of the land.[9]

In the early 1950s the Noble Foundation implemented another aspect of its research efforts and embarked on an extremely successful program in cancer research. With the main thrust of the foundation's program centering around the cause and effect of cancer, a number of its discoveries appeared in *Tissue Culture: Methods and Applications* by Dr. Paul F. Kruse, Jr., and M. K. Patterson, Jr. The most comprehensive work ever printed in the field, it was published by Academic Press of New York City in 1973. Countless papers on the research conducted by the Noble Foundation have been presented to scientific groups throughout the world, and its researchers are respected both nationally and internationally for their contributions.[10]

As the foundation grew so did its facilities. By 1975, seventy-five researchers were on the staff, nine of them with doctor-of-philosophy degrees, and by 1977 the institution included a 405-acre farm as headquarters, as well as demonstration and research facilities totaling 3,370 acres. Some of the more important contributions of the Samuel Roberts Noble Foundation include the development of a bull performance testing program for weight gainability; the development of Elbon Rye, the first rye to be developed and released in Oklahoma that produced outstanding forage production; the creation of Bonel Rye, which surpassed Elbon in productability; work in chemical pond weed control; initiation of the intensified winter pasture program; pioneering agricultural extension work; seasonal forage production of small research grains; development of a culture medium for the growth of cells; and "a prominent role in the development of the drug L-asparaginase." John March, president of the Noble Foundation, estimated that through June, 1978, the organization had spent in excess of $35,000,000 on research and grants. All of this was made possible by the farsightedness of Lloyd Noble and his success as one of Oklahoma's pioneer oilmen.[11]

Higher education has also been a great benefactor of the Oklahoma petroleum industry. This applied not only to physical facilities, such as the Lloyd Noble Arena and the Goddard Health Center at the University of Oklahoma, but to academic excellence as well. The Merrick Foundation, established by the pioneering oil family headed by Ward Merrick, provided a $750,000 endowment to the History Department of the University of Oklahoma for the establishment of the Merrick Chair in Western American History. This was the first fully-endowed academic chair at the university, and according to the department chairman, Dr. Norman Crockett, was a "tremendous shot in the arm" for the study of Western History at the institution. Affording a senior historian the opportunity to provide a great stimulus for the study of Oklahoma's western heritage, the Merrick Chair was first filled in the summer of 1978 by Dr. Paul Glad.[12]

However, the preservation of Oklahoma's heritage was not the only endeavor of the Merrick Foundation. Both Ward Merrick, Sr., and Ward Merrick, Jr., were greatly interested in the establishment of a modern health facility for the south-central portion of the state, and as a result, the foundation contributed heavily toward the creation of the Southern Oklahoma Memorial Hospital Association, which operates a modern health care facility at Ardmore. Opened in the 1950s, the hospital was valued at $1,500,000 and offered one of the most advanced health programs in the state. The Merrick Foundation also made grants to local and state schools, other hospitals across Oklahoma and a number of religious projects. Although the organization was not limited to Oklahoma in its philanthropic endeavors, Ward Merrick, Jr., estimated that ninety percent of its funds were spent within the state. Among the most successful programs of the foundation was the presentation of an annual teaching award to the instructor judged by his or her peers as having contributed the most to the advancement of the ideals of the American free enterprise system at the University of Oklahoma, Oklahoma State University, and Oklahoma Christian College. In addition to the recognition accorded the individual, the award carried a cash stipend of $1,000. The Merrick Foundation was also heavily involved in the concept of the "college without walls program" and the instigation of vocational training facilities in the Ardmore area. Originating in 1947 with a $2,000,000 endowment, the Merrick Foundation has since spent in excess of $2,000,000 in grants and contributions, and in 1978, had accumulated assets valued at $2,400,000.[13]

An outstanding contribution to world literature owes its creation to the Oklahoma petroleum industry. The Walter Neustadt family of Ardmore, Oklahoma, involved in oil production for five generations, endowed the establishment of the Neustadt International Prize for Literature through the University of Oklahoma. The award was originally conceived in 1969 as the *Books Abroad* International Prize for Literature; however, in 1972, Walter Neustadt, Jr., agreed to provide a two hundred thousand dollar endowment for the project from moneys accumulated through the family's involvement in petroleum production, and the name of the award was changed to the Neustadt International Prize for Literature. Second in prestige only to the Nobel Prize, the Neustadt-endowed presentation is given biennially to the most outstanding international author selected by an international jury of twelve. The award carries a $10,000 cash stipend. Very few other international prizes of this magnitude exist within the United States, according to Ivar Ivask, the editor of *World Literature Today*. The Neustadts had always been among the cultural leaders of the state, and they felt that the University of Oklahoma's international literary quarterly (formerly *Books Abroad*), which sponsors the award the prize was the highest contribution they could make to bring further international recognition to the state.[14]

John E. Kirkpatrick, having entered the oil business in Oklahoma after returning from naval service in World War II, established the Kirkpatrick Foundation with the announced purpose of supporting "religious, charitable, scientific, literary, or educational undertakings within the United States, or any of its possessions." Although the foundation was given a broad base of operations, by far the majority of its efforts have been in Oklahoma. Such a project had been in the minds of John and Eleanor Kirkpatrick for many years, and in 1955, the time seemed opportune for such an endeavor. To finance the effort, the Kirkpatricks approached the board of directors of

Eleanor Blake Kirkpatrick works with her husband, John E. Kirkpatrick, and other family members in overseeing the Kirkpatrick Foundation. *Courtesy of the Oklahoma Heritage Association.*

the Kirkpatrick Oil Company to "establish a program in recognition of its financial position, its community obligations and public relations needs." As a result, the board agreed that the firm would "contribute substantial sums of property to the foundation in order that it would be able to carry on its praiseworthy activities." Kirkpatrick gave the foundation stock in the Kirkpatrick Oil Company and announced that he planned to continue a "program of making substantial donations to the foundation in order for it to be endowed ultimately with money or property with a value in excess of $1,500,000.00."[15]

Adopting a declaration of support, the board of

directors of Kirkpatrick Oil Company resolved that the firm's initial contribution to the foundation would be a number of its most promising oil and gas leases. When the transfer took place on November 1, 1956, the property given to the foundation was valued at $500,000; however, this was just the beginning. Over the next twenty years, the Kirkpatrick Foundation received a total of $1,617,621 in contributions and $2,803,900 in income from investment dividends, interest, the sale of securities, and royalties. Out of this total, the organization distributed $1,966,812 during its first two decades of existence. Throughout the history of the foundation, John and Eleanor Kirkpatrick remained active members of the board of trustees. With less than $2,000,000 in contributions, the organization has given away nearly $2,000,000 while maintaining a working capital of an additional $2,000,000 for future grants.[16]

Under the Kirkpatrick's leadership, the foundation had a notable impact on Oklahoma City, the state, and nation. It played a major role in the development of the Oklahoma City Community Foundation and was active in the formulation of the "Great Plan," which created the ongoing relationship between Oklahoma City University and the Massachusetts Institute of Technology in Boston. It was responsible for many of the cultural developments within the Oklahoma City area: the Kirkpatrick Fine Arts Building on the Oklahoma City University campus, the Lyric Theater, the Oklahoma Science and Arts Foundation, the Kirkpatrick Planetarium, the Paine-Kirkpatrick Wing of the National Cowboy Hall of Fame and Western Heritage Center, and the Oklahoma Center for Arts and Sciences. However, the Kirkpatrick Foundation also did much work on an individual basis by providing scholarship loans to the Naval Reserve Officers Training Corp Midshipmen at the University of Oklahoma and funding the Saint Anthony Hospital Dental Clinic for indigent patients. The Oklahoma City Zoo, one of Kirkpatrick's favorite projects, has the reputation of being one of the finest in the world. In addition, thousands of people have benefited from the work of the Oklahoma Medical Research Foundation, much of which was made possible by grants from the Kirkpatrick Foundation.[17]

The Kirkpatrick's daughter, Joan, and their grandson, Christian, have also been active in the organization, which is described as "a family foundation." In this way, the foundation "was able to accomplish a great deal with a minimum of bureaucracy." In addition, both John and Eleanor Kirkpatrick have devoted a large amount of their time to the organization to insure that "the maximum benefit was received for every dollar." Moreover, according to one friend, "he [John Kirkpatrick] probably worries more over the way the funds of the Kirkpatrick Foundation were utilized than he did his own personal fortune." Because much of the money put into the organization had originated in Oklahoma, the Kirkpatricks believed that it should be spent in the state, and more than 276 different beneficiaries had received grants as of 1975. Of this number 46 grantees were either medically-related or hospitals, 28 were involved in cultural development or heritage, 19 were religiously oriented, 19 were dealing with higher education, 11 were libraries or centers of learning, and 5 were related to patriotic themes. Any Oklahoman who has visited the Oklahoma City Zoo, benefited from the facilities of the Oklahoma Medical Research Foundation, attended the Oklahoma City Symphony, or been fascinated with the exhibits of the Oklahoma Center for the Arts and Science has enjoyed a portion of the state's petroleum legacy.[18]

One of the most beautiful gifts to the people of Oklahoma was provided by the family of Judge Robert A. Hefner. Hefner, an Ardmore attorney when the oil boom swept the south-central portion of the state, took advantage of the opportunity and entered the petroleum business. Later, he was elected a justice of the Oklahoma Supreme Court in 1926. Moving his family to Oklahoma City, Judge Hefner purchased a stately mansion at 201 Northwest Fourteenth Street. The structure had originally been constructed in 1917, and during his lifetime, Judge Hefner traveled throughout the world seeking elegant furnishings and works of art for the house.[19]

In 1933, Hefner retired from the bench to resume private law practice, and in 1939, he was elected mayor of Oklahoma City. During this period, the Hefners continued to acquire material for the home. As the Hefner's collection of art, china, and other furnishings grew, the family searched for a way the material could be enjoyed by the citizens of the state. As a result, in 1970, Judge Hefner and his family announced their gift of the home, its furnishings, and attendant properties to the Oklahoma Heritage Association. Judge Hefner died in 1971, and on August 6, 1972, the Oklahoma Heritage House was formally "dedicated to the people of Oklahoma."[20]

Since that time, the facilities have become the headquarters of the Oklahoma Heritage Association, which "nurtures a progressive concept of honoring

Robert A. Hefner, *left*, and Frank Buttram, *right*, two well-known Oklahoma petroleum philanthropists. *Courtesy of the Oklahoma Heritage Association.*

builders of the future heritage of the state of Oklahoma as well as the preservation of the state's past." Judge Hefner's children, Robert A., Evelyn, and William, continue to play an active role in the Heritage Association and its efforts for a "general program of the enhancement of the on-going heritage and quality of life in Oklahoma." The home presents an opportunity for all citizens of the state to view the rare china collection, beautiful art work, the elaborate furniture from throughout the world, and the impressive architecture of the residence. In addition, the gift furnished the Oklahoma Heritage Association with the facilities necessary to offer seminars and institutes on the state's heritage and to expound on "the details of the history of Oklahoma." It also provided a home for the Oklahoma Hall of Fame Galleria, the Shepherd Oklahoma Heritage Library, and the Oklahoma Heritage Archives.[21]

Another legacy of the petroleum industry to the people of the state was given by Dean A. McGee, chairman of the board of the Kerr-McGee Corporation, who established the McGee Foundation. McGee, who first became involved in the oil business in 1927, funded the construction of the $2 million Dean A. McGee Eye Institute, which has served hundreds of Oklahomans with vision problems. The five-story structure was built to house the Lions Clubs of Oklahoma Eye Bank, an optical shop, an animal research division, and several individual patient-care and research rooms. It had been estimated that "blindness costs Oklahoma taxpayers over two million dollars a year," and the facilities provided by McGee allowed for the treatment of twenty-five thousand patients annually.[22]

The three-fold purpose of the Dean A. McGee Eye Institute was to screen and treat persons for eye

The Oklahoma Heritage House, formerly the home of Judge Robert A. Hefner. *Courtesy of the Oklahoma Heritage Association.*

disease, to teach and train physicians to become specialists in eye diseases, and to research eye disorders. In order that the facilities might be available to all citizens of the state, a mobile unit was also provided to serve the fifty-seven Oklahoma counties lacking the services of ophthalmologists. In addition, the institute became an educational affiliate of the Oklahoma University College of Medicine's Department of Ophthalmology in the training of resident physicians and paramedics in eye disease. Countless Oklahomans have benefited from the treatment facilities and research services provided by McGee, made possible through his involvement in the petroleum industry.[23]

Following the same vein as McGee in supporting medical research and treatment, was W. K. Warren, the originator of the William K. Warren Foundation. From an initial capital expenditure of $300, Warren built his oil business into an organization valued in excess of $300,000,000, and in 1948 he chartered the William K. Warren Foundation. Since that time Warren, his wife, and their seven children have contributed large sums of money to be utilized in grants to hospitals, churches, and schools. According to Warren, "We decided in 1953 the best way to distribute some of our earthly benefits was to build an outstanding hospital," and the result of this decision was the construction of Saint Francis Hospital in Tulsa. Taking a great personal interest in the project, Warren visited sixty-three hospitals throughout the country to ensure that the most feasible architectural plans and specifications were incorporated in the structure.[24]

A religious man, Warren readily admitted that "I prayed a lot," while establishing his oil business, and "still do." His religious beliefs were reflected in his philanthropic endeavors, and by the time the first patient was admitted in October, 1960, the "hospital represented the largest single gift ever made by an individual to a Catholic Order." A lay leader in the Roman Catholic Church, Warren expressed his idea of charity on a plaque placed near the main entrance to the facility: "This hospital is given to the Sister Adorers of the Most Precious Blood for the people of Tulsa and vicinity with the hope, wish and prayer that life may be happier, healthier and longer."[25]

In addition to providing the hospital, Warren served as president and chairman of the William K. Warren Medical Research Center, which began work on a twenty million dollar addition in 1973, and as president and chairman of the board of the Tulsa Vianney Girls School, having financed the construction. He also contributed heavily toward the financing of St. John's Hospital in Tulsa. For these contributions, he was invested as a Knight of Malta, an order devoted "to the care of the sick and needy in hospitals since the Middle Ages," by the Catholic Church.[26]

Another pioneer Oklahoma oilman who left an oil legacy to the state was John Elmer Mabee. Born in 1879, Mabee was a self-made individual whose "program of self-education made him proficient in finance, economics and business and his application of these skills in oil drilling, [and] lease trading . . . made possible his ability to spread gifts and endowments with a lavish hand." In 1945, Mabee and his wife, Lottie J., established the J. E. and L. J. Mabee Foundation, Inc., which granted "millions of dollars

John E. Mabee, who formed the Tulsa-based Mabee Foundation. *Courtesy of the Oklahoma Heritage Association.*

. . . to schools, hospitals, churches, orphanages and character-building organizations." Based in Tulsa, the organization contributed heavily to the University of Tulsa's building program and helped finance construction of two dormitories, a reading clinic, and a speech and hearing aid clinic at the institution. In addition, the foundation aided the Young Men's Christian Association, the Salvation Army, the Red Cross, and the International Petroleum Exposition. A major factor in any philanthropic effort in the Tulsa area, Mabee's goal was "To aid Christian religious organization, charitable organizations, preparatory, vocational and technical schools, institutions of higher learning, and scientific research; [and] to support hospitals and other agencies and institutions engaged in the discovery, treatment, and care of

Oklahoma art legacy. Alfred E. Aaronson, *left*, and Thomas Gilcrease, *right*, whose love of art left a legacy for all Oklahomans. *Courtesy of the Oklahoma Heritage Association.*

diseases." After his death in January, 1961, the foundation continued to serve the people of Oklahoma in much the same way as "Mr. Philanthropy" (a name frequently applied to Mabee by area newspapers) intended.[27]

Among the cultural leaders of Oklahoma is Alfred E. Aaronson, who came to the state in 1913 and entered the oil industry. Serving as vice-president of both the Mid-Continent Petroleum Company and the Mid-Continent Gas Company from 1913 to 1918, then as president of the Tuloma Oil Company from 1915 to 1926, he participated in the most exciting era of the Oklahoma oil industry before turning his talents to other investments. Aaronson had developed a keen cultural interest in the Tulsa area when, in 1954, he learned that the state was in danger of losing the renowned art collection accumulated by Thomas Gilcrease. The collection consisted of a "fabulous and irreplaceable collection of paintings, etchings, folios, sketches, books, sculptures in bronze, wood and stone, pre-Columbian treasures, and artifacts of Americana." Aaronson, appalled that such a treasure might be lost to Oklahomans, led the fight to raise the necessary funding to keep it within the state. Acutely aware of the tremendous economic growth of the region, Aaronson believed that "it is vital that our cultural development keep pace with this growth," and "Tulsa's acquisition of the Gil-

The Thomas Gilcrease Institute of American History and Art in Tulsa. *Courtesy of the Beryl D. Ford Collection.*

crease collection was an important step in this direction." To a great extent, it was the effort of Aaronson that made the Thomas Gilcrease Institute of American History and Art a reality.[28]

Another interest of Aaronson's was an adequate library system in Tulsa, and he worked untiringly toward this goal. He spearheaded "the drive for a metropolitan library system for Tulsa and served as the first chairman of the Tulsa City-County Library Commission in 1960. Later, as he developed an interest in human relations, he started a special collection in the library devoted to this aspect of man's behavior. It was Aaronson's belief that "In the growth of a community, satisfactory human relations between various groups of its citizens becomes a matter of prime importance. Libraries provide a place to rest, be quiet, step off the moving platform of the moment and think," he stated. "It is difficult to imagine a world without a public library. In the modern world, the library is not only for the learned and the scholarly, it is literally for everybody. . . .

The public library is a common man's university." Because of the ideals of a man such as Alfred Aaronson, a better understanding of mankind has been among the most valuable of Oklahoma's petroleum legacies.[29]

No discussion of the oil legacy inherited by the people of Oklahoma would be complete without an acknowledgment of the efforts of the two Phillips brothers, Frank and Waite. Both were pioneers in the petroleum industry of Oklahoma and heavily involved in the development of the state's oil deposits. In addition, both were well-known philanthropists and left the citizens of the state cultural and historical facilities that otherwise would not have been a reality.[30]

Frank Phillips is perhaps best known for his legacy of the Woolaroc Museum, located in Osage County, fourteen miles southwest of Bartlesville. Named from a combination of the words *wood*, *lake*, and *rock*, it contains a wildlife refuge, lodge, and museum. Initially it served as a source of recreation to Phillips and his friends and business associates in the 1920s, but its 55,000 exhibits were later opened to the state's citizens. Waite Phillips also left a museum filled with rare and valuable works of art for the people of Oklahoma to enjoy: Philbrook Art Center in Tulsa. Originally Phillips' home, the Italian Renaissance-styled mansion was constructed in 1926 at 2727 South Rockford Road. Surrounded by twenty-three acres of formally landscaped gardens, Philbrook houses an internationally recognized collection of art, including the famed Clark Field collection of American Indian baskets and pottery, the Samuel H. Kress collection of Italian Renaissance paintings and sculpture, the George H. Taber collection of Chinese jades, the Standard Oil Company of New Jersey's collection of oil industry paintings, and many others. In addition, the institution offers a variety of programs and educational workshops for Oklahomans to enjoy.[31]

Oilmen were not alone in giving generously to the improvement of the welfare and well-being of the citizens of the state. Many of the major oil companies in Oklahoma have also contributed. Atlantic Richfield Company, one of the state's producing firms, gave $8,000,000 to the Atlantic Richfield Foundation in 1977, and of this amount, $7,250,000 was distributed nationwide to "educational institutions, the United Fund, humanities and the arts, community programs, health and medical services organizations, environmental research and protection programs, and public information and other projects." Although not all of this money was spent in Oklahoma, a substantial grant was made to the Baptist Medical Center's Burn Center in Oklahoma City. In the cities of Ardmore, Shawnee, Bartlesville, and Woodward which house the firm's subdistrict offices, Atlantic Richfield employees have also given of their time and money to aid various civic projects.[32]

The Cities Service Company of Tulsa has channeled money into civic, cultural, and educational projects in the state. The Cities Service Foundation, a corporate institution involved in activities throughout the United States, is administered through the Tulsa offices of the corporation with an annual budget of $1,000,000. Contributing to the University of Oklahoma, Oklahoma State University, and the University of Tulsa, it has made unrestricted gifts to individual departments. In addition, the foundation has contributed heavily to the United Fund. The Henry L. Doherty Education Foundation, named for the founder of the company, was established in 1939 to "help sons and daughters of Cities Service employees finance their college education." More than 2,100 students have been awarded scholarships in this program.[33]

The petroleum industry in Oklahoma has always taken its responsibilities toward education seriously. John G. McLean, the chief executive officer of Conoco from 1969 to 1974, described this commitment when he stated that "The value of corporate support to education is even more important today than any other time in recent years," because "There is an ever-increasing need throughout society for breadth of understanding and specialized intellectual skills." Conoco responded by an outlay of almost $6,000,000 over a fifteen-year period for educational grants, and in 1978, approximately 207 colleges and universities received a total of approximately $1,500,000 from the company. Phillips Petroleum Company also has a major scholarship program, in which sixty-six children of employees of the firm receive individual $4,000 grants. These and other contributions to the educational process of Oklahoma's youth were one of the oil industry's greatest gifts and have been of tremendous benefit to the state's colleges and universities. Literally thousands of young people in the state have been able to acquire college educations as a result of Oklahoma's petroleum resources.[34]

These cultural, medical, and educational benefits were made possible by the crude pumped from beneath the earth of Oklahoma, as well as the interest of oilmen in the welfare of the state's citizens. Although countless other examples of Oklahoma's petroleum legacy are available, the Kerr-McGee Foundation, Inc.; the Phillips Petroleum Foundation, Inc.; the Skelly Oil Company Foundation; the Thomas Gilcrease Foundation; the Kerr Foundation, Inc.; the McCasland Foundation, the Vivian Bilby Foundation, Inc.; and the Frank Phillips Foundation offer a graphic illustration of the benefits reaped from the state's underground wealth. When these efforts are added to the taxes paid by the petroleum industry, $318,000,000 in 1977, the value of Oklahoma's oil legacy is staggering.[35]

## Chapter 21.
# Outstanding Oklahoma Oilmen

THROUGHOUT THE OIL HISTORY of the state, the citizens of Oklahoma have been fortunate to benefit from the many petroleum philanthropists who, after making their fortunes, gave away vast sums of money. The American free enterprise system embodied in the Oklahoma oil business has left a legacy for all the people of the state. Individuals such as E. W. Marland, Lew Wentz, Frank Phillips, Waite Phillips, James A. Chapman, Robert McFarlin, Joe Champlin, Lloyd Noble, W. G. Skelly, Harry F. Sinclair, John E. Kirkpatrick, Judge Robert Hefner, Robert Hefner, Jr., William Hefner, Evelyn Hefner Combs, and many others have contributed greatly to civic, educational, cultural, scientific, and charitable programs throughout the state; few Oklahomans have not benefited from their efforts. In an attempt to recognize the many contributions made by state oilmen, the Oklahoma Petroleum Council (OPC) initiated an "Outstanding Oklahoma Oil Man Award" in 1961. Presented at the annual meeting of the OPC, it was intended to reflect the recipient's contribution to the Oklahoma petroleum industry, as well as his civic and philanthropic endeavors.[1]

At a dinner held in the Mayo Hotel in Tulsa on October 12, 1961, Wirt Franklin of Ardmore received the OPC's first "Outstanding Oklahoma Oil Man Award." A longtime Oklahoma oilman, Franklin was born in Missouri. He moved to Muskogee, Indian Territory, as a stenographer for the Dawes Commission to the Five Civilized Tribes in 1902. Later, he was promoted to law clerk and eventually was given charge of the preparation of the Choctaw and Chickasaw rolls. In 1905 he formed a law partnership with S. A. Apple, and in June of 1906, Franklin moved to Ardmore, where he practiced law until the opening of the Healdton Pool in 1913.[2]

Franklin's first venture into the petroleum industry was with the Crystal Oil Company during the heyday of the Healdton field. When the firm was sold in 1916, he operated independently until the organization of the Wirt Franklin Petroleum Corporation in 1927, of which he served as president until 1937. Afterward, he once again operated as an independent. During his career, Franklin was the first president of the Independent Petroleum Association of America, serving from 1929 until 1935. He was also an honorary life member of the board of directors of the American Petroleum Institute and a director of the Kansas-Oklahoma Division of the Mid-Continent Oil and Gas Association. Between August of 1941 and July of 1943, Franklin was the Director in Charge and Director of Production for District Two, comprised of fifteen Midwestern states, including Oklahoma, with the Petroleum Administration for War; afterward, until March of 1944, he served as special assistant to the deputy administrator. Throughout his life, Franklin remained active in the civic affairs of Ardmore, and he was an elder and trustee of the First Presbyterian Church of Ardmore. In 1932, he was a candidate for the United States Senate.[3]

As the second recipient of the "Outstanding Oklahoma Oil Man Award" in 1962, the Oklahoma Petroleum Council honored Herbert R. Straight of Bartlesville, the former president and chairman of the board of Cities Service Oil Company and Cities Service Pipe Line Company. Known as the "Dean of Mid-Continent Petrolecrats," he was born in Tidioute, Pennsylvania, in 1874, and was the son of R. J. Straight, a prominent Pennsylvania oil-producer and operator. At the age of sixteen, Straight began work in the oil fields as a roustabout for $1.50 per day. After graduation from Leland Stanford University in Stanford, California, in 1897, he returned to the East Coast and became an independent pro-

Clarence H. Wright, the 1964 recipient of the Oklahoma Petroleum Council's Outstanding Oklahoma Oil Man Award. *Courtesy of the Oklahoma Heritage Association.*

ducer in partnership with his father in Bradford, Pennsylvania. Moving to Oklahoma in 1911 as a representative of the T. N. Barnsdall interest, Straight soon joined with a group of petroleum firms that later became the Cities Service Oil Company. It was Straight who drilled the discovery well in the El Dorado field in Butler County, Kansas, in 1915, and in 1920, he was named vice-president and manager of what became Cities Service Petroleum Company. He gave up active participation in company affairs at the end of 1947; however, he remained on the board of directors.[4]

Among the "firsts" which Straight contributed to the petroleum industry was his use of an electric motor to drill a well in the Bradford field of Pennsylvania and the utilization of wire lines in drilling operations. He also was a pioneer in the use of a pipe mast as a derrick substitute, the use of a steel shaft for the bull wheel, and the flowing of oil wells with compressed air. In civic affairs, Straight helped organize the Bartlesville YMCA and served as the organization's first president. For many years he was the senior warden of St. Luke's Episcopal Church of Bartlesville, a member of the town's Park Board, president of the local Chamber of Commerce, vice-president of the Cherokee Area Boy Scout Council, and a member of the Bartlesville Rotary Club. In addition, he served as a director of the First National Bank of Bartlesville and the First National Bank of Tulsa.[5]

Dr. A. I. Levorsen, regarded as one of the world's most prominent petroleum geologists and responsible for the location of many Oklahoma oil fields, was the third recipient of the OPC award in 1963. Born at Fergus Falls, Minnesota, on July 5, 1894, Levorsen graduated from the University of Minnesota and came to Oklahoma during the 1920s. Serving as a geologist for several oil companies, he consistently maintained that there was an abundance of oil to be found "in the most unlikely places and he encouraged geologists everywhere to use their imagination and to be unorthodox in their thinking." To prove his theory, Levorsen discovered the Fitts Pool in 1931 by "unconventional geological reasoning." It was his thinking that "all oil is not on structure but great reserves lie in big stratigraphic traps."[6]

The author of several leading petroleum geology textbooks, Levorsen, beginning in 1935, served as a consulting geologist to many petroleum companies and foreign governments. He was professor of geology and dean of the school of mineral sciences at Stanford University from 1945 to 1951. Serving as president of the American Association of Petroleum Geologists, Levorsen also was president of the Geological Society of America and the American Geological Institute. In addition, he was awarded the "Sydney Powers Memorial Medal Award" by the American Association of Petroleum Geologists—the highest recognition presented by the group.[7]

In 1964, Clarence H. Wright of Tulsa was honored with the "Outstanding Oklahoma Oil Man Award." Wright had a varied background prior to entering the petroleum industry, and his experiences in private enterprise aided him a great deal in his future efforts. In 1920, he invested in the newly organized Sunray Oil Company and was named a director and vice-president of the firm in 1921. Becoming president in 1930, he began to play an even more active role in the Oklahoma oil industry after the general offices of Sunray were established in Tulsa in 1934. Very concerned with the development of youth organizations and hospitals, Wright was constantly involved in civic affairs.[8]

K. S. Adams of Bartlesville received the 1965 citation. Known throughout the petroleum industry as "Boots," Adams moved to Bartlesville in 1920 after his graduation from the University of Kansas. Although his first job was on an ice wagon, he soon joined Phillips Petroleum Company as a warehouse clerk, and by the time he was thirty-two, he had been named an assistant to the firm's president, Frank Phillips. When Phillips became chairman of the board in 1938, Adams was elected president. He became chief executive officer in 1949, and chairman of the board in 1951. Retiring as chief executive officer in April of 1964, Adams remained active in the company as board chairman and chairman of the finance committee.[9]

Very involved in civic and industry affairs, Adams was elected to the Helms Athletic Foundation's Basketball Hall of Fame in 1958; was a director of the Tenth District Federal Reserve Bank of Kansas City; was a director and member of the executive committee of the American Petroleum Institute; and commander of the local American Legion Post. The recipient of the "Oklahoma Outstanding Legionnaire Award," Adams was awarded an honorary Doctor of Laws degree by both Drury College of Springfield, Missouri, and Oklahoma Baptist University. In 1958, he was inducted into the Oklahoma Hall of Fame. A thirty-third degree Mason, Adams was a member of the DeMolay Supreme Council and was presented both the "General Grand Chapter Award" and the "Distinguished Service Medal" of Royal Arch Masons.[10]

The chairman of the Big Chief Drilling Company, William T. Payne, was given the sixth annual "Outstanding Oklahoma Oil Man Award" at a banquet in the Persian Room of the Skirvin Tower in Oklahoma City on October 4, 1965. Payne, who was chairman of the Oklahoma Regents for Higher Education, was presented the citation by Oklahoma Governor Henry Bellmon. Born in Tecumseh, Nebraska, Payne moved to a farm near Shawnee. Working his way through college, he was graduated from Oklahoma A&M in 1915, and he enrolled in graduate school at

K. S. ("Boots") Adams, the Oklahoma Petroleum Council's 1965 honoree. *Courtesy of the Oklahoma Heritage Association.*

Massachusetts State College the following year. Payne was employed as a chemist and bacteriologist for the City of Detroit, Michigan, before serving as a second lieutenant in World War I. In 1919, he entered the petroleum industry as a scout for the North American Oil Company, and two years later, he joined Walter Helmerich. In 1926 the two men formed Helmerich and Payne Oil Company, of which Payne was vice-president until 1936. During his time in the oil fields, he was instrumental in developing the method of extracting oil from Mississippi Limestone and, as a result, opened vast new areas of the state to petroleum production. Payne also served

Ward S. Merrick, who was selected by the Oklahoma Petroleum Council as their honoree in 1968. *Courtesy of the Oklahoma Heritage Association.*

as president of the American Association of Oilwell Drilling Contractors, the general Mid-Continent Oil and Gas Association, and the Kansas-Oklahoma Division of the Mid-Continent Oil and Gas Association.[11]

Continuously active in oil business affairs, Payne "held some of the highest offices in national petroleum associations" and, at the same time, provided "truly outstanding" leadership for many of the state's civic and philanthropic causes. He was the first chairman of the Oklahoma Oil Industry Information Committee, the predecessor of the Oklahoma Petroleum Council, and a director of the Independent Petroleum Association of America, the Oklahoma City Chamber of Commerce, and the OPC. In addition, Payne was a trustee of the United Fund of Oklahoma City and the National United Presbyterian Foundation. He was awarded the Mercy Hospital "Award for Outstanding Service"; the "Silver Beaver," "Silver Antelope," and the "Man with a Heart" citations from the Boy Scouts; and in 1960, the United Fund's "Award for Outstanding Citizen of the Year." He was a member of the hall of fame of both the Oklahoma University Medical Sciences and the Oklahoma State University Alumni. Payne also served as director of the National Association of Manufacturers, the Frontiers of Science Foundation, and the American Petroleum Institute.[12]

In 1967, the OPC honored W. K. Warren, the founder of Warren Petroleum Corporation and president of the International Petroleum Exposition. Called "one of the truly outstanding leaders of our state," Warren was born at Nashville, Tennessee, on December 3, 1897, and moved to Tulsa in 1916. Organizing the Warren Petroleum Corporation in 1922, he served as its chairman of the board and chief executive officer until his retirement in 1961. Warren was cited for his service as president of the IPE, which "focused world-wide attention upon Oklahoma," and for "his generous gifts to worthy causes," which "have proved to be of great benefit to the citizens of Oklahoma."[13]

Warren's many civic duties included the presidency of Saint Francis Hospital and of the William K. Warren Medical Research Center, both in Tulsa. Serving as chairman of the advisory council of the college of commerce of the University of Notre Dame, he was also an honorary life member of the board of trustees of the University of Tulsa. Warren was chairman of the board of the Transwestern Pipeline Company, director of the American Petroleum Institute, president of the Natural Gas Processors Association, the recipient of the NGPA's "Hanlon Award," president of the Twenty-five Year Club of the Petroleum Industry, and a member of the Mid-Continent Oil and Gas Association.[14]

Ward S. Merrick, Ardmore independent oil-producer and community leader, was selected as the 1968 honoree at the OPC's tenth anniversary celebration. Praised "for his numerous contributions toward the development of the southern part of Oklahoma and for his philanthropic activities," Merrick was the third generation of his family to enter the oil business. Born on June 27, 1895, at Randolph, New York, he moved to Chicago in 1904, and, after graduation from high school, served with the army in France during World War I. After military service, Merrick joined his parents, who were living in Ardmore, Oklahoma, at the time, and went to work for his father in the Healdton field.[15]

In honor of his father, Merrick established the

Merrick Foundation in 1948 and later joined other area leaders in forming the Southern Oklahoma Memorial Association, which operated the Memorial Hospital of Southern Oklahoma. Merrick served as chairman of the board of trustees for the first ten years of the institution. Awarded the "University of Oklahoma Distinguished Service Citation," he was a longtime member of the University of Oklahoma Foundation Board, and made possible the construction of the Merrick Computer Building at the school. Merrick also was a member of the board of directors of the Oklahoma Historical Society, a charter trustee of the Chickasaw Historical Society, and chairman of the Ardmore City Planning Commission. One of the originators of Ardmore's "Community Activities, Inc." Merrick was always prominent in civic improvement. In addition, he was a member of the Independent Petroleum Association of America, the Oklahoma Petroleum Council, and served as vice-president of the Oklahoma Independent Petroleum Association.[16]

The ninth recipient of the "Outstanding Oklahoma Oil Man Award" was P. C. Lauinger, the chairman of the board of the Petroleum Publishing Company which produced the *Oil and Gas Journal* and five other oil-trade publications with a world-wide circulation. Involved in the petroleum industry throughout his life, Lauinger was born in Oil City, Pennsylvania, on August 8, 1900. In 1922, he was graduated from Georgetown University in Washington, D.C., and that same year, he became a staff representative for the Petroleum Publishing Company. He was named president of the firm in 1931, and held that position until 1968, when he became chairman of the board.[17]

Under Lauinger's direction, the Petroleum Publishing Company, in addition to the *Oil and Gas Journal*, also produced *Petroleo Interamericano*, *Oil and Gas International*, *Oil and Gas Equipment*, *Offshore*, and *Ocean Oil Weekly*. He was also a director of the Continental Oil Company, Williams Brothers Company, the National Bank of Tulsa, Home Federal Savings and Loan Association of Tulsa, the Texas Mid-Continent Oil and Gas Association, and the Derrick Publishing Company of Oil City. A member of the board of directors of Georgetown University, Lauinger was awarded a Knight of Malta and a Knight Grand Cross in the Equestrian Order of the Holy Sepulchre of Jerusalem; he was also a member of Sigma Delta Chi.[18]

Dean A. McGee, the chairman of the board and chief executive officer of Kerr-McGee Corporation, was selected to receive the 1970 presentation. Born in Humboldt, Kansas, on March 20, 1904, McGee received a B.S. degree in Mining Engineering from the University of Kansas in 1926. After teaching at the University of Kansas School of Engineering Geology for a year, he went to work for Phillips Petroleum Company in 1927 as a petroleum geologist. Eventually, McGee became the firm's chief geologist. He joined Kerlyn Oil Company as the vice-president in charge of production and exploration in 1937 and became executive vice-president in 1942. Four years later, he was named executive vice-president and director of Kerr-McGee Oil Industries. He became president and chief executive officer in 1954 and was named chairman of the board of directors in 1963. In 1965, the firm was reorganized as Kerr-McGee Corporation, and two years later, McGee was again elected chairman of the board.[19]

In addition, he was a member of the board of directors of the General Electric Company, the Oklahoma Natural Gas Company, Fidelity National Bank and Trust Company of Oklahoma City, the First National Bank and Trust Company of Muskogee, and Chairman of the Board of Directors of Central Plains Enterprises. McGee's contacts with the petroleum industry allowed him to become a member of the American Petroleum Institute, the National Petroleum Council, and the American Association of Petroleum Geologists. He also was the author of several papers on energy, petroleum geology, and production practices. A Fellow of the Association for the Advancement of Science and the Oklahoma Academy of Science, he was a member of the boards of trustees of the California Institute of Technology, Oklahoma City University, and the National Cowboy Hall of Fame and Western Heritage Center. Serving as a elder in the Westminster Presbyterian Church, McGee was also a member of the Oklahoma Hall of Fame and the Oklahoma Medical Sciences Hall of Fame.[20]

In 1971 the OPC honored William Wayne Keeler as the "Outstanding Oklahoma Oil Man." Although of a pioneer Bartlesville family, Keeler was born on April 5, 1908, in Dalhart, Texas, where his parents were temporarily living. Graduating from Bartlesville High School in 1926, he entered the University of Kansas Engineering School. While continuing his studies at the institution, he took a job with the Phillips Petroleum Company in Kansas City, Kansas, in 1926. Working his way upward in the Phillips organization, Keeler was chosen as vice-president of the firm's executive department in 1947, and at the

Six recipients of the Outstanding Oklahoma Oil Man Award. *Left to right:* William W. Keeler (1971), Dean A. McGee (1970), John E. Kirkpatrick (1974), Jack H. Abernathy (1978), P. C. Lauinger, Sr. (1969), and William T. Payne (1966). *Courtesy of the Oklahoma Petroleum Council.*

same time, elected to the board of directors; in 1954 he became a member of the executive committee. Named executive vice-president in 1956, Keeler was made chairman of the executive committee six years later. During World War II, he worked with a Phillips refinery construction project in Mexico, and served with the Office of Petroleum Administration for War. In 1952 Keeler was selected as director of refining for the Petroleum Administration for Defense, and he served as chairman of the Military Petroleum Advisory Board from 1954 to 1962.[21]

In addition, he was chairman of the board of directors of the National Association of Manufacturers, a trustee of the National Petroleum Refiners Association, and a director of the American Petroleum Institute, the Independent Petroleum Association of America, and the International Petroleum Exposition. In the civic field, Keeler held numerous offices, including memberships on the board of directors of the Thomas Gilcrease Institute of American History and Art, the White House Youth Conference, and the board of directors of the Bartlesville Chamber of

Commerce. A Cherokee Indian, he was appointed principal chief of the Cherokee Nation in 1949. Keeler was also a thirty-third-degree Scottish Rite Mason, a Shriner and a Royal Jester. In 1966, he was inducted into the Oklahoma Hall of Fame, and in 1969, he was awarded the University of Oklahoma's highest honor, the "Distinguished Service Citation." In addition, he received the American Academy of Achievement's "Golden Plate Award" as one of the nation's "giants of accomplishment."[22]

T. H. McCasland, the chairman of the board of Mack Oil Company in Duncan and the man "responsible for numerous discoveries of oil and gas . . . in southwestern Oklahoma," was selected by the OPC in 1972. Born on February 21, 1895, near Duncan, still a part of the Chickasaw Nation at that time, McCasland was graduated from the University of Oklahoma in 1916. After serving in World War I, he entered the petroleum industry as a lease trader; however, by the mid-1930s, McCasland had acquired a small rotary rig and started his own drilling venture. Operating as an individual until he organized the Mack Oil Company in 1946, he served first as the organization's president and later as its chairman of the board.[23]

Among the many discoveries McCasland made in Oklahoma were: the Gas City field in Stephens County, the Woods Sand Horizon of the Knox Pool in Grady County; the Southeast Knox Pool in Stephens and Grady counties; the Cruce Hoxbar Sand Pool in Stephens County; and the Extension North Area of the Markham Sand Pool in Stephens County. He also served as president of the Duncan Chamber of Commerce, the Duncan Rotary Club, the University of Oklahoma Dad's Association, and the University of Oklahoma's Student Association. He was presented the University of Oklahoma's "Distinguished Service Award," and served as a member of the Academy of University Fellows, as well as a director of the Independent Petroleum Association of America and the Oklahoma Independent Petroleum Association.[24]

Receiving the "Outstanding Oklahoma Oil Man Award" in 1973 was D. D. Bovaird, the chairman of the board of the Bovaird Supply Company of Tulsa. A pioneer in the oil and gas equipment supply business, Bovaird was born in Bradford, Pennsylvania, in 1896, but moved with his family to Independence, Kansas, at the age of ten. After graduation from high school, he attended the College of Emporia in Kansas, before being graduated with a B.A. degree from the University of Michigan in Ann Arbor in 1918;

T. H. McCasland, Who received the Oklahoma Petroleum Council citation as Outstanding Oklahoma Oil Man in 1972. *Courtesy of the Oklahoma Heritage Association.*

two years later he received a B.S. degree in Mechanical Engineering from the same institution. In 1920, he went to work for the Bovaird Supply Company and in 1927 was named treasurer of the firm. At the death of his brother, William M., in April, 1949, he became president of both Bovaird Supply Company and Bovaird, Inc.; in 1961, he was selected as chairman of the board, and his son, William J. Bovaird, was elected president.[25]

Although Bovaird was deeply involved in building one of the largest independent supply store companies, he also was very active in many civic and charitable efforts. He was a member of the board of trustees of the University of Tulsa, the Hillcrest Medical Center, and the First Presbyterian Church. In addition, Bovaird participated in the Tulsa Community Chest, the Tulsa Boys Home, the YMCA, the Family and Children's Welfare Service, and the Philbrook Art Center. In the oil business, he served as a director of the American Petroleum Institute, a member of the National Petroleum Council, a director and president of the Petroleum Equipment Suppliers Association, and a director of the Executive Committee of the International Petroleum Exposition. Bovaird was also a member of the Independent Petroleum Association, the Mid-Continent Oil and Gas Association, the Texas Mid-Continent Oil and Gas Association, the Texas Independent Producers and Royalty Owners Association, and the Panhandle Producers and Royalty Owners Association.[26]

John E. Kirkpatrick, a prominent Oklahoma City oil executive and civic leader, was honored by the OPC in 1974. Born in Oklahoma City in 1908, Kirkpatrick attended the United States Military Academy, Marion Institute, and the United States Naval Academy, from which he was graduated in 1931. After serving four years in the navy, he entered the Harvard Graduate School of Business Administration, and later moved to Tulsa to organize the Allied Steel Products Corporation. Recalled to active duty with the navy in 1941, he served in both the Atlantic and Pacific aboard the U.S.S. *North Carolina*, *Alaska*, and *Oklahoma City*. Remaining active in the naval reserve, Kirkpatrick eventually rose to the rank of rear admiral. Returning to Oklahoma City after World War II, he entered the oil business with Hubert E. Bale and formed Kirkpatrick and Bale, which became the Kirkpatrick Oil Company in 1950.[27]

Taking a strong interest in the civic and cultural development of Oklahoma City and the state, Kirkpatrick, together with his wife, Eleanor, did much in the field of philanthrophy, donating the Fine Arts Auditorium of Oklahoma City University, the Oklahoma Art Center, and the Oklahoma Science and Arts Foundation building. He also served as president of the Lyric Theater, the Oklahoma City Community Foundation, and the Oklahoma Zoological Society. In addition, Kirkpatrick was national vice-chairman of the National Conference of Christians and Jews, president of the Oklahoma City Symphony Society, chairman of the board of the Oklahoma Theater Center, vice-chairman of the board of trustees of Mercy Hospital, chairman of the board of the Oklahoma Science and Arts Foundation, and vice-president of the Economic Club of Oklahoma. A trustee of the National Cowboy Hall of Fame and Western Heritage Center, Kirkpatrick served as the Honorary Consul of the Republic of Korea in Oklahoma City and as vice-chairman, director, and member of the executive committee of the Liberty National Corporation. In addition, he was a director of the May Avenue Bank, the Southwest Title and Trust Company, and the Stockyards Bank of Oklahoma City.[28]

Joseph A. LaFortune, Sr., of Tulsa was the posthumous honoree of the Oklahoma Petroleum Council in 1975. A longtime civic leader and philanthropist, LaFortune was a native of South Bend, Indiana, and moved to Tulsa in 1920, where he worked for the *Tulsa Daily World* and the *National Petroleum News*. In 1924, he joined the Warren Petroleum Corporation as a stockholder and secretary. Elected vice-president in 1929, he became executive vice-president a short time later, and was selected as vice-chairman in 1952. LaFortune retired in 1954 and was engaged in independent oil operations and investments. During the Korean War, he served as Deputy Petroleum Administrator for Defense.[29]

An outstanding civic leader, LaFortune donated the necessary land for LaFortune Park in Tulsa, funds for an athletic stadium at Memorial High School, and the financing for an athletic dormitory at the University of Tulsa. A lifelong member of the Roman Catholic Church, LaFortune contributed heavily to the construction program at Cascia Hall Prep School in Tulsa and Notre Dame University. He was a supporter of the Tulsa Psychiatric Foundation, the Tulsa Education Foundation, the Boy Scouts of America, Tulsa Opera, and the Tulsa Philharmonic Society. The holder of the "Fillius Ordinis" bestowed by the Augustinian Order, LaFortune also was a Knight Commander of St. Gregory, a Knight of Malta, a member of the Oklahoma Hall of Fame, and the 1970 recipient of the "Brotherhood Award" of the National Conference of Christians and Jews. In addition, he was a president and director of the Natural Gasoline Association of America and a director of the Public Service Company of Oklahoma. LaFortune died on August 5, 1975, prior to receiving the award.[30]

In 1976, Robert A. Hefner, Jr., was presented the "Outstanding Oklahoma Oil Man Award" by the OPC. Born in Beaumont, Texas, in 1907, Hefner moved to Ardmore in 1909. Attending Ardmore High School and Culver Military Academy, he was graduated from Stanford University in 1928, cum laude with a B.A. degree. Hefner then attended the Harvard University School of Law, and he was awarded a LL.B. and a J.D. degree from the University of Oklahoma in 1930. Athletically inclined, Hefner won college letters in boxing, basketball, and golf. In addition, he received the Gold Medal for Oklahoma in violin in 1921, and played with the San Francisco Symphony Orchestra from 1924 to 1928. Opening a law practice in Oklahoma City in 1931, Hefner served as an assistant to the attorney general of the United States in 1934–1935, before moving to Illinois in 1936, where he specialized in legal questions pertaining to the petroleum industry. Returning to Oklahoma City in 1946 as the manager of Hefner Production Company and the managing partner of the Hefner Company, he built the latter into a holder of more than 12,000,000,000 cubic feet of natural gas reserves and 3,000,000 barrels of oil reserves.[31]

One of the donors of the Hefner family home and furnishings to the Oklahoma Heritage Association for use as the Oklahoma Heritage Center, Hefner also took a great interest in the development of the Baptist Memorial Hospital, the National Cowboy Hall of Fame and Western Heritage Center, and the Last Frontier Council of the Boy Scouts. In addition, he was an organizer and member of the board of directors of the Oklahoma Independent Petroleum Association, vice-president of the Independent Petroleum Association of America, and a member of the Oklahoma Hall of Fame. He was also active in the All Souls' Episcopal Church, the Oklahoma Bar Association, and the Chamber of Commerce.[32]

Receiving the acknowledgment in 1977 was Robert W. McDowell, a prominent Tulsa oil executive and civic leader. Born in Anthony, Kansas, on August 9, 1895, McDowell moved to Oklahoma when he was two years of age. After serving as a first lieutenant in the army during World War I, he began his oil career in Tulsa in 1918 with the Producers and Refiners Corporation. In 1924, McDowell became a vice-president and general manager of Hawkeye Oil Company in Waterloo, Iowa, and when Mid-Continent Petroleum Corporation acquired the Hawkeye firm, he was later transferred to the company's general office in Tulsa. Named sales vice-president in 1930, executive vice-president in 1938, and a director in 1944, McDowell was elected president of Mid-Continent in 1949. When Sunray merged with Mid-Continent in 1955, he was named president of Sunray's marketing and refining sub-

Robert A. Hefner, Jr., who was honored by the Oklahoma Petroleum Council in 1976. *Courtesy of the Oklahoma Heritage Association.*

sidiary, and in 1960, he was chosen as chairman of the board of Sunray DX Oil Company.[33]

Well-known in oil industry affairs, McDowell was director of the American Petroleum Institute, president and director of the National Petroleum Refiners Association, and a member of the National Petroleum Council. During World War II, he served as vice-chairman of the Petroleum Industry Marketing Committee for the Second District. In addition, McDowell was president of the Western Petroleum Refiners Association and the Twenty-five Year Club of the Petroleum Industry. Serving as a member of the board of trustees of the University of Tulsa, he also was a trustee of the First Presbyterian Church of

Tulsa and a member of the Board of Directors of Tulsa Opera. Active in the Tulsa Chamber of Commerce and the Tulsa Hurricane Club, McDowell was a member of the Shriners, Jesters, Masons, and Tulsa's Community Chest.[34]

Honored in 1978 was Jack H. Abernathy, who was graduated with a B.S. degree in Petroleum Engineering from the University of Oklahoma in 1932. Entering the oil business first as a petroleum engineer for Sinclair Oil and Gas Company and then as chief engineer and general production superintendent for Sunray DX Oil Company, he later became chief engineer of the Oklahoma City Wilcox Pool Engineering Association, president and chairman of the Board of Big Chief Drilling Company, president of Seneca Oil Company, and president and chairman of Post Oak Oil Company. Eventually he rose to vice-chairman of Entex, Inc., as well as director and member of the executive committee of Liberty National Bank and Trust Company and Southwestern Bank and Trust Company of Oklahoma City.[35]

In addition, Abernathy served as a member of the National Petroleum Council and president of both the International Association of Drilling Contractors and the Mid-Continent Oil and Gas Association. Previously presented the Texas Mid-Continent Oil and Gas Association's "Distinguished Service Award," he also received the "Distinguished Alumni Award" of the University of Oklahoma School of Petroleum Engineers and was inducted into the Oklahoma Hall of Fame. During his long career of public service, Abernathy served as director of the Oklahoma City Chamber of Commerce, the Oklahoma City Zoological Society, the Oklahoma State Chamber of Commerce, and the United States Chamber of Commerce, as well as a trustee for the University of Oklahoma Foundation and the Oklahoma City Community Foundation. He also acted as a member of the Oklahoma Governor's Council for Petroleum Development, a trustee for the Midwest Research Institute, and a director of the Independent Petroleum Association of America.[36]

The men honored by the Oklahoma Petroleum Council with the "Outstanding Oklahoma Oil Man Award" are but a few of the countless individuals involved in the state's petroleum industry who have contributed greatly to the growth of both Oklahoma and the nation. However, they are an accurate representation of what the American free enterprise system, when combined with the oil business, can accomplish in civic, cultural, and charitable endeavors. They are carrying on the legacy of oilmen throughout Oklahoma's history, and they graphically illustrate what the petroleum industry means to the state.

## Chapter 22.
# The Future

DURING THE DECADE of the 1950s the petroleum industry in Oklahoma experienced a drastic decline, and only a few men (Julius Livingston was one) entered the oil arena during this troubled era with any measure of success. Much of the blame for the low price of crude was laid on the federal government for its policy of insisting on artificially low energy costs. The action was not only to provide lower prices for the consumers but also represented a commitment to economic growth. Many politicians strongly believed in the relationship between cheap energy supplies and business expansion. However, the policy created a dependence on foreign imports, the consequences of which were felt in the 1970s, when the energy crisis struck the nation full force.[1]

Following World War II, Oklahoma showed a general increase in total oil production: in 1949 the state produced 151,902,000 barrels of crude, and ten years later the amount had increased to 198,090,000 barrels of oil. Although state regulation was generally slanted to encourage new production during this time—with the exception of the years from 1949 through 1956 and 1967—the rate of depletion actually exceeded discoveries. In 1967 the Oklahoma Corporation Commission set a new policy which provided for extra allowables ranging from 20 barrels of oil daily for shallow wells to 3,050 barrels per day for wells that exceeded fourteen thousand feet in depth. Nevertheless, virtually no significant new fields were found. One exception was the discovery of the Sooner Trend in Logan, Kingfisher, and Garfield counties in 1965.[2]

With very few new pools located, a number of oilmen believed that perhaps there were no new discoveries to be made within Oklahoma. This idea was reinforced in 1968, when only twenty percent of the 419 wells drilled proved to be producers. Seemingly in recognition of such a fact, the OCC ended its forty-five-year-old policy of holding regular monthly proration hearings and instead opted for bi-monthly meetings. This belief was also reflected in the amount of acreage under lease for oil and gas in Oklahoma: from 20,830,000 acres in 1959, the amount dipped to 19,850,000 in 1962, to 18,175,000 in 1965, to 16,785,000 in 1966, and to 16,380,000 in 1967. While the actual number of producing wells increased from 75,210 in 1958 to 80,583 in 1966, the total number of wells drilled dropped from 6,354 to 4,112 during the same period. The number of exploratory wells decreased as well: 803 wildcat wells were drilled in 1962, but that number was reduced to 521 four years later.[3]

During the post-World War II era, Oklahoma continued to lead in the conservation of petroleum reserves. A Compulsory Unitization Law, "House Bill No. 339," was adopted by the state legislature in 1945. Granting the Oklahoma Corporation Commission jurisdiction over unitization, the act was applied to all "common sources of supply of oil, oil and gas, or gas distillate" and prescribed that each "unit and unit area shall be limited to all or a portion of a single common source of supply." The announced goal of the legislation was "the efficient . . . management or control of the further development and operation of the unit" to permit "persons . . . entitled to share in or benefit by the production . . . to produce or receive . . . their fair, equitable and reasonable share"; to provide a method of apportioning the cost of further development; to create "an operating committee to have general over-all management and control of the unit and the conduct of its business and affairs"; and other minor points. Although an attempt was later made to repeal the unitization law, it was unsuccessful. Nevertheless, it was altered to some extent by the 1951 legislature. That same gathering also enacted a Standard Gas

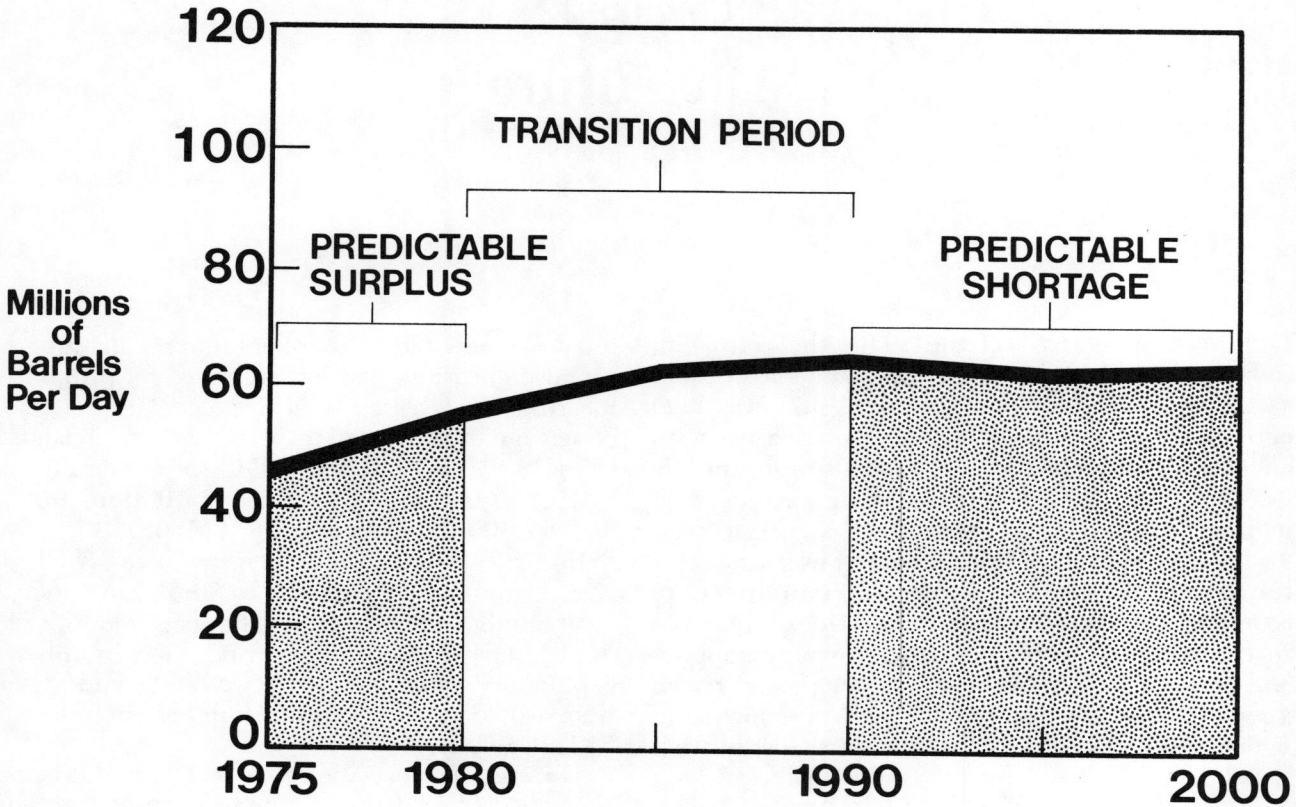

The oil supply outlook. It is generally assumed that unless the United States increases its development of energy sources the free world will face a serious shortage near the end of the twentieth century. *Courtesy of Continental Oil Company.*

Measurement Law establishing the "standard pressure base of 14.65 pounds per square inch and the standard temperature base of sixty degrees Fahrenheit."[4]

Thus the general drift of the petroleum industry within Oklahoma during the years following the great oil booms was lethargic. This changed, however, with the advent of the Arab oil embargo and the associated energy crisis of 1973. The renewed energy consciousness resulted in an increased interest in Oklahoma crude production. Although there was a modest decline in energy demand resulting from government regulations and voluntary cutbacks in 1974 and 1975, in the year 1976 the nation was again using energy at an unprecedented rate. Nonetheless, the price of domestic crude remained artificially low because of government regulations.[5]

By 1975, Oklahoma had five oil fields ranked among the top hundred producing pools in the United States: Sho-Vel-Tum field in Carter and Stephens counties was in twelfth place, with a proved reserve of 297,651,000 barrels; Sooner Trend in Kingfisher County was in sixty-third place, with reserves totaling 64,145,000 barrels; Golden Trend in Garvin County was numbered sixty-sixth, with reserves amounting to 61,780,000 barrels; Healdton field in Carter County boasted reserves of 58,119,000 barrels and was ranked seventy-third; and Postle field in Texas County was number ninety-four, with reserves totaling 44,021,000 barrels. Even so, the earnings of many oil companies were severely restricted because of the government's policy. Advanced technology, greatly increasing the cost of exploration, contributed to the economic decline, as did the elimination of the percentage depletion allowance by the federal government for all but small independent oil companies. Nonetheless, Oklahoma began to experience another oil boom as the renewed demand for energy generated a more vigorous search for oil. Thus in the first quarter of 1975 the state ranked second in the nation in the number of wells "drilled in search for oil and gas."[6]

What was the solution to the seemingly contradictory policy in which the government encouraged the search for new energy sources and, at the same time, hamstrung the oil industry with bureaucratic regulations? Several Oklahoma newspapers discussed the question and were nearly unanimous in their views. The *Claremore Daily Progress* commented that "There may be a ton of gas underneath Oklahoma, but the basic law of economics is that nothing can be sold for less than it costs to produce unless someone pays the difference." The result of governmental control of the petroleum industry has been to "virtually eliminate the incentive to find and exploit new sources of gas to be sold in interstate commerce because the risks are not financially stable," declared the *Duncan Daily Banner*. Echoing the same sentiments, the *Marietta Monitor* stated that the "last couple of years the tendency in Congress has seemed to be that if the energy problem is ignored long enough it will go away." Facts seemed to bear out these statements for, by the end of 1976, Oklahoma's proved crude oil reserves totaled 1,186,553,000 barrels, a decrease of 53,134,000 barrels, or 4.3 percent. The message was clear: although there were "considerable quantities of oil and gas" remaining to be found, the minerals were generally located in "inaccessible places where production costs will be high," and no oilman was going to "pump oil out of the ground if he could not sell it at a price" that would allow him to recover his expenses.[7]

For the Oklahoma petroleum industry, the year 1977 brought "an upsurge in drilling activity that . . . dotted the landscape with active rigs." However, at the same time, D. W. Calvert, president of the Oklahoma Petroleum Council, declared that "the costs of probing into the deepest areas, such as the Anadarko Basin, are rising so rapidly that such enormous expenditures cannot be justified unless the rate of return on natural gas and/or oil is sufficient to enable the wells to pay off if they do turn out to be . . . [producers]." In continuing, Calvert maintained that further "incentives are absolutely necessary to encourage producers to take the greater financial risks involved in drilling more wells and deeper wells in difficult terrains." "Our national energy policy must ensure an economic climate which maximizes the development of the natural resources within our own borders," Calvert concluded, because "Either the United States will take the steps to explore for and produce more oil and gas domestically or we must face up to the fact that our nation will continue to become more and more dependent upon unreliable sources of imported oil."[8]

Most Oklahoma oilmen echoed these sentiments. Edward A. Smith, active in state oil fields for more than sixty years, argued that government regulation actually hindered continued development of the nation's oil reserves. "Less than five percent of the potential gas-producing and oil-producing areas in the United States have been tested," Smith maintains; however, the remaining ninety-five percent is generally found in inaccessible and remote locations, often at great depths; therefore the cost of extraction is higher. As a result, oil companies are unwilling to embark on many new exploration programs because government regulations prevent them from recovering their initial investment. "It's simple," according to William B. Osborn, Jr., a longtime state petroleum producer. "Oilmen are not going to pay twenty dollars to get a barrel of oil from the earth which, under government regulations, they can sell for ten dollars." Many oilmen such as Donald Thompson, contend that such policy of regulation is "Oppressive . . . [and] contrary to the best interests of the United States.[9]

The facts seem to confirm these accusations of a governmental bureaucracy smothering the oil industry. For example, the United States Department of Energy had an annual budget of $10,600,000,000 in 1978, which amounted to "$266,871 for each of the 39,763 wells drilled . . . $58.35 for each of the 181,855,700 feet drilled . . . $3.59 for each barrel of domestic crude oil . . . $1.67 for every barrel of petroleum products consumed . . . and 10 cents for every gallon of gasoline consumed in 1976." In addition, this represented a budget that "exceeds 1975 capital and exploration expenditures by the petroleum industry to explore for and produce domestic crude oil, natural gas and natural gas liquids . . . and it is almost three-fifths of all U.S. capital expenditures in all U.S. petroleum sectors that year." The huge bureaucratic budget of the Department of

In order for America to increase its production of petroleum, it will be necessary to explore regions that are accessible only by air, and such operations as this Halliburton Services airmobile cementing unit are expensive. *Courtesy of Halliburton Services.*

Energy "exceeds the 1974 profits of the seven largest oil companies." With such stifling governmental regulation of the oil industry, it is little wonder that Smith complains that he "cannot plan ahead in the oil business because of bureaucracy," and that such regulatory practices make "it impossible for someone to accomplish what he did again."[10]

In their opposition to government regulations, Oklahoma oilmen do not advocate a return to the unrestricted boom era of the 1920s and 1930s when huge amounts of crude and natural gas were wasted because of a lack of conservation. Few, if any, wish to regress to a period when the industry was governed by the "law of capture." They are, however, concerned by the use of cheap energy as a political tool to woo voters while artificially low prices do irreparable damage to America's petroleum industry. Cheap energy is a thing of the past; advancing exploratory technology, unimagined a decade ago, has so increased the cost of recovering crude that low prices have been maintained only by federal intervention. "Now, there's no reason for anyone to think that they're going to have cheap energy again," Smith declared. "They are not!" In addition, the cost of complying with bureaucratic paperwork continues to mount, and it is no longer feasible for the oil companies to absorb the rising expenditures; therefore they pass such costs on to the consumer. Thus, as governmental regulations grow, the price of oil rises accordingly. R. M. Tillman of Conoco estimates that his company spends approximately five million dollars annually to comply with federal paperwork, and that figure is passed on to the public. Paradoxically, in its goal to provide cheap energy, it is obvious that federal regulations—instead of holding down the prices of petroleum products—actually contribute to the price increases.[11]

Not only is federal intervention and meddling in the petroleum industry driving up the price paid by consumers, it is also hampering the exploration for domestic oil. Much of the crude in Oklahoma has already been located and pumped for several years. Therefore, much of the remaining petroleum is found in marginal pools, and until the price is raised to a level where it is profitable, oil companies are not going to develop methods of recovering these marginal deposits. Stanley Learned, former chairman of the board of Phillips Petroleum Company, predicts, "As the price rises more marginal producing areas will be tapped"; thus the oil shortage will be curtailed to some extent.[12]

While most state oilmen are unanimous in their cry for relief from bureaucratic federal restrictions, which they claim are "stifling" the petroleum industry, they are generally satisfied with state regulatory practices and believe they are both equal and fair. Most regard the pioneering efforts of Oklahoma oilmen to bring order to the chaos of the 1920s and 1930s as being responsible for the state's progressive oil regulations. Because of the role played by oilmen in developing Oklahoma's conservation laws, the state enjoys perhaps the most productive relationship between petroleum producers and regulatory agencies in the nation.[13]

The adverse regulations of the federal government have made the tremendous boom periods which the state experienced in the early twentieth century a thing of the past. Oilmen disagree, however, on whether the process of accumulating vast fortunes overnight can still be duplicated. Many of those who lived through the fast-paced booms believe that it "couldn't be done now—taxes—can't just go out trading [leases] again." The "self-made oil man is about to go out of style," predicts Smith as he argues that the "personal contacts" which in the early days made the "wheeling and dealing" possible are "just not here anymore." Walter Neustadt, Jr., disagrees, however, and firmly believes that the opportunities are still there. Nontheless, he concedes that the days of his father's and grandfather's era are long past.[14]

Regardless of who is right, the state is still a mecca for farsighted oilmen who believe that persistence and advancing technology will locate additional supplies of oil and gas. Edward H. Leede, a native of New York and a long longtime resident of Midland, Texas, has been active in the Mid-Continent region since the 1950s. Leede and a partner, geologist Clyde Pine, have been convinced of the potential offered by deep well exploration in the Anadarko Basin area for some time. Organizing an office in Oklahoma City in 1977, Leede Exploration employed an Oklahoma geologist, David G. Campbell, considered by many industry observers as an authority on the Mid-Continent. Campbell, in the 1960s, helped develop the geologic concepts that predicted the potential of the Watonga Trend in western Oklahoma, as well as other plays on the Anadarko Shelf and in the Arkoma Basin. Leede Exploration—a partnership consisting of Leede, Pine, Robert H. Frazier, Tom A. Morgan, Charles D. Ray, and Jerry Hudgeons—embarked on an extensive exploration program in western Oklahoma utilizing the most up-to-date technology for deep hole drilling. Both Leede and Campbell are optimistic about the potential of the region because

Many petroleum geologists, such as David G. Campbell, the Mid-Continent Division exploration manager for Leede Exploration, remain convinced of the future significance of Oklahoma's role in the oil industry. *Courtesy of Leede Exploration.*

in 1978, of the sixty-five wildcat wells sunk in the region, fifteen were completed as discovery wells, for a 65 percent success ratio. As a result, both men believe that the future of the state's petroleum industry will be determined by the discovery of other deep fields in the Anadarko Basin.[15]

Most oilmen acknowledge that there is still a large amount of petroleum in Oklahoma. Also, they generally agree with Ward Merrick, Jr., that it will be found in smaller deposits and will be expensive to extract. Although most of the great pools—Healdton, Greater Seminole, Glenn Pool, and others—have already been found, producers are now reexamining these areas. The majority of these fields were opened at a time when oilmen did not consider a well a success unless it produced several hundred barrels daily, and many holes were abandoned when production decreased. With the increased petroleum demand many of these were reopened as stripper wells. Contrary to the popular connotation, the term *stripper well* does not mean those "at the economic edge"; rather it applies to wells producing "ten barrels [of crude] per day or less." With the oil squeeze, these wells became very important; and in 1979 "the vast majority of all oil-producers in the state" were engaged in stripper-well production. In addition, in

their haste to reach the tremendous reservoirs of crude, the early-day oilmen often overlooked smaller deposits at shallower depths, and in many instances new holes are being drilled in the older fields to tap these pools. Other men, such as Dean A. McGee, believe that advancing geophysics and technology will unlock vast new underground reservoirs where many oilmen once thought no petroleum existed.[16]

Increasing prices also open another source of crude to Oklahoma oilmen. This petroleum, which was previously not within reach because of technological limitations, is taken from areas that were formerly believed to be depleted. Within the last few decades many older oil fields have been reopened by either secondary or tertiary recovery methods. Secondary recovery relies on the injection of water or gas into the oil-bearing formations to force accumulations of crude, which then can be brought to the surface. The more recent development of tertiary recovery involves a highly complicated chemical reaction to accomplish the same goal. Special chemicals are pumped into holes drilled in propitious locations within a known oil field from which most of the recoverable oil has already been pumped. The chemicals work their way through the formation, forcing the crude before them until the oil accumulates in sufficient quantity to be removed. Actually, it has been estimated that only approximately twenty-five percent of the crude has been removed from pools which were previously depleted by primary production techniques. Oilmen hope to increase this amount by another twenty-five percent with secondary methods and ultimately increase the total to seventy-five percent with tertiary recovery.[17]

At the other end of the spectrum are the oilmen who believe the future lies in drilling even deeper wells in the search for oil. Perhaps the best example of this is the Anadarko Basin in western Oklahoma, where wells of between seventeen thousand and twenty-one thousand feet are needed. Such wells are a far cry from those of the 1920s and 1930s, when a hole of two thousand or three thousand feet was considered deep, and they graphically illustrate the great advancement in petroleum technology that has taken place in a relatively short period of time. Such deep holes are, however, extremely expensive, and despite the great technological advances that have been made, no one can guarantee that oil will be found until the well is drilled. Of the one in ten wells that find oil, the majority are abandoned before payout, and only one in fifty discovers significant oil or gas. Therefore, price incentives must make the gam-

The offshore operations of Oklahoma-based companies must also be increased to offset the increased demand for petroleum. *Courtesy of Reading & Bates Offshore Drilling Company.*

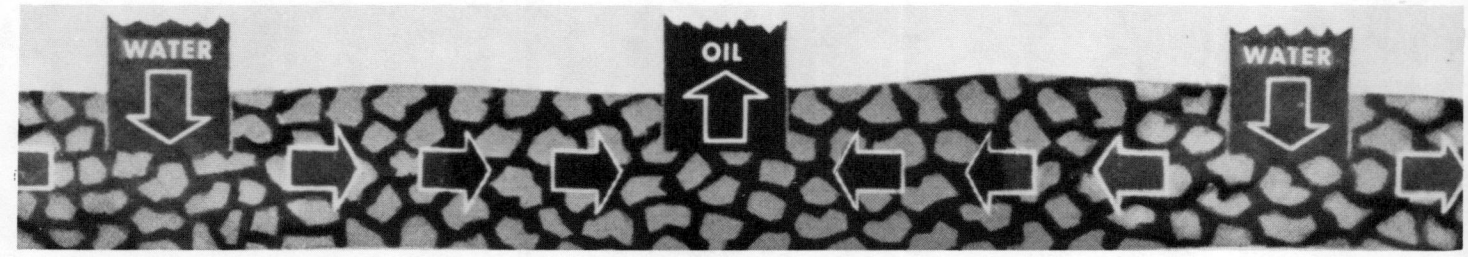

Secondary recovery through the use of water flooding will increase the output of many of the state's older fields. In this operation, as water flows away from the injection wells, it loosens oil clinging to particles of sand so that pumps can lift the water and oil to the surface. *Courtesy of the Interstate Oil Compact Commission*.

ble worthwhile for the region to be fully developed.[18]

Wherever the future of Oklahoma oil lies, all oilmen agree on one aspect of the industry: luck. To the pioneer oilman, luck was of extreme importance. T. H. McCasland, Sr., who first entered the petroleum industry prior to America's entry into World War I, estimates only one out of every one hundred individuals who embarked on an oil career "made it," and their success can be traced to a single factor: they were blessed by fate to drill the right hole at the right time, while countless others sank dry holes or brought in producers at a time when there was no market available. Not only did most pioneer oilmen believe that luck plays a tremendous part in the success or failure of those in the petroleum industry, many contemporary oilmen share the same belief. Robert J. Collins of Eason Oil Company, who holds a doctorate in geology, maintains that luck has a lot to do with the search for crude, inasmuch as no one can guarantee that a well will strike oil. Only by investing a large amount of money and actually drilling a hole can it be determined if petroleum is present in paying quantities.[19]

Repeatedly oilmen relate the role that luck plays in the success or failure of an oil well. On one occasion William T. Payne and Walter Helmerich planned to drill a well in the Braman area in Kay County during that region's boom period. The two men had only a nine-acre tract that had been leased in an area where Comar Oil Company was drilling several other wells. Because their acreage was located within Braman's boundaries, Payne and Helmerich were deluged with pleas to drill the well on the property of various individuals. To avoid the dispute, the two men arbitrarily selected a site on the edge of the town. Brought in at 100 barrels of oil per day, the well continued to increase in output until 5,500 barrels flowed from it daily. Neither man realized the extent of the "luck" involved in their strike until none of the other wells in the area found oil, nor did offsets to the discovery well. The peculiar geological formation of the region required that a well be drilled in precisely the spot chosen by Payne and Helmerich: a site selected to end a dispute between the property owners involved. Much the same story was told by Jack T. Conn and Charles Urschel regarding Slick City about six miles north of Ada. A well was drilled in the region in the 1920s that "blew off the top of the crown block when it came in." The well set off a rush to the area by nearby residents, and the discovery well eventually produced "more than one million barrels of oil," but all four of the offsets which were drilled on ten-acre locations, were dry. Luck![20]

Although it is difficult to forecast the future of the Oklahoma petroleum industry, by far the majority of the state's oilmen express great optimism over the prospects, and few, if any, indicate a willingness to abandon the profession. In fact, there are many Okahoma families who have passed their involvement in the industry from generation to generation, and in some instances three, four, and five generations have followed one another into the oil fields. Most oilmen agree that there is still a place in the business for bright young men who are willing to work hard and take risks. Still, more will fail than succeed.[21]

One factor frequently favoring new independents in the oil business is their familiarity with the area. Inasmuch as most of the crude located in Oklahoma has been found by independents operating in a region where they are well known, trusted, and not considered as "outsiders," they enjoy an advantage over their counterparts in the larger corporations. This has

It is hoped that research facilities such as those of Phillips Petroleum Company just to the west of Bartlesville can develop even more revolutionary means to increase Oklahoma's petroleum production. *Courtesy of the Petroleum Publishing Company.*

been especially true since the 1950s and 1960s, when many of the major producers ceased to search actively for new petroleum deposits within the state. Now, however, the Arab oil embargo has rekindled the interest of major oil companies in Oklahoma, and it is difficult to judge what role they will have in the future.[22]

It is difficult to conclude whether the future of the state's oil industry will be determined by shallow wells, secondary recovery, tertiary methods, or deep wells. The adherents of all these methods present very good arguments. In all likelihood, the future will rest on a combination of these predictions, as well as new techniques only dreamed of today. Improved exploratory technology may well be the real key to the future. One thing is clear: only adequate tax and price incentives will encourage and stimulate continued exploration and technological advancement.[23]

It is certain that Oklahoma will continue to play a dominant role in the nation's petroleum industry. In 1978 alone the state witnessed 5,899 well completions according to figures provided by the Petroleum Information Corporation. This was a gain of eighteen percent over 1977 and placed Oklahoma second only to Texas in wells completed throughout the United States. Osage County alone had 602 well completions in 1978 and ranked second among all the nation's counties. In addition, of the top ten geologic provinces in America the Anadarko Basin was fourth in well completions, and the Sooner Trend was fifth among the ten leading fields ranked by completion. Nonetheless, the state did experience a slight drop in crude-oil and condensate production from 1977 to

1978—an estimated 4,175,000 barrels as compared to 3,972,000 barrels, for a decrease of 4.9 percent. Gas production was up 3.1 percent with an estimated production for 1978 amounting to 5,027.4 million cubic feet as compared to 4,874.4 million cubic feet in 1977. Oklahoma continued to supply the United States with 4.58 percent of its total crude oil production and 8.97 percent of all gas production.[24]

The price of continued production and exploration continues to rise at an unprecendented rate for Oklahoma oilmen, especially those involved in deep-hole production. Inflation has caused the cost of completing wells to skyrocket. While Oklahoma had 97 wells drilled to 15,000 feet or deeper in 1978, out of a nationwide total of 604, sinking a hole below 20,000 feet cost an average of $240.26 per foot. The cost of shallower wells was pegged at $50.05 per foot. For Oklahoma to continue its leadership role in the nation's energy production, state oilmen have expended vast amounts of capital.[25]

The state's contribution to the national oil industry is unchallenged. For years some of the world's largest energy companies—Phillips Petroleum, Cities Service, Continental Oil Company, Kerr-McGee Corporation, and Getty Refining and Marketing Company—have been headquartered in the Sooner State. Many other companies, Atlantic Richfield Company, Texaco, Inc., Getty Oil Company, Sun Company, Champlin, and others—have maintained important division or regional offices in the state. Much of the work undertaken to ease America's energy shortage is being conducted at the Amoco Research Center in Tulsa, the laboratories of Halliburton Services in Duncan, the Continental Oil facilities in Ponca City, Cities Service's research center in Tulsa, Phillips Petroleum's offices in Bartlesville, and Kerr-McGee's headquarters in Oklahoma City. Thus it is unlikely that Oklahoma's role in the nation's petroleum industry will diminish.[26]

It is obvious to anyone with a cursory knowledge of history that oil played a tremendous role in the development of Oklahoma. Its legacy to the citizens of the state is difficult to calculate in full. There are few Oklahomans whose lives it has not touched in some way—directly, or in medicine, education, economics, or cultural affairs. Petroleum catapulted the new state into the twentieth century; the vast natural resource financed its economic development. Perhaps modern Oklahoma is the greatest legacy of this elusive treasure, the discovery and utilization of which have made the words *Oklahoma* and *oil* synonymous throughout the world.[27]

# Notes

## Chapter 1

1. Carl Coke Rister, *Oil! Titan of the Southwest*, pp. 12–13; Interview, Louisa Rider, April 26, 1938, Grant Foreman, ed., "Indian-Pioneer History," 113 vols., Indian Archives, Oklahoma Historical Society, Oklahoma City, vol. 113, pp. 424–25. (Oklahoma Historical Society hereafter cited as OHS.)
2. Rister, *Oil!*, pp. 12–13; Muriel H. Wright, "First Oklahoma Oil Was Produced in 1859," *Chronicles of Oklahoma*, vol. 4, no. 4 (December, 1926), p. 322.
3. Wright, "First Oklahoma Oil," pp. 322–23; Rister, *Oil!*, p. 13.
4. Rister, *Oil!*, p. 14; Harold F. Williamson, Ralph L. Andreano, Arnold R. Daum, and Gilbert C. Klose, *The American Petroleum Industry*, vol. 1, p. 3.
5. Interview, Rider, April 26, 1938; Rister, *Oil!*, p. 14.
6. Rister, *Oil!*, pp. 14–15; Wright, "First Oklahoma Oil," pp. 323–24; Carl B. Glasscock, *Then Came Oil: The Story of the Last Frontier*, pp. 112–13.
7. Rister, *Oil!*, pp. 14–15; Wright, "First Oklahoma Oil," p. 324.
8. Wright, "First Oklahoma Oil," p. 324; Glasscock, *Then Came Oil*, p. 113; Rister, *Oil!*, pp. 15–16.
9. Rister, *Oil!*, p. 16; Williamson et al., *American Petroleum Industry*, vol. I, p. 549.
10. Glasscock, *Then Came Oil*, p. 113; Wright, "First Oklahoma Oil," p. 324; Rister, *Oil!*, p. 16.
11. Rister, *Oil!*, pp. 16–17; "Acts of the Choctaw Nation, 1884–1889," Indian Archives, OHS, Choctaw vol. 304, pp. 9–10.
12. "Acts of the Choctaw Nation," vol. 304, pp. 9–11.
13. *Ibid.*, pp. 10–11.
14. "Laws of the Cherokee Nation, 1884," Indian Archives, OHS, Cherokee vol. 261, pp. 89–93.
15. Rister, *Oil!*, p. 17; Wright, "First Oklahoma Oil," p. 325.
16. Wright, "First Oklahoma Oil," pp. 325–26; Rister, *Oil!*, pp. 17–18.
17. Rister, *Oil!*, p. 18; Charles J. Kappler, *Indian Affairs: Laws and Treaties*, vol. 2, pp. 443, 929; Wright, "First Oklahoma Oil," p. 326.
18. Wright, "First Oklahoma Oil," pp. 326–27.
19. *Ibid.*, p. 327; Rister, *Oil!*, p. 18.
20. Rister, *Oil!*, p. 18; "Laws of the Cherokee Nation, 1884," Cherokee vol. 261, p. 92; Wright, "First Oklahoma Oil," p. 327.
21. Wright, "First Oklahoma Oil," pp. 327–28; Rister, *Oil!*, pp. 18–19.
22. Rister, *Oil!*, p. 19; Wright, "First Oklahoma Oil," pp. 327–28.
23. Wright, "First Oklahoma Oil," p. 328; "Well Log, 1888, on Clear Boggy 14 miles from Atoka, drilled by National Oil Co. of St. Louis, H. W. Faucett, Mgr.," Museum, OHS.
24. Edward Allison Hill, *Story of the Mid-Continent Oil and Gas Field*, p. 15; Rister, *Oil!*, p. 19.
25. Rister, *Oil!*, p. 20; Glasscock, *Then Came Oil*, pp. 114–15.
26. Glasscock, *Then Came Oil*, p. 114; Rister, *Oil!*, pp. 20–21; Hill, *Mid-Continent Oil and Gas Field*, p. 15; Frank F. Finney, Sr., "The Indian Territory Illuminating Oil Company," *Chronicles of Oklahoma*, vol. 37, no. 2 (Summer, 1959), pp. 153–54.
27. Finney, "Indian Territory Illuminating Oil Company"; Hill, *Mid-Continent Oil and Gas Field*, p. 14; D. P. Killian, "Henry Vernon Foster, 1875–1939," *Chronicles of Oklahoma*, vol. 20, no. 4 (December, 1942), p. 441; Rister, *Oil!*, p. 21.
28. Rister, *Oil!*, p. 21; Finney, "Indian Territory Illuminating Oil Company," pp. 152–53.
29. Finney, "The Indian Territory Illuminating Oil Company," p. 153; Rister, *Oil!*, p. 21.
30. Rister, *Oil!*, p. 21; Finney, "Indian Territory Illuminating Oil Company," p. 153.
31. Finney, "Indian Territory Illuminating Oil Company," pp. 153–54; Roy M. McClintock, "Oil in Oklahoma," Roy M. McClintock Collection, Oklahoma State Archives, Oklahoma City, p. 2; Rister, *Oil!*, pp. 21–22.

## Chapter 2

1. *Bartlesville Enterprise*, August 15, 1906; Margaret W. Teague, *History of Washington County and Surrounding Area*, vol. 2, pp. 6–7.
2. Teague, *History of Washington County*, vol. 2, p. 7;

*Bartlesville Enterprise*, August 15, 1906; Rister, *Oil!*, p. 22; Glasscock, *Then Came Oil*, p. 115.

3. Glasscock, *Then Came Oil*, p. 115; *Bartlesville Enterprise*, August 15, 1906; Teague, *History of Washington County*, vol. 2, pp. 7–8.

4. Teague, *History of Washington County*, vol. 2, pp. 8–9; Rister, *Oil!*, p. 23; Finney, "Indian Territory Illuminating Oil Company," p. 154; *Bartlesville Enterprise*, August 15, 1906.

5. *Bartlesville Enterprise*, August 15, 1906; Interview, Andrew Jackson Fugate, June 18, 1937, Foreman, ed., "Indian-Pioneer History," vol. 25, p. 237; Glasscock, *Then Came Oil*, pp. 116–17; Teague, *History of Washington County*, vol. 2, p. 8–9.

6. Teague, *History of Washington County*, vol. 2, pp. 9–10; *Bartlesville Enterprise*, August 15, 1906.

7. *Bartlesville Enterprise*, August 15, 1906; Rister, *Oil!*, p. 22; Finney, "Indian Territory Illuminating Oil Company," p. 154; Glasscock, *Then Came Oil*, p. 117; Teague, *History of Washington County*, vol. 2, p. 10.

8. Teague, *History of Washington County*, vol. 2, pp. 11–12; *Bartlesville Enterprise*, August 15, 1906; Glasscock, *Then Came Oil*, pp. 117–19; Rister, *Oil!*, p. 22.

9. Rister, *Oil!*, p. 22; Teague, *History of Washington County*, vol. 2, p. 12.

10. Williamson et al., *American Petroleum Industry*, vol. 2, pp. 90–91; Hill, *Mid-Continent Oil and Gas Field*, p. 16; Rister, *Oil!*, pp. 412–13.

11. Rister, *Oil!*, p. 80; *Bartlesville Enterprise*, August 15, 1906; Finney, "Indian Territory Illuminating Oil Company," p. 155.

12. Finney, "Indian Territory Illuminating Oil Company," p. 115; *Bartlesville Enterprise*, August 15, 1906.

13. *Bartlesville Enterprise*, August 15, 1906; Finney, "Indian Territory Illuminating Oil Company," p. 155; Rister, *Oil!*, p. 80.

14. Rister, *Oil!*, p. 80.

15. *Ibid.*, pp. 80–81; "Act of June 28, 1898, Including Atoka Agreement," Seth K. Corden and W. B. Richards, comps., *The Oklahoma Red Book*, vol. 1, p. 532.

16. "The Curtis Act," Corden and Richards, *Oklahoma Red Book*, vol. 1, pp. 544–45; Rister, *Oil!*, p. 81.

17. Rister, *Oil!*, p. 81; "The Curtis Act," p. 545.

18. "The Curtis Act," p. 548; Rister, *Oil!*, p. 80.

19. Rister, *Oil!*, p. 81; *Bartlesville Enterprise*, August 15, 1906.

20. James L. Allhands, "History of the Construction of the Frisco Railway Lines in Oklahoma," *Chronicles of Oklahoma*, vol. 3, no. 3 (September, 1925), pp. 229–39; John W. Morris, Charles R. Goins, and Edwin C. McReynolds, *Historical Atlas of Oklahoma*, p. 64; Donovan L. Hofsommer, ed., *Railroads in Oklahoma*, pp. 1–74; Finney, "Indian Territory Illuminating Oil Company," p. 155.

21. Finney, "Indian Territory Illuminating Oil Company," p. 155; Teague, *History of Washington County*, vol. 2, p. 18.

22. LeRoy H. Fischer, ed., *Territorial Governors of Oklahoma*, pp. 3–6; *Bartlesville Enterprise*, August 15, 1906.

23. Williamson et al., *American Petroleum Industry*, vol. 1, pp. 525–731; *ibid.*, vol. 2, pp. 184–86.

24. *Ibid.*, pp. 185–90.

25. *Ibid.*, pp. 192–95.

26. *Ibid.*, pp. 19–20; Rister, *Oil!*, p. 81.

27. Rister, *Oil!*, p. 81; Cassius M. Barnes, *Report of the Governor of Oklahoma to the Secretary of the Interior for the Fiscal Year Ended June 30, 1898*, p. 51.

28. Cassius M. Barnes, *Annual Report of the Governor of Oklahoma to the Secretary of the Interior for the Fiscal Year Ended June 30, 1899*, p. 80; Rister, *Oil!*, p. 81.

29. Rister, *Oil!*, pp. 80–81; Glasscock, *Then Came Oil*, pp. 132–35; Fred S. Clinton, "First Oil and Gas Well in Tulsa County," *Chronicles of Oklahoma*, vol. 30, no. 3 (Autumn, 1952), p. 312; *Kansas City Times*, June 26, 1901, as quoted in Rister, *Oil!*, p. 85.

## Chapter 3

1. Rister, *Oil!*, pp. 80–94; Joe D. Roberts, "Ten-Barrel 'Whodunit' at Red Fork," Publications Department, OHS, *passim*; Glasscock, *Then Came Oil*, pp. 131–35; Clinton, "First Oil and Gas Well in Tulsa County," pp. 312–32.

2. Clinton, "First Oil and Gas Well in Tulsa County," pp. 312–13; Glasscock, *Then Came Oil*, pp. 131–33.

3. Glasscock, *Then Came Oil*, pp. 131–33; Clinton, "First Oil and Gas Well in Tulsa County," pp. 312–13.

4. Clinton, "First Oil and Gas Well in Tulsa County," pp. 312–14, 322–23; Glasscock, *Then Came Oil*, pp. 131–34.

5. Glasscock, *Then Came Oil*, p. 133; Clinton, "First Oil and Gas Well in Tulsa County," p. 314.

6. Clinton, "First Oil and Gas Well in Tulsa County," pp. 314–15; Glasscock, *Then Came Oil*, p. 134.

7. Glasscock, *Then Came Oil*, pp. 134–35; Clinton, "First Oil and Gas Well in Tulsa County," pp. 314–15.

8. Clinton, "First Oil and Gas Well in Tulsa County," p. 315; Glasscock, *Then Came Oil*, p. 315.

9. Glasscock, *Then Came Oil*, p. 315; Clinton, "First Oil and Gas Well in Tulsa County," p. 316.

10. Rister, *Oil!*, pp. 81–82; Glasscock, *Then Came Oil*, pp. 131–32; F. C. Hubbard, "Affadavit," Heydrick Collection, Western History Collections, University of Oklahoma, Norman; Roberts, "Ten-Barrel 'Whodunit' at Red Fork," pp. 3–4; Gaston Litton, *History of Oklahoma at the Golden Anniversary of Statehood*, vol. 2, p. 177.

11. Litton, *History of Oklahoma*, pp. 177–78; Rister, *Oil!*, pp. 81–82; Glasscock, *Then Came Oil*, p. 132.

12. Glasscock, *Then Came Oil*, pp. 133–34; Rister, *Oil!*, p. 83; Litton, *History of Oklahoma*, vol. 2, p. 178.

13. Litton, *History of Oklahoma*, vol. 2, p. 178; Rister, *Oil!*, p. 83; Glasscock, *Then Came Oil*, p. 134.

14. Glasscock, *Then Came Oil*, p. 135; Litton, *History of Oklahoma*, vol. 2, p. 178; Rister, *Oil!*, pp. 83–84.

15. Rister, *Oil!*, pp. 83–84; Glasscock, *Then Came Oil*, pp. 134–35; Clinton, "First Oil and Gas Well in Tulsa County," pp. 314–16, 322.

16. *Kansas City Times*, June 26, 1901, as quoted in Rister,

*Oil!*, p. 85; *Muskogee Weekly Phoenix*, July 4, 1901; *Tulsa Democrat*, June 28, 1901.

17. *Tulsa Democrat*, June 28, 1901; *Vinita Daily Chieftain*, June 28, 1901; Rister, *Oil!*, p. 84.

18. Rister, *Oil!*, pp. 84–85; Clinton, "First Oil and Gas Well in Tulsa County," pp. 316–17; Glasscock, *Then Came Oil*, pp. 135–36.

19. Rister, *Oil!*, pp. 85–86; *Daily Chieftain*, June 27, 1901; United States Department of the Interior, *Report, 1901, Indian Affairs* pp. 225–70.

20. United States Department of the Interior, *Report, 1901, Indian Affairs*, pp. 33, 232.

21. *Daily Chieftain*, June 28, 1901; Rister, *Oil!*, pp. 86–87.

22. "Original Creek Agreement, March 1, 1901," Corden and Richards, *Oklahoma Red Book*, vol. 1, pp. 562, 570; *Tulsa Democrat*, July 5, 1901.

23. *Tulsa Democrat*, July 5, 1901; Rister, *Oil!*, p. 86.

24. Rister, *Oil!*, p. 86; Clinton, "First Oil and Gas Well in Tulsa County," p. 317; Glasscock, *Then Came Oil*, p. 136.

25. *Tulsa Democrat*, August 2, 1901, and August 23, 1901; Rister, *Oil!*, p. 86.

26. Rister, *Oil!*, pp. 86–87; "Original Creek Agreement," pp. 559–72; *Tulsa Democrat*, August 2, 1901.

27. *Tulsa Democrat*, August 2, 1901; Litton, *History of Oklahoma*, vol. 2, p. 178.

28. Litton, *History of Oklahoma*, vol. 2, p. 178; *Tulsa Democrat*, August 23, 1901.

29. Rister, *Oil!*, pp. 86–87; Litton, *History of Oklahoma*, vol. 2, p. 78.

30. Litton, *History of Oklahoma*, vol. 2, pp. 178–79; Rister, *Oil!*, pp. 87–88; *Oklahoma State Capital*, July 18, 1901; Thompson B. Ferguson, *Annual Report of the Governor of Oklahoma to the Secretary of the Interior for the Fiscal Year Ended June 30, 1902*, pp. 65–66.

31. Ferguson, *Annual Report of the Governor*, p. 66; Rister, *Oil!*, p. 88.

32. Litton, *History of Oklahoma*, vol. 2, p. 179; *Tulsa Democrat*, August 23, 1901; Thompson B. Ferguson, *Report of the Governor of Oklahoma to the Secretary of the Interior, 1904*, p. 83.

33. Ferguson, *Report of the Governor 1904*, pp. 83–84; Rister, *Oil!*, pp. 88–89; Litton, *History of Oklahoma*, vol. 2, pp. 178–80.

# Chapter 4

1. Teague, *History of Washington County*, vol. 2, pp. 19–22; Finney, "Indian Territory Illuminating Oil Company," pp. 156–58.

2. Finney, "Indian Territory Illuminating Oil Company," pp. 158–59; Rister, *Oil!*, p. 88.

3. Teague, *History of Washington County*, vol. 2, pp. 18–22; Finney, "Indian Territory Illuminating Oil Company," pp. 155–57.

4. Finney, "Indian Territory Illuminating Oil Company," pp. 157–59; Thompson B. Ferguson, *Report of the Governor of Oklahoma to the Secretary of the Interior for the Year Ended June 30, 1905*, p. 23; Frank Frantz, *Report of the Governor of Oklahoma to the Secretary of the Interior for the Year Ended June 30, 1906*, p. 39; Teague, *History of Washington County*, vol. 2, p. 22.

5. Phillips Petroleum Company, *Some Highlights of Phillips Petroleum Company*, Oklahoma Petroleum Collection, Oklahoma Heritage Association, Oklahoma City, p. 2 (Oklahoma Petroleum Collection hereafter cited OPC); Interview, Dr. O. S. Sommerville, May 7, 1963, Archives, Phillips Petroleum Company, Bartlesville, Oklahoma (Phillips Petroleum Company hereafter cited PPC); "History of Phillips Petroleum Company to 1951," June 5, 1963, Archives, PPC; *Bartlesville Philnews*, June, 1967.

6. *Philnews*, June, 1967; Interview, Earnest E. Traywick, July 20, 1960, Archives, PPC; Phillips Petroleum Company, *Some Highlights*, p. 3.

7. Phillips Petroleum Company, *Some Highlights*, p. 3; Interview, R. C. Alden, June 15, 1959, Archives, PPC; *Philnews*, June, 1967.

8. *Philnews*, June, 1967; Phillips Petroleum Company, *Some Highlights*, pp. 3–5.

9. Finney, "Indian Territory Illuminating Oil Company," p. 158; Teague, *History of Washington County*, vol. 2, p. 23.

10. Litton, *History of Oklahoma*, vol. 2, p. 179; Rister, *Oil!*, p. 88; *The Triangle*, May 27, June 3, and June 10, 1904.

11. *The Triangle*, June 3 and July 1, 1904.

12. *The Triangle*, July 8, 1904; Rister, *Oil!*, p. 88.

13. Rister, *Oil!*, pp. 88–89; *The Triangle*, July 22 and July 29, 1904; Litton, *History of Oklahoma*, vol. 2, p. 179.

14. Litton, *History of Oklahoma*, vol. 2, p. 179; Franz, *Report of the Governor* (1906), p. 39; *The Triangle*, July 22 and July 29, 1904; Ferguson, *Report of the Governor 1905*, p. 23.

15. Ferguson, *Report of the Governor 1905*, p. 23; L. C. Snider, *Oil and Gas in the Mid-Continent Fields*, pp. 145–46; Litton, *History of Oklahoma*, vol. 2, p. 179.

16. Litton, *History of Oklahoma*, vol. 2, p. 179; Rister, *Oil!*, p. 89; Clarence B. Douglas, *The History of Tulsa, Oklahoma: A City with a Personality*, vol. 2, p. 206; J. Vere Frazier, Jr., "History of the Glennpool Oil Field," master's thesis, University of Oklahoma, 1951, pp. 12–14.

17. Frazier, "History of the Glennpool Oil Field," pp. 12–13; Rister, *Oil!*, p. 89; Douglas, *History of Tulsa*, vol. 1, p. 206.

18. Douglas, *History of Tulsa*, vol. 1, p. 206; *Tulsa Daily World*, November 26, 1935; Frazier, "History of the Glennpool Oil Field," p. 13.

19. Frazier, "History of the Glennpool Oil Field," p. 13; Douglas, *History of Tulsa*, vol. 1, p. 206; Litton, *History of Oklahoma*, vol. 2, p. 179.

20. Litton, *History of Oklahoma*, vol. 2, p. 179; Rister, *Oil!*, p. 89; *Tulsa Democrat*, December 22, 1905; *Tulsa Daily World*, February 17, 1907; Frazier, "History of Glennpool Oil Field," pp. 11–14.

21. Frazier, "History of Glennpool Oil Field," pp. 11, 15; Douglas, *History of Tulsa*, vol. 1, p. 209.

22. Douglas, *History of Tulsa*, vol. 1, pp. 206–209; Rister, *Oil!*, pp. 89–90; *Tulsa Daily World*, February 17, 1907;

Frazier, "History of the Glennpool Oil Field," pp. 14–15.

23. Frazier, "History of the Glennpool Oil Field," pp. 14–17; Douglas, *History of Tulsa*, vol. 1, pp. 206–209; *Mounds Enterprise*, January 19, March 16, and March 23, 1906.

24. *Mounds Enterprise*, February 9, 1906; Litton, *History of Oklahoma*, vol. 2, p. 179; Frazier, "History of the Glennpool Oil Field," p. 16.

25. Frazier, "History of the Glennpool Oil Field," pp. 18–20; Litton, *History of Oklahoma*, vol. 2, p. 180; Rister, *Oil!*, p. 91.

26. Rister, *Oil!*, p. 90; Frazier, "History of the Glennpool Oil Field," pp. 20–22.

27. Frazier, "History of the Glennpool Oil Field," pp. 24, 99; Douglas, *History of Tulsa*, vol. 1, p. 209; Snyder, *Oil and Gas*, p. 190; *Oil Investor's Journal*, vol. 5, no. 3 (August 3, 1906), p. 14.

28. Sinclair Oil and Gas Company, *A Great Name in Oil: Sinclair Through Fifty Years*, pp. 13–15.

29. Ibid., pp. 14–33.

30. Douglas, *History of Tulsa*, vol. 1, p. 209; Rister, *Oil!*, pp. 90–94; Frazier, "History of the Glennpool Oil Field," pp. 26–27.

31. Frazier, "History of the Glennpool Oil Field," pp. 28–29; *Oil Investor's Journal*, vol. 5, no. 11 (November 18, 1906), pp. 1–2; *Oil Investor's Journal*, vol. 5, no. 13 (December 3, 1906), p. 12; Rister, *Oil!*, pp. 92–93.

32. Rister, *Oil!*, pp. 91–94; *Oil Investor's Journal*, vol. 5, no. 13 (December 3, 1906), p. 12; *Mounds Enterprise*, December 7, 1906; Frazier, "History of the Glennpool Oil Field," pp. 29–44.

33. Frazier, "History of the Glennpool Oil Field," pp. 44–54; Rister, *Oil!*, pp. 92–94; *Oil Investor's Journal*, vol. 6, no. 6 (September 5, 1907), p. 8.

34. United States Geological Survey, *Mineral Resources of the United States, Part II, Nonmetals, 1906–1915*, p. 300; Frazier, "History of the Glennpool Oil Field," pp. 52–54.

35. Frazier, "History of the Glennpool Oil Field," pp. 53–55; Snyder, *Oil and Gas*, pp. 191–92.

36. Snyder, *Oil and Gas*, pp. 188–93; Frazier, "History of the Glennpool Oil Field," pp. 55–57; Rister, *Oil!*, pp. 92–94.

37. Rister, *Oil!*, pp. 93–94; Williamson et al., *American Petroleum Industry*, vol. 2, pp. 22–23; Litton, *History of Oklahoma*, vol. 2, p. 180.

38. Rister, *Oil!*, pp. 93–94; Frazier, "History of the Glennpool Oil Field," pp. 57, 71–72; Snyder, *Oil and Gas*, pp. 189–93.

## Chapter 5

1. Alvin O. Turner, "The Regulation of the Oklahoma Oil Industry," Ph.D. dissertation, Oklahoma State University, 1977, pp. 50–62; Litton, *History of Oklahoma*, vol. 2, p. 180.

2. Litton, *History of Oklahoma*, vol. 2, p. 180; Bess Mills-Bullard, comp., "Digest of Oklahoma Oil and Gas Fields," *Oklahoma Geological Survey Bulletin No. 40*, vol. 1 (July, 1928), p. 185; Charles G. Forbes, "The Origin and Early Development of the Oil Industry in Oklahoma," Ph.D. dissertation, University of Oklahoma, 1939, p. 158.

3. Forbes, "Origin and Early Development," p. 158; Mills-Bullard, "Oklahoma Oil and Gas Fields," p. 208.

4. Mills-Bullard, "Oklahoma Oil and Gas Fields," pp. 109, 132, 239, 255; Forbes, "Origin and Early Development," pp. 158–59.

5. Forbes, "Origin and Early Development," pp. 158–59; Mills-Bullard, "Oklahoma Oil and Gas Fields," pp. 112–13, 164–65, 168, 174, 236.

6. Mills-Bullard, "Oklahoma Oil and Gas Fields," pp. 121, 157–58, 198, 273–74; Forbes, "Origin and Early Development," p. 159.

7. Forbes, "Origin and Early Development," p. 158; Litton, *History of Oklahoma*, vol. 2, p. 180.

8. Blue Clark, "The Beginning of Oil and Gas Conservation in Oklahoma, 1907–1931," *Chronicles of Oklahoma*, vol. 55, no. 4 (Winter, 1977–78), pp. 375–77; Turner, "Regulation of the Oklahoma Oil Industry," pp. iii, 3–5.

9. Turner, "Regulation of the Oklahoma Oil Industry," pp. 3, 9–12; "An Act Relating to Minerals, November 3, 1847," Cherokee Minerals File, Indian Archives, OHS; "Amendment to the Minerals Act of 1847, November 29, 1873," Cherokee Minerals File, Indian Archives, OHS; H. Craig Miner, *The Corporation and the Indian*, pp. 144–46.

10. Miner, *The Corporation and the Indian*, pp. 143–63; Turner, "Regulation of the Oklahoma Oil Industry," pp. 12–13.

11. Turner, "Regulation of the Oklahoma Oil Industry," pp. 14–18.

12. Ibid., p. 18; "The Curtis Act," pp. 544–45; Miner, *The Corporation and the Indian*, pp. 151–53.

13. Miner, *The Corporation and the Indian*, pp. 152–56; Turner, "Regulation of the Oklahoma Oil Industry," pp. 21–25.

14. Turner, "Regulation of the Oklahoma Oil Industry," pp. 28–30; Miner, *The Corporation and the Indian*, pp. 156–63; Forbes, "Origin and Early Development," pp. 27–29.

15. Forbes, "Origin and Early Development," pp. 30–32; Turner, "Regulation of the Oklahoma Oil Industry," pp. 31–32.

16. Turner, "Regulation of the Oklahoma Oil Industry," pp. 35–47; Forbes, "Origin and Early Development," pp. 32–33; "Constitution of the State of Sequoyah," Corden and Richards, *The Oklahoma Red Book*, vol. 1, pp. 623–74.

17. Turner, "Regulation of the Oklahoma Oil Industry," pp. 50–51; Blue Clark, "Oil and Gas Conservation," pp. 375–76; Territory of Oklahoma, *Session Laws of 1895*, pp. 174–76.

18. Territory of Oklahoma, *Session Laws of 1895*, pp. 176–77.

19. Territory of Oklahoma, *Session Laws of 1899*, pp. 186–89; Territory of Oklahoma, *Session Laws of 1903*, pp. 180–95; Turner, "Regulation of the Oklahoma Oil Industry," p. 51.

20. Turner, "Regulation of the Oklahoma Oil Industry," pp. 52–54; Territory of Oklahoma, *Session Laws of 1905*, pp. 309–10.

21. Territory of Oklahoma, *Session Laws of 1905*, pp. 310–12; Turner, "Regulation of the Oklahoma Oil Industry," pp. 53–54; Blue Clark, "Oil and Gas Conservation," pp. 375–77.

22. Blue Clark, "Oil and Gas Conservation," pp. 375–77; Turner, "Regulation of the Oklahoma Oil Industry," pp. 50–55; Forbes, "Origin and Early Development," pp. 48–49.

23. Forbes, "Origin and Early Development," p. 33; "Enabling Act," Corden and Richards, *The Oklahoma Red Book*, vol. 1, pp. 33–34, 36–39.

24. Turner, "Regulation of the Oklahoma Oil Industry," pp. 54–56; "Constitution of Oklahoma," Corden and Richards, *The Oklahoma Red Book*, vol. 1, pp. 59, 66, 75.

25. *Harlow's Weekly* (Oklahoma City), January 31, 1914; Turner, "Regulation of the Oklahoma Oil Industry," pp. 56–57; State of Oklahoma, *Emergency Laws Passed by First Legislature State of Oklahoma 1907–1908*, pp. 25–31, 55–61.

26. State of Oklahoma, *Emergency Laws*, pp. 57–58; Turner, "Regulation of the Oklahoma Oil Industry," p. 56.

27. Forbes, "Origin and Early Development," pp. 34–37; *Mounds Enterprise*, October 18, 1907.

28. Turner, "Regulation of the Oklahoma Oil Industry," pp. 59–60; State of Oklahoma, *Session Laws of 1909*, pp. 425–30; "Oklahoma Constitution," p. 69; Henry G. Snyder, comp., *The Compiled Laws of Oklahoma 1909*, pp. 1096–99.

29. Henry G. Snyder, *Compiled Laws*, pp. 431–33; Turner, "Regulation of the Oklahoma Oil Industry," pp. 60–61.

30. Turner, "Regulation of the Oklahoma Oil Industry," pp. 60–61; State of Oklahoma, *Compiled Laws of Oklahoma 1909*, pp. 433–36.

31. Turner, "Regulation of the Oklahoma Oil Industry," pp. 61–62; Blue Clark, "Oil and Gas Conservation," pp. 375–79; Forbes, "Origin and Early Development," pp. 35–49.

32. Forbes, "Origin and Early Development," pp. 112–13, 116; Mills-Bullard, "Oklahoma Oil and Gas Fields," pp. 140–41, 211–12; Charles E. Bowles, "Oklahoma Petroleum: An Industrial Survey," *Oklahoma Geological Survey Bulletin No. 40*, vol. 1 (July, 1928), pp. 80–83.

## Chapter 6

1. John J. Mathews, *The Osages: Children of the Middle Waters*, pp. 771–73; Charles J. Kappler, comp., *Indian Affairs: Laws and Treaties*, vol. 3, pp. 256–58; "An Act Establishing the Territory of Oklahoma," Corden and Richards, *The Oklahoma Red Book*, vol. 1, p. 428; "Constitution of Oklahoma," p. 102; Forbes, "Origin and Early Development," p. 114.

2. Forbes, "Origin and Early Development," p. 114; Rister, *Oil!*, pp. 21–23; H. T. Beckwith, "Osage County," *Oklahoma Geological Survey Bulletin No. 40*, vol. 3 (July, 1930), p. 245.

3. *Osage Journal* (Pawhuska), April 10, 1902; Forbes, "Origin and Early Development," pp. 114–15.

4. Forbes, "Origin and Early Development," pp. 115–16; Finney, "Indian Territory Illuminating Oil Company," pp. 156–57; *Osage Journal*, April 10, 1902.

5. Forbes, "Origin and Early Development," p. 116; Beckwith, "Osage County," pp. 245–46.

6. Beckwith, "Osage County," pp. 245–46; Mills-Bullard, "Oklahoma Oil and Gas Fields," pp. 108, 117–18, 206, 210, 269–70.

7. Forbes, "Origin and Early Development," pp. 116–17; Beckwith, "Osage County," pp. 245–46.

8. Interview, Lois Straight Johnson, June 20, 1978, OPC; John Steiger to Kenny A. Franks, March 8, 1979, OPC; John Steiger, "Brief History of Cities Service Oil Company," OPC; "Speech by Lois Straight Johnson, April 16, 1975," OPC; John Steiger, "Relationship Between ITIO and Cities Service," OPC.

9. Steiger, "ITIO and Cities Service," OPC; "Speech by Lois Straight Johnson, April 16, 1975"; Williamson et al., *American Petroleum Industry*, vol. 2, p. 100; Interview, Johnson, June 20, 1978.

10. Mills-Bullard, "Oklahoma Oil and Gas Fields," pp. 115–16, 244, 247; Beckwith, "Osage County," pp. 245–46.

11. Beckwith, "Osage County," pp. 245–46; Forbes, "Origin and Early Development," pp. 116–18.

12. Forbes, "Origin and Early Development," pp. 43–46; Kappler, *Indian Affairs*, vol. 3, pp. 137–38, 254–56; *Osage Journal*, October 12 [19], 1911.

13. United States Department of the Interior, *Regulations to Govern the Leasing of Lands in the Osage Reservation, Oklahoma, for Oil and Gas Mining Purposes*, pp. 1–13; Forbes, "Origin and Early Development," pp. 46–47, 118–20; Gerald Forbes, "History of the Osage Blanket Lease," *Chronicles of Oklahoma*, vol. 19, no. 1 (March, 1941), pp. 78–79.

14. Gerald Forbes, "Osage Blanket Lease," pp. 78–79; *Osage Journal*, November 14, 1912; Beckwith, "Osage County," p. 245.

15. Beckwith, "Osage County," p. 246; Mills-Bullard, "Oklahoma Oil and Gas Fields," p. 269; Forbes, "Origin and Early Development," pp. 120–21.

16. Forbes, "Origin and Early Development," p. 120; Beckwith, "Osage County," p. 246; Mills-Bullard, "Oklahoma Oil and Gas Fields," pp. 182–83, 222, 225–26.

17. Forbes, "Origin and Early Development," p. 49; *Osage Journal*, February 11, 1915; *Oil and Gas Journal* (Tulsa), February 11, 1915; Gerald Forbes, "Osage Blanket Lease," pp. 79–80.

18. Gerald Forbes, "Osage Blanket Lease," p. 80; United States Department of the Interior, *Reports of the Department of the Interior for the Fiscal Year Ended June 30, 1915, Administrative Reports*, vol. 2, pp. 27–28.

19. United States Department of the Interior, *Reports of . . . Fiscal Year Ended June 30, 1915*, pp. 28–29; Gerald Forbes, "Osage Blanket Lease," p. 80.

20. Gerald Forbes, "Osage Blanket Lease," pp. 80–81; United States Department of the Interior, *Reports of . . . Fiscal Year Ended June 30, 1915*, vol. 2, p. 29; Beckwith, "Osage County," p. 246.

21. Beckwith, "Osage County," p. 246; Forbes, "Origin and Early Development," pp. 121–22; Litton, *History of Oklahoma*, vol. 2, p. 184; Mills-Bullard, "Oklahoma Oil and Gas Fields," pp. 110, 172, 197–98, 246.

22. Mills-Bullard, "Oklahoma Oil and Gas Fields," pp. 106–107, 147, 198–99, 270–71; Beckwith, "Osage County," p. 246; Litton, *History of Oklahoma*, vol. 2, pp. 183–84.

23. Litton, *History of Oklahoma*, p. 183; Mills-Bullard, "Oklahoma Oil and Gas Fields," pp. 214–16, 218–19, 271–72.

24. Forbes, "Origin and Early Development," pp. 122–26; *Osage Journal*, February 25, 1905; Gerald Forbes, "Osage Blanket Lease," pp. 74–76.

25. Interview, H. E. Koopman and Melvin B. Heine, August 12, 1960, Archives, PPC; Interview, Ralph L. Stewart, August 1, 1960, Archives, PPC.

26. Kappler, *Indian Affairs*, vol. 3, pp. 252–58; Mathews, *The Osages*, pp. 772–73.

27. Mathews, *The Osages*, pp. 772–78; Litton, *History of Oklahoma*, vol. 2, p. 184; Forbes, "Origin and Early Development," pp. 120–21.

28. Forbes, "Origin and Early Development," pp. 121–22; Mathews, *The Osages*, pp. 774–81; Kate Pearson Burwell, "The Richest People in the World," *Sturm's Statehood Magazine*, vol. 2, no. 4 (June, 1906), pp. 91–93; Robert Gregory, *Oil in Oklahoma*, pp. 53–58.

29. Teague, *History of Washington County*, vol. 2, pp. 23–24; George E. Tinker, "Who Gets the Oil?" *The Osage Magazine*, vol. 1 (May, 1910), pp. 53–60; Litton, *History of Oklahoma*, vol. 2, p. 184.

30. Litton, *History of Oklahoma*, vol. 2, pp. 184–85; Fred G. Hollman, "The Richest Pagans on Earth," *Indian Sketches*, Library, OHS; Mathews, *The Osages*, pp. 780–81; Beckwith, "Osage County," pp. 246–47.

31. Beckwith, "Osage County," pp. 246–47; Mathews, *The Osages*, p. 775; Litton, *History of Oklahoma*, vol. 2, p. 185; Mills-Bullard, "Oklahoma Oil and Gas Fields," pp. 124–25; Forbes, "Origin and Early Development," p. 221.

32. Forbes, "Origin and Early Development," pp. 100–101; Litton, *History of Oklahoma*, vol. 2, pp. 180–81; Mills-Bullard, "Oklahoma Oil and Gas Fields," pp. 140–41.

## Chapter 7

1. Rister, *Oil!*, p. 119; Mills-Bullard, "Oklahoma Oil and Gas Fields," pp. 140–41; Glasscock, *Then Came Oil*, pp. 216–18.

2. Glasscock, *Then Came Oil*, pp. 218–19; Forbes, "Origin and Early Development," pp. 100–101; Odie B. Faulk, Carl N. Tyson, and James H. Thomas, *The McMan: The Lives of Robert M. McFarlin and James A. Chapman*, pp. 20–21.

3. Faulk et al., *The McMan*, pp. 20–22; Glasscock, *Then Came Oil*, pp. 218–19; Litton, *History of Oklahoma*, vol. 2, pp. 180–81; Forbes, "Origin and Early Development," pp. 100–101.

4. Forbes, "Origin and Early Development," p. 101; Glasscock, *Then Came Oil*, pp. 218–19; Faulk et al., *The McMan*, pp. 22–23.

5. Faulk et al., *The McMan*, pp. 23–24; Glasscock, *Then Came Oil*, pp. 218–19; Forbes, "Origin and Early Development," pp. 101–102.

6. Forbes, "Origin and Early Development," pp. 100–101; Faulk et al., *The McMan*, p. 24; Glasscock, *Then Came Oil*, pp. 219–20; Mills-Bullard, "Oklahoma Oil and Gas Fields," pp. 140–41.

7. *Cushing Independent*, February 29, 1912; *Cushing Democrat*, February 29, 1912.

8. Mills-Bullard, "Oklahoma Oil and Gas Fields," pp. 140–41; Faulk et al., *The McMan*, pp. 24–25; Glasscock, *Then Came Oil*, p. 220; Forbes, "Origin and Early Development," pp. 101–102.

9. Forbes, "Origin and Early Development," p. 101; Glasscock, *Then Came Oil*, pp. 220–21.

10. Glasscock, *Then Came Oil*, pp. 220–21; *Cushing Independent*, March 21, 1912; *Cushing Democrat*, March 21, 1912; Forbes, "Origin and Early Development," p. 101.

11. Forbes, "Origin and Early Development," pp. 101–102; Glasscock, *Then Came Oil*, p. 221.

12. *Cushing Independent*, March 21, 1912.

13. *Ibid.*, March 28, 1912; Rister, *Oil!*, p. 119; Forbes, "Origin and Early Development," p. 102.

14. "Origin and Early Development," pp. 103–104; *Cushing Democrat*, April 4, April 11, and April 18, 1912.

15. *Cushing Independent*, April 18, 1912; Glasscock, *Then Came Oil*, pp. 222–23; Forbes, "Origin and Early Development," pp. 103, 110–12.

16. Forbes, "Origin and Early Development," pp. 103–104; *Cushing Democrat*, May 9, 1912; Morris, Goins, and McReynolds, *Historical Atlas of Oklahoma*, p. 70; Mills-Bullard, "Oklahoma Oil and Gas Fields," pp. 140–41; Snyder, *Oil and Gas*, pp. 215–17.

17. Snyder, *Oil and Gas*, pp. 220–21; Faulk et al., *The McMan*, p. 37; Forbes, "Origin and Early Development," p. 104.

18. Forbes, "Origin and Early Development," p. 105; Glasscock, *Then Came Oil*, pp. 224–25.

19. *Cushing Independent*, January 1, 1915; Forbes, "Origin and Early Development," pp. 104–105; "Creek County," *Oklahoma Geological Survey Bulletin No. 19*, Part 2 (April, 1917), p. 175; Snyder, *Oil and Gas*, pp. 220–21.

20. Snyder, *Oil and Gas*, pp. 215–22; Mills-Bullard, "Oklahoma Oil and Gas Fields," pp. 149, 204–205, 242–43, 272.

21. Forbes, "Origin and Early Development," pp. 104–106; Snyder, *Oil and Gas*, pp. 220–22.

22. Snyder, *Oil and Gas*, p. 221; Rister, *Oil!*, p. 123; Forbes, "Origin and Early Development," pp. 106–107.

23. *Cushing Independent*, January 8, 1915, and January 15, 1915.

24. Forbes, "Origin and Early Development," pp. 106–107; Turner, "Regulation of the Oklahoma Oil Industry," pp. 81–84.

25. Turner, "Regulation of the Oklahoma Oil Industry," pp. 84–85; Forbes, "Origin and Early Development," pp. 106–107; Rister, *Oil!*, p. 123.

26. *Harlow's Weekly*, July 4, 1914; Turner, "Regulation of the Oklahoma Oil Industry," pp. 84–85.

27. Turner, "Regulation of the Oklahoma Oil Industry,"

pp. 84–85; Forbes, "Origin and Early Development," p. 106; *Harlow's Weekly*, July 4, 1914.

28. *Harlow's Weekly*, July 14, 1914; Turner, "Regulation of the Oklahoma Oil Industry," pp. 85–86; "Creek County," pp. 175–77.

29. "Creek County," pp. 164–71; Forbes, "Origin and Early Development," pp. 107–109.

30. Forbes, "Origin and Early Development," pp. 105–10; Rister, *Oil!*, pp. 123–24; "Creek County," p. 177.

31. "Creek County," pp. 177–78; Forbes, "Origin and Early Development," pp. 104–12; *Harlow's Weekly*, July 4, 1913.

32. Litton, *History of Oklahoma*, vol. 2, pp. 180–81; Rister, *Oil!*, pp. 124–25; "Creek County," pp. 172–79; Mills-Bullard, "Oklahoma Oil and Gas Fields," pp. 140–41.

33. Mills-Bullard, "Oklahoma Oil and Gas Fields," pp. 166–67; Rister, *Oil!*, p. 125; Litton, *History of Oklahoma*, vol. 2, pp. 181–82.

## Chapter 8

1. Litton, *History of Oklahoma*, vol. 2, p. 181; C. W. Tomlinson, "Carter County," *Oklahoma Geological Survey Bulletin No. 40*, vol. 2 (July, 1930), p. 269; Gilbert L. Robinson, "History of the Healdton Oil Field" (master's thesis, University of Oklahoma, 1937), pp. 1–2; Forbes, "Origin and Early Development," p. 136.

2. Forbes, "Origin and Early Development," pp. 136–37; *Tulsa Democrat*, May 24, 1901.

3. Forbes, "Origin and Early Development," p. 137; L. L. Hutchison, "Preliminary Report on the Rock Asphalt, Asphaltite, Petroleum, and Natural Gas in Oklahoma," *Oklahoma Geological Survey Bulletin No. 2* (March, 1911), pp. 245–49.

4. Hutchison, "Preliminary Report," p. 249; Forbes, "Origin and Early Development," p. 137.

5. Forbes, "Origin and Early Development," pp. 137–38; Hutchison, "Preliminary Report," pp. 249–50.

6. Hutchison, "Preliminary Report," pp. 255–56; Forbes, "Origin and Early Development," pp. 137–38.

7. Forbes, "Origin and Early Development," pp. 137–38; Litton, *History of Oklahoma*, vol. 2, p. 181; Tomlinson, "Carter County," p. 269; Robinson, "Healdton Oil Field," pp. 2–4.

8. Robinson, "Healdton Oil Field," pp. 2–4; Litton, *History of Oklahoma*, vol. 2, pp. 181–82; Tomlinson, "Carter County," p. 269; Forbes, "Origin and Early Development," p. 139.

9. Forbes, "Origin and Early Development," pp. 139–40.

10. *Ibid.*, p. 141; Tomlinson, "Carter County," pp. 269–70; *Daily Ardmoreite*, August 12, 1913, and August 14, 1914; Rister, *Oil!*, pp. 126–27; Litton, *History of Oklahoma*, vol. 2, p. 182; "Carter County," *Oklahoma Geological Survey Bulletin No. 19*, Part 2 (April, 1917), pp. 91–92.

11. "Carter County," p. 92; Rister, *Oil!*, pp. 126–27; Tomlinson, "Carter County," pp. 269–70. Forbes, "Origin and Early Development," pp. 140–41.

12. Forbes, "Origin and Early Development," pp. 141–43; Rister, *Oil!*, pp. 128–29; Litton, *History of Oklahoma*, vol. 2, p. 182; *Daily Ardmoreite*, August 11, 1913, and October 8, 1914; Tomlinson, "Carter County," p. 92.

13. Tomlinson, "Carter County," p. 90; Robinson, "Healdton Oil Field," pp. 142–43; Forbes, "Origin and Early Development," pp. 143–44.

14. Forbes, "Origin and Early Development," pp. 142–43; *Daily Ardmoreite*, March 15, 1914; Robinson, "Healdton Oil Field," p. 14; Rister, *Oil!*, pp. 129–31; Turner, "Regulation of the Oklahoma Oil Industry," pp. 72–74.

15. Turner, "Regulation of the Oklahoma Oil Industry," pp. 73–78; Rister, *Oil!*, pp. 130–31; Robinson, "Healdton Oil Field," pp. 14–15.

16. Robinson, "Healdton Oil Field," pp. 46–53; Turner, "Regulation of the Oklahoma Oil Industry," pp. 81–85; Rister, *Oil!*, pp. 131–33; *Daily Ardmoreite*, August 16, 1914.

17. Interview, Ward Merrick, Jr., June 8, 1978, OPC.

18. Litton, *History of Oklahoma*, vol. 2, p. 182; *Daily Ardmoreite*, August 27, 1914, August 28, 1914, and August 30, 1914.

19. *Daily Ardmoreite*, August 13, 1914.

20. *Ibid.*, August 27, 1914, and August 30, 1914.

21. *Ibid.*, August 27, 1914, August 28, 1914, and August 30, 1914.

22. *Ibid.*

23. *Ibid.*

24. *Ibid.*

25. *Ibid.*

26. *Ibid.*, August 30, 1914.

27. *Ibid.*

28. *Ibid.*, August 27, 1914, August 28, 1914, and August 30, 1914.

29. *Ibid.*, August 28, 1914, and August 30, 1914.

30. *Ibid.*, August 30, 1914; Litton, *History of Oklahoma*, vol. 2, p. 182.

31. *Daily Ardmoreite*, August 30, 1914, and August 31, 1914.

32. *Ibid.*, September 2, 1914; Rister, *Oil!*, pp. 130–33.

33. Rister, *Oil!*, pp. 138–39; Robinson, "Healdton Oil Field," pp. 48–57; Turner, "Regulation of the Oklahoma Oil Industry," pp. 81–107; Forbes, "Origin and Early Development," pp. 150–56.

34. Forbes, "Origin and Early Development," pp. 153–54; Tomlinson, "Carter County," pp. 269–70; "Carter County," pp. 92–102.

35. "Carter County," pp. 92–96; Robinson, "Healdton Oil Field," p. 21; Litton, *History of Oklahoma*, vol. 2, p. 182.

36. Interview, Robert J. Collins, July 10, 1978, OPC; "The Company: Its History," OPC.

37. "The Company: Its History, OPC,; Interview, Collins, July 10, 1978.

38. Forbes, "Origin and Early Development," pp. 155–56; Rister, *Oil!*, p. 119; Litton, *History of Oklahoma*, vol. 2, p. 198.

39. Gerald Forbes, "Southwestern Oil Boom Towns," *Chronicles of Oklahoma*, vol. 17, no. 4 (December, 1939), p. 393; Drue Lemel DeBerry, "The Ethos of the Oklahoma Oil Boom Frontier, 1905–1929" (Master's thesis, University of

Oklahoma, 1970), pp. 1–5.

## Chapter 9

1. Gerald Forbes, "Southwestern Oil Boom Towns," p. 393; DeBerry, "Ethos" p. 9.
2. DeBerry, "Ethos," p. 10; Litton, *History of Oklahoma*, vol. 2, pp. 197–98; Gerald Forbes, "Southwestern Oil Boom Towns," pp. 393–95.
3. Gerald Forbes, "Southwestern Oil Boom Towns," p. 395; DeBerry, "Ethos," pp. 18–19; Rister, *Oil!*, p. 120.
4. Interview, William M. Beard, July 10, 1978, OPC; Beard Oil Company, *1977 Annual Report*, O.P.C.
5. "The Machinist," Oil in Oklahoma Series, Oklahoma Writers Project, Library, OHS, pp. 6–7.
6. Ibid., pp. 7–8.
7. "The Rig Builder," Oil in Oklahoma Series, Oklahoma Writers Project, Library, OHS, pp. 4–5.
8. Ibid., pp. 7–8.
9. "The Switcher," Oil in Oklahoma Series, Oklahoma Writers Project, Library, OHS, p. 4.
10. Interview, Stanley Learned, June 9, 1978, OPC.
11. Interview, William M. Barber, July 1, 1959, Archives, PPC.
12. Interview, Ralph L. Stewart, August 1, 1960, Archives, PPC.
13. "The Oil-Field Cook," Oil in Oklahoma Series, Oklahoma Writers Project, Library, OHS, pp. 1–2.
14. "The Pumper," Oil in Oklahoma Series, Oklahoma Writers Project, Library, OHS, pp. 1–5.
15. Ibid., pp. 1–3; "The Rig Builder," p. 8.
16. Gerald Forbes, "Southwestern Oil Boom Towns," pp. 395–96; DeBerry, "Ethos," pp. 25–26.
17. DeBerry, "Ethos," pp. 26–28; Rister, *Oil!*, p. 120; Barbara Shirley, "An Interview with Ed A. Smith of Service Drilling Company . . . On June 14, 1977," OPC; Gerald Forbes, "Southwestern Oil Boom Towns," p. 396.
18. DeBerry, "Ethos," p. 24; James L. Gilbert, "Three Sands: Oklahoma Oil Field and Community of the 1920's" (master's thesis, University of Oklahoma, 1967), pp. 87–88.
19. Gilbert, "Three Sands," pp. 105–106; Frazier, "History of the Glennpool Oil Field," pp. 61–62; DeBerry, "Ethos," pp. 62–63.
20. DeBerry, "Ethos," p. 67; *Daily Ardmoreite*, November 21, 1915; Robinson, "Healdton Oil Field," p. 58.
21. Robinson, "Healdton Oil Field," p. 58; Gerald Forbes, "Southwestern Oil Boom Towns," pp. 396–97; DeBerry, "Ethos," pp. 66–67; Rister, *Oil!*, pp. 120–21.
22. Interview, William B. Osborn, Jr., July 17, 1978, OPC.
23. Interview, William T. Payne, August 5, 1978, OPC.
24. Ibid.
25. Ibid.
26. DeBerry, "Ethos," pp. 73–75; Robinson, "Healdton Oil Field," pp. 63–68; Glasscock, *Then Came Oil*, pp. 253–54.
27. Earle E. Emerson, *Playing My Part*, pp. 60–61.
28. Interview, Jack Conn, July 11, 1978, OPC; Roy P. Stewart and Pendleton Woods, *Born Grown: An Oklahoma City History*, p. 239.
29. Stewart and Woods, *Born Grown*, p. 239; Interview, Conn, July 11, 1978; Interview, Charles Urschel, September 20, 1978, OPC.
30. Robinson, "Healdton Oil Field," pp. 58–70; DeBerry, "Ethos," pp. 66–70; "Supply Salesman," Oil in Oklahoma Series, Oklahoma Writers Project, Library, OHS, p. 2; "Spudder Man," Oil in Oklahoma Series, Oklahoma Writers Project, Library, OHS, pp. 3, 8–9; "Constitution of Oklahoma," Corden and Richards, *The Oklahoma Red Book*, vol. 1, p. 41; Gerald Forbes, "Southwestern Oil Boom Towns," p. 397.
31. Gerald Forbes, "Southwestern Oil Boom Towns," p. 397; Rister, *Oil!*, p. 122; Faulk et al., *The McMan*, p. 34; "Casinghead Gasoline Plant," Oil in Oklahoma Series, Oklahoma Writers Project, Library, OHS, pp. 30–31.
32. Gerald Forbes, "Southwestern Oil Boom Towns," p. 398; Gilbert, "Three Sands," pp. 106–108.
33. Gilbert, "Three Sands," p. 108; Rister, *Oil!*, p. 202; DeBerry, "Ethos," pp. 81–83.
34. DeBerry, "Ethos," pp. 70–71; Glasscock, *Then Came Oil*, pp. 255–56; Gilbert, "Three Sands," pp. 93–94.
35. Gilbert, "Three Sands," pp. 108–109; DeBerry, "Ethos," pp. 84–85.
36. DeBerry, "Ethos," pp. 109–11; Gerald Forbes, "Southwestern Oil Boom Towns," pp. 398–99; DeBerry, "Ethos," pp. 398–99; Gilbert, "Three Sands," pp. 59, 62, 64–66.
37. Gilbert, "Three Sands," p. 111; Gerald Forbes, "Southwestern Oil Boom Towns," pp. 398–400.
38. Gerald Forbes, "Southwestern Oil Boom Towns," pp. 399–400; DeBerry, "Ethos," 88–106.

## Chapter 10

1. Litton, *History of Oklahoma*, vol. 2, pp. 198–200; Rister, *Oil!*, pp. 143, 231; Williamson et al., *American Petroleum Industry*, vol. 2, pp. 299–300.
2. American Petroleum Institute, *Petroleum Facts and Figures*, pp. 14, 23, 34, 102; Mills-Bullard, "Oklahoma Oil and Gas Fields," pp. 103–274.
3. Rister, *Oil!*, p. 196; Beckwith, "Osage County," pp. 246–47; Litton, *History of Oklahoma*, vol. 2, p. 183.
4. Litton, *History of Oklahoma*, vol. 2, p. 183; Rister, *Oil!*, pp. 196–97; Mills-Bullard, "Oklahoma Oil and Gas Fields," pp. 124–25; Beckwith, "Osage County," pp. 246–47.
5. Beckwith, "Osage County," pp. 246–47; Rister, *Oil!*, pp. 196–98; Litton, *History of Oklahoma*, vol. 2, p. 183.
6. Litton, "History of Oklahoma," vol. 2, pp. 183–84; Beckwith, "Osage County," pp. 246–47.
7. Beckwith, "Osage County," pp. 246–47; "Our Government's Biggest Gambling Center at Pawhuska" and "More Than $3,000,000 Was Paid for Leases at Recent Osage Sale," as quoted in Rister, *Oil!*, pp. 200–202; United States Department of the Interior, *Report of the Commissioner of Indian Affairs for the*

*Fiscal Year Ended June 30, 1922*, pp. 23–24; Mills-Bullard, "Oklahoma Oil and Gas Fields," pp. 124–25; Litton, *History of Oklahoma*, vol. 2, p. 184.

8. Interview, Paul Endacott, June 20, 1978, OPC.

9. *Ibid*.

10. Rister, *Oil!*, pp. 197–98; Litton, *History of Oklahoma*, vol. 2, pp. 183–84; Beckwith, "Osage County," pp. 246–47.

11. Beckwith, "Osage County," pp. 246–47; American Petroleum Institute, *Petroleum Facts and Figures*, p. 112.

12. John Steiger to Kenny A. Franks, March 14, 1978, OPC; Clark, "The Beginning of Oil and Gas Conservation in Oklahoma, 1907–1931," p. 377; Interview, Stanley Learned, June 9, 1978, OPC.

13. Interview, Stanley Learned, June 9, 1978, OPC.

14. *Ibid*.; Steiger to Franks, March 14, 1978; Williamson et al., *American Petroleum Industry*, vol. 2, pp. 426–27.

15. Rister, *Oil!*, p. 198; Litton, *History of Oklahoma*, vol. 2, pp. 183–84; Beckwith, "Osage County," pp. 246–47; Mills-Bullard, "Oklahoma Oil and Gas Fields," p. 124.

16. Mills-Bullard, "Oklahoma Oil and Gas Fields," pp. 107–108, 155–57, 188, 190, 224–25; Beckwith, "Osage County," p. 246.

17. Rister, *Oil!*, p. 231; Litton, *History of Oklahoma*, vol. 2, p. 185; Tomlinson, "Carter County," pp. 274–76; Mills-Bullard, "Oklahoma Oil and Gas Fields," pp. 168–69.

18. *Daily Ardmoreite*, June 6 and June 7, 1919.

19. Rister, *Oil!*, pp. 231–32; Tomlinson, "Carter County," pp. 275–76; Mills-Bullard, "Oklahoma Oil and Gas Fields," pp. 168–69; Litton, *History of Oklahoma*, vol. 2, p. 185.

20. Litton, *History of Oklahoma*, pp. 185–86; Tomlinson, "Carter County," p. 278; Rister, *Oil!*, p. 232.

21. Interview, T. H. McCasland, June 7, 1978, OPC.

22. *Ibid*.

23. Rister, *Oil!*, p. 203; State of Oklahoma, *Directory of Oklahoma, 1975*, p. 63; Litton, *History of Oklahoma*, vol. 2, p. 185; Gilbert, "Three Sands," pp. 4–5.

24. Continental Oil Company, *Conoco: The First One Hundred Years*, p. 63; "Pioneering in Oil: Conoco's First Century," OPC.

25. "Pioneering in Oil"; Continental Oil Company, *Conoco*, pp. 63–67.

26. Continental Oil Company, *Conoco*, pp. 67–68; "Pioneering in Oil."

27. "Pioneering in Oil"; Continental Oil Company, *Conoco*, pp. 11–13.

28. Continental Oil Company, *Conoco*, pp. 13–77; "Pioneering in Oil."

29. Continental Oil Company, "Pioneering in Oil"; Continental Oil Company, *Conoco*, pp. 75–119. Interview, Keely Marshall, June 19, 1978, OPC.

30. Mills-Bullard, "Oklahoma Oil and Gas Fields," pp. 192–93, 200; G. C. Clark and C. L. Cooper, "Kay, Grant, Garfield and Noble Counties," *Oklahoma Geological Survey Bulletin No. 40*, vol. 2 (July, 1930), p. 83.

31. Clark and Cooper, "Kay, Grant, Garfield and Noble Counties," pp. 86–92; Litton, *History of Oklahoma*, vol. 2, p. 185; Mills-Bullard, "Oklahoma Oil and Gas Fields," pp. 114–17, 143, 158–59.

32. *Enid Cycler*, Spring, 1978; Henry B. Bass, "Herbert Hiram Champlin," *Chronicles of Oklahoma*, vol. 33, no. 1 (Spring, 1955), pp. 43–46.

33. Bass, "Herbert Hiram Champlin," p. 45.

34. *Ibid*., pp. 45–46.

35. Rister, *Oil!*, p. 203; *Daily Oklahoman* (Oklahoma City), September 23, 1923; Gilbert, "Three Sands," pp. 11–15.

36. Gilbert, "Three Sands," pp. 15–16; *Daily Oklahoman*, September 23, 1923; *Tonkawa News*, June 23, 1921; Clark and Cooper, "Kay, Grant, Garfield and Noble Counties," p. 93.

37. Clark and Cooper, "Kay, Grant, Garfield and Noble Counties," p. 93; Rister, *Oil!*, p. 203; Mills-Bullard, "Oklahoma Oil and Gas Fields," p. 256; Gilbert, "Three Sands," p. 16.

38. Gilbert, "Three Sands," pp. 16–17; *Tonkawa News*, June 23, 1921.

39. *Tonkawa News*, June 23, 1921; Rister, *Oil!*, p. 204; *Daily Oklahoman*, September 23, 1923; Gilbert, "Three Sands," p. 17.

40. Gilbert, "Three Sands," pp. 17–20; *Tonkawa News*, June 30, 1921; *Daily Oklahoman*, September 23, 1923.

41. *Daily Oklahoman*, September 23, 1923; Rister, *Oil!*, p. 204; Gilbert, "Three Sands," pp. 20–27.

42. Gilbert, "Three Sands," pp. 21–26; Mills-Bullard, "Oklahoma Oil and Gas Fields," p. 256; *Daily Oklahoman*, September 23, 1923; Clark and Cooper, "Kay, Grant, Garfield and Noble Counties," pp. 93–96.

43. Clark and Cooper, "Kay, Grant, Garfield and Noble Counties," pp. 93–96; Rister, *Oil!*, p. 203; Mills-Bullard, "Oklahoma Oil and Gas Fields," p. 256; *Daily Oklahoman*, September 23, 1923; Gilbert, "Three Sands," pp. 30–31.

44. Gilbert, "Three Sands," pp. 27–34; Litton, *History of Oklahoma*, vol. 2, p. 185; *Daily Oklahoman*, September 23, 1923; Rister, *Oil!*, pp. 204–05.

45. Rister, *Oil!*, p. 204; Gilbert, "Three Sands," pp. 11–12, 28–29.

46. Interview, V. D. Peters, November 17, 1960, Archives, PPC; *Philnews* (Bartlesville), November 28, 1939.

47. Bill Kelder, ed., *A History of Apco Oil Corporation and Its Predecessor Company Anderson-Prichard Oil Corporation*, pp. 1–11.

48. *Ibid*., pp. 11–150.

49. A. I. Levorsen, "Geology of Seminole County," *Oklahoma Geological Survey Bulletin No. 40*, vol. 3 (July, 1930), pp. 289, 321; Litton, *History of Oklahoma*, vol. 2, pp. 185–87; Mills-Bullard, "Oklahoma Oil and Gas Fields," pp. 120, 173, 254, 260; Clark and Cooper, "Kay, Grant, Garfield and Noble Counties," pp. 96–104.

## Chapter 11

1. Max W. Ball, Douglas Ball, and Daniel S. Turner, *This Fascinating Oil Business*, pp. 359, 369–70; Levorsen, "Geology of Seminole County," p. 289; Litton, *History of Oklahoma*, vol. 2, pp. 185–86.

2. Litton, *History of Oklahoma*, vol. 2, p. 186; Rister, *Oil!*, p. 235.

3. Rister, *Oil!*, pp. 231–33; Mills-Bullard, "Oklahoma Oil and Gas Fields," pp. 105, 156, 171, 192; J. Phillip Boyle, "Hughes County," *Oklahoma Geological Survey Bulletin No. 40*, vol. 3 (July, 1930), pp. 624–25.

4. Rister, *Oil!*, p. 234; *Daily Oklahoman*, June 24, 1923; Litton, *History of Oklahoma*, vol. 2, p. 186.

5. Litton, *History of Oklahoma*, vol. 2, p. 186; *Wewoka Capital-Democrat*, March 22, 1923; Levorsen, "Geology of Seminole County," p. 322; Rister, *Oil!*, p. 234; Mills-Bullard, "Oklahoma Oil and Gas Fields," pp. 256–66; *Seminole County News*, March 22, 1923.

6. Interview, William B. Osborn, Jr., July 17, 1978, OPC.

7. Rister, *Oil!*, pp. 234–35; *Wewoka Capital-Democrat*, March 22 and April 10, 1923.

8. *Wewoka Capital-Democrat*, April 19, 1923; Rister, *Oil!*, pp. 234–35.

9. Rister, *Oil!*, p. 235; Levorsen, "Geology of Seminole County," pp. 322–24; Mills-Bullard, "Oklahoma Oil and Gas Fields," pp. 265–66.

10. Mills-Bullard, "Oklahoma Oil and Gas Fields," pp. 139–40; Rister, *Oil!*, pp. 236–37; Litton, *History of Oklahoma*, vol. 2, p. 187; Levorsen, "Geology of Seminole County," pp. 324–25; *Wewoka Capital-Democrat*, September 27, 1923.

11. *Wewoka Capital-Democrat*, March 20, 1924, and March 27, 1924; Mills-Bullard, "Oklahoma Oil and Gas Fields," pp. 139–40; Rister, *Oil!*, pp. 236–37.

12. Rister, *Oil!*, pp. 236–37; *Wewoka Capital-Democrat*, March 20, 1924, and March 27, 1924.

13. Rister, *Oil!*, p. 235; Mills-Bullard, "Oklahoma Oil and Gas Fields," pp. 139–40; Levorsen, "Geology of Seminole County," p. 325.

14. Levorsen, "Geology of Seminole County," pp. 326–28; Rister, *Oil!*, p. 237.

15. Rister, *Oil!*, p. 237; *Wewoka Capital-Democrat*, January 15, 1925, and December 31, 1925, and March 25, 1926; Litton, *History of Oklahoma*, vol. 2, pp. 186–87.

16. Rister, *Oil!*, pp. 235, 241–42; Levorsen, "Geology of Seminole County," pp. 328–40.

17. Levorsen, "Geology of Seminole County," p. 336; Rister, *Oil!*, p. 237; Mills-Bullard, "Oklahoma Oil and Gas Fields," p. 151; T. E. Weirich, "Pottawatomie County," *Oklahoma Geological Survey Bulletin No. 40* (July, 1930), vol. 3, p. 591.

18. Weirich, "Pottawatomie County," pp. 391–94; Rister, *Oil!*, pp. 237–42; Levorsen, "Geology of Seminole County," pp. 336–40.

19. Levorsen, "Geology of Pottawatomie County," p. 328; Mills-Bullard, "Oklahoma Oil and Gas Fields," p. 241; American Association of Petroleum Geologists, *Trek of the Oil Finders: A History of Exploration for Petroleum*, p. 129; Rister, *Oil!*, p. 328.

20. Rister, *Oil!*, pp. 238–39; *Seminole County News*, March 11, 1929; Levorsen, "Geology of Seminole County," p. 328.

21. Interview, Osborn, July 17, 1978.

22. Rister, *Oil!*, pp. 237–39; *Seminole County News*, March 11, 1926, and March 18, 1926; Levorsen, "Geology of Seminole County," pp. 328–29.

23. Levorsen, "Geology of Seminole County," pp. 328–29; Rister, *Oil!*, pp. 238–39; *Seminole County News*, July 22, 1926, and July 29, 1926, and August 5, 1926; Mills-Bullard, "Oklahoma Oil and Gas Fields," p. 241.

24. Mills-Bullard, "Oklahoma Oil and Gas Fields," p. 241; Rister, *Oil!*, pp. 238–39; *Seminole County News*, August 5, 1926; Levorsen, "Geology of Seminole County," pp. 328–34.

25. Levorsen, "Geology of Seminole County," p. 334; Mills-Bullard, "Oklahoma Oil and Gas Fields," p. 240; *Seminole County News*, April 22, 1926; Rister, *Oil!*, pp. 235–41.

26. Rister, *Oil!*, pp. 240–41; *Wewoka Capital-Democrat*, October 14, 1926; Levorsen, "Geology of Seminole County," p. 334.

27. Levorsen, "Geology of Seminole County," pp. 334–36; Mills-Bullard, "Oklahoma Oil and Gas Fields," p. 240; Rister, *Oil!*, 240–41.

28. Rister, *Oil!*, p. 241; Weirich, "Pottawatomie County," p. 594; *Tecumseh County Democrat*, July 23, 1926; *Shawnee Weekly Herald*, July 22, 1926.

29. Levorsen, "Geology of Seminole County," pp. 342–43; Rister, *Oil!*, pp. 235, 241–42; Mills-Bullard, "Oklahoma Oil and Gas Fields," p. 186.

30. Mills-Bullard, "Oklahoma Oil and Gas Fields," p. 105; R. A. Conkling, "Pontotoc County," *Oklahoma Geological Survey Bulletin No. 40* (July, 1930), vol. 3, pp. 127–29; Boyle, "Hughes County," p. 624; *Oil and Gas Journal*, January 17, 1929; Rister, *Oil!*, p. 235.

31. Rister, *Oil!*, pp. 242–44; Blue Clark, "Oil and Gas Conservation," pp. 387–88; Litton, *History of Oklahoma*, vol. 2, pp. 186–87.

32. Litton, *History of Oklahoma*, p. 187; Williamson et al., *American Petroleum Industry*, vol. 2, p. 323; Rister, *Oil!*, pp. 243–44.

33. Rister, *Oil!*, pp. 242–45; Litton, *History of Oklahoma*, vol. 2, p. 187; Interview, Urschel, September 20, 1978.

34. Interview, Learned, June 9, 1978; Interview, Harry A. Trower, September 9, 1960, Archives, PPC; Williamson et al., *American Petroleum Industry*, vol. 2, pp. 575–76.

35. Williamson et al., *American Petroleum Industry*, vol. 2, pp. 576–78; Interview, Learned, June 9, 1978.

36. Interview, Learned, June 9, 1978; Williamson et al., *American Petroleum Industry*, vol. 2, p. 578.

# Chapter 12

1. Stewart and Woods, *Born Grown*, pp. 212–24; Rister, *Oil!*, pp. 248–69; Litton, *History of Oklahoma*, vol. 2, pp. 187–91.

2. Litton, *History of Oklahoma*, vol. 2, pp. 187–88; Stewart and Woods, *Born Grown*, p. 212; Rister, *Oil!*, pp. 248–49.

3. Rister, *Oil!*, pp. 248–49; A. Travis, "Oklahoma County," *Oklahoma Geological Survey Bulletin No. 40*, vol. 2 (July, 1930), p. 434; George H. Shirk, *Oklahoma Place Names*, p. 225; "Oklahoma County," *Oklahoma Geological Survey Bulle-*

*tin No. 19*, Part 2 (April, 1917), p. 365.

4. "Oklahoma County," pp. 364–65; Litton, *History of Oklahoma*, vol. 2, p. 188; Shirk, *Oklahoma Place Names*, p. 169; Travis, "Oklahoma County," p. 434.

5. Travis, "Oklahoma County," p. 434; Rister, *Oil!*, pp. 249; Stewart and Woods, *Born Grown*, p. 212.

6. Stewart and Woods, *Born Grown*, p. 212; Rister, *Oil!*, pp. 249–50; Travis, "Oklahoma County," pp. 434–35.

7. Travis, "Oklahoma County," p. 435; Stewart and Woods, *Born Grown*, p. 212; Litton, *History of Oklahoma*, vol. 2, p. 188.

8. Litton, *History of Oklahoma*, vol. 2, p. 188; Travis, "Oklahoma County," p. 435; Stewart and Woods, *Born Grown*, pp. 212–13; Rister, *Oil!*, p. 250.

9. Rister, *Oil!*, p. 250; *Daily Oklahoman*, November 30, 1926.

10. *Daily Oklahoman*, November 30, 1926; Rister, *Oil!*, p. 250; Travis, "Oklahoma County," p. 435; Stewart and Woods, *Born Grown*, p. 213.

11. Stewart and Woods, *Born Grown*, p. 213; Travis, "Oklahoma Country," p. 435; Rister, *Oil!*, p. 250; *Daily Oklahoman*, June 6, 1928, and 13, 1928.

12. *Daily Oklahoman*, December 5, 1928; Stewart and Woods, *Born Grown*, p. 213; "Oil Discovery Anniversary Marked," Vertical Files, Library, OHS; Rister, *Oil!*, p. 250.

13. Rister, *Oil!*, p. 250; Stewart and Woods, *Born Grown*, p. 213; *Daily Oklahoman*, December 5, 1928.

14. *Daily Oklahoman*, December 5, 1928; Rister, *Oil!*, p. 250; Stewart and Woods, *Born Grown*, p. 213.

15. *Daily Oklahoman*, December 5, 1928.

16. *Ibid.*

17. Rister, *Oil!*, p. 251; Stewart and Woods, *Born Grown*, p. 213; *Daily Oklahoman*, December 5, 1928, and December 6, 1928.

18. *Daily Oklahoman*, December 5, 1928.

19. *Ibid.*, December 6, 1928; Stewart and Woods, *Born Grown*, p. 213; Rister, *Oil!*, p. 255.

20. Rister, *Oil!*, p. 255; *Daily Oklahoman*, December 6, 1928, December 7, 1928, and December 8, 1928; *Oil and Gas Journal*, December 13, 1928, December 20, 1928, December 27, 1928, and January 3, 1929; Stewart and Woods, *Born Grown*, pp. 213–14.

21. Stewart and Woods, *Born Grown*, pp. 213–14; *Daily Oklahoman*, December 6, 1928, and December 7, 1928; Litton, *History of Oklahoma*, vol. 2, p. 188.

22. Rister, *Oil!*, p. 255; Stewart and Woods, *Born Grown*, p. 214; Interview, Dean A. McGee, OPC; "Oklahoma City Discovery Well," Vertical File, Library, OHS.

23. "Oklahoma City Discovery Well" (OHS); Stewart and Woods, *Born Grown*, pp. 214–19; Rister, *Oil!*, pp. 255–69; Litton, *History of Oklahoma*, vol. 2, pp. 188–90; *Daily Oklahoman*, March 5, 1930.

24. Jack Duncan, "History of the Harper Oil Company," OPC; Interview, Jack Duncan, July 13, 1978, OPC.

25. Interview, Duncan, July 13, 1978; Duncan, "History of the Harper Oil Company."

26. Duncan, "History of the Harper Oil Company"; Interview, Duncan, July 13, 1978.

27. Rister, *Oil!*, pp. 258–59; Steiger to Franks, March 8, 1979; *Daily Oklahoman*, March 30, 1930; Interview, James O. Kemm, March 14, 1979, OPC; Stewart and Woods, *Born Grown*, p. 215.

28. Stewart and Woods, *Born Grown*, p. 215; *Daily Oklahoman*, March 30, 1930, and March 31, 1930; Rister, *Oil!*, pp. 259–60.

29. Rister, *Oil!*, p. 259; Stewart and Woods, *Born Grown*, p. 215; *Daily Oklahoman*, March 30, 1930, March 31, 1930, April 1, 1930, and April 2, 1930.

30. *Daily Oklahoman*, April 3, 1930, and April 4, 1930; Litton, *History of Oklahoma*, vol. 2, p. 189; Rister, *Oil!*, p. 259.

31. Rister, *Oil!*, p. 260; *Daily Oklahoman*, April 3, 1930, April 4, 1930, and April 5, 1930; Stewart and Woods, *Born Grown*, p. 215.

32. Stewart and Woods, *Born Grown*, p. 215; *Daily Oklahoman*, April 4 and 5, 1930; Rister, *Oil!*, p. 261.

33. Rister, *Oil!*, pp. 259–61; *Daily Oklahoman*, April 4, 1930, and April 5, 1930; Stewart and Woods, *Born Grown*, pp. 215–16; Litton, *History of Oklahoma*, vol. 2, p. 189.

34. Litton, *History of Oklahoma*, vol. 2, pp. 187–90, 194–97; Interview, J. Sam Williams, May 7, 1963, Archives, PPC; Rister, *Oil!*, pp. 255–69; Stewart and Woods, *Born Grown*, pp. 214–18; Blue Clark, "Oil and Gas Conservation," pp. 379–91; Williamson et al., *American Petroleum Industry*, vol. 2, pp. 303–306, 541–43.

## Chapter 13

1. Turner, "Regulation of the Oklahoma Oil Industry," p. 65; Richard Hays Powell, "The Oil Industry and the Depression from the Development of Greater Seminole Through the Passage of the Oil Code" (master's thesis, University of Oklahoma, 1968), pp. 1–16; Williamson et al., *American Petroleum Industry*, vol. 2, pp. 49–51; Blue Clark, "Oil and Gas Conservation," pp. 378–79.

2. Blue Clark, "Oil and Gas Conservation," pp. 381–82; *Harlow's Weekly*, April 26, 1913; Turner, "Regulation of the Oklahoma Oil Industry," pp. 65–68.

3. Turner, "Regulation of the Oklahoma Oil Industry," pp. 67–69; State of Oklahoma, *Session Laws of 1913*, pp. 166–79, 439–41; *Harlow's Weekly*, March 8, 1913; Blue Clark, "Oil and Gas Conservation," pp. 381–82.

4. Blue Clark, "Oil and Gas Conservation," p. 379; *Daily Oklahoman*, April 24, 1914; Williamson et al., *American Petroleum Industry*, vol. 2, p. 321; Turner, "Regulation of the Oklahoma Oil Industry," pp. 69–81.

5. Turner, "Regulation of the Oklahoma Oil Industry," pp. 81–84; *Harlow's Weekly*, May 9, 1914; Oklahoma Corporation Commission, *Seventh Annual Report*, pp. 503–508; Blue Clark, "Oil and Gas Conservation," pp. 382–84.

6. Blue Clark, "Oil and Gas Conservation," pp. 384–85; Williamson et al., *American Petroleum Industry*, vol. 2, pp. 321–27; State of Oklahoma, *Session Laws of 1915*, pp. 28–31, 326–29; Turner, "Regulation of the Oklahoma Oil Industry,"

pp. 91–98.

7. Turner, "Regulation of the Oklahoma Oil Industry," pp. 98–99; Oklahoma Corporation Commission, *Eighth and Ninth Annual Reports*, pp. 252–65; Blue Clark, "Oil and Gas Conservation," pp. 384–85.

8. Blue Clark, "Oil and Gas Conservation," pp. 384–86; Turner, "Regulation of the Oklahoma Oil Industry," pp. 98–101; Oklahoma Corporation Commission, *Eighth and Ninth Annual Reports*, pp. 252–62.

9. Oklahoma Corporation Commission, *Eighth and Ninth Annual Reports*, pp. 261–62; Blue Clark, "Oil and Gas Conservation," pp. 385–86; Turner, "Regulation of the Oklahoma Oil Industry," pp. 100–101.

10. Turner, "Regulation of the Oklahoma Oil Industry," pp. 101–104; Oklahoma Corporation Commission, *Eighth and Ninth Annual Reports*, pp. 287–323.

11. Oklahoma Corporation Commission, *Eighth and Ninth Annual Reports*, pp. 319–21; Turner, "Regulation of the Oklahoma Oil Industry," p. 103.

12. Turner, "Regulation of the Oklahoma Oil Industry," p. 103; Oklahoma Corporation Commission, *Eighth and Ninth Annual Reports*, pp. 321–23.

13. Blue Clark, "Oil and Gas Conservation," pp. 385–86; Turner, "Regulation of the Oklahoma Oil Industry," pp. 103–107; American Petroleum Institute, *Petroleum Facts and Figures* (1928), pp. 28–29; Williamson et al., *American Petroleum Industry*, vol. 2, pp. 321–22; Oklahoma Corporation Commission, *Tenth Annual Report*, pp. 525–41.

14. Oklahoma Corporation Commission, *Tenth Annual Report*, pp. 533–39; Blue Clark, "Oil and Gas Conservation," p. 386; Oklahoma Corporation Commission, *Eighth and Ninth Annual Reports*, pp. 318–23; Turner, "Regulation of the Oklahoma Oil Industry," pp. 104–107.

15. Turner, "Regulation of the Oklahoma Oil Industry," pp. 105–106; Oklahoma Corporation Commission, *Tenth Annual Report*, pp. 533–38.

16. Oklahoma Corporation Commission, *Tenth Annual Report*, pp. 538–39; Turner, "Regulation of the Oklahoma Oil Industry," pp. 105–106.

17. Turner, "Regulation of the Oklahoma Oil Industry," pp. 106–107; Blue Clark, "Oil and Gas Conservation," pp. 386–88; Williamson et al., *American Petroleum Industry*, vol. 2, pp. 321–23.

18. Williamson et al., *American Petroleum Industry*, pp. 316–17, 326; Steiger to Franks, March 14, 1978.

19. Steiger to Franks, March 14, 1978; Williamson et al., *American Petroleum Industry*, vol. 2, pp. 316–17.

20. Williamson et al., *American Petroleum Industry*, pp. 316–17, 326; Steiger to Franks, March 14, 1978; Interview, Charles E. Beecher, June 20, 1978, OPC.

21. Interview, Beecher, June 20, 1978; Steiger to Franks, March 14, 1978; Williamson et al., *American Petroleum Industry*, vol. 2, pp. 316–17, 326.

22. Rister, *Oil!*, pp. 255–69; Litton, *History of Oklahoma*, vol. 2, pp. 188–90; *Daily Oklahoman*, March 5, 1930; Stewart and Woods, *Born Grown*, pp. 214–19.

23. Stewart and Woods, *Born Grown*, pp. 214–19; *Daily Oklahoman*, May 6, 1932, and May 7, 1932; Rister, *Oil!*, pp. 260–62.

24. Rister, *Oil!*, pp. 260–62; *Daily Oklahoman*, May 6, 1932, and May 7, 1932; Stewart and Woods, *Born Grown*, pp. 214–19.

25. Stewart and Woods, *Born Grown*, pp. 212–16; Blue Clark, "Oil and Gas Conservation," pp. 387–89; Rister, *Oil!*, pp. 239–64; Williamson et al., *American Petroleum Industry*, vol. 2, pp. 321–25; Turner, "Regulation of the Oklahoma Oil Industry," pp. 196–218.

26. Turner, "Regulation of the Oklahoma Oil Industry," pp. 168–93; Interview, Edward A. Smith, July 21, 1978, OPC; Check Mid-Kansas Oil & Gas Co. to E. A. Smith, July 20, 1931, OPC; Interview, Oscar L. Cordell, November 17, 1960, Archives, PPC; Blue Clark, "Oil and Gas Conservation," pp. 387–89; Stewart and Woods, *Born Grown*, pp. 214–17; Williamson et al., *American Petroleum Industry*, vol. 2, pp. 323–27; Rister, *Oil!*, pp. 261–66.

27. Rister, *Oil!*, pp. 263–65; *Daily Oklahoman*, July 11, 1931, July 25, 1931, and July 28, 1931; Williamson et al., *American Petroleum Industry*, vol. 2, pp. 540–43; Turner, "Regulation of the Oklahoma Oil Industry," pp. 181–93; Blue Clark, "Oil and Gas Conservation," pp. 388–89.

28. Blue Clark, "Oil and Gas Conservation," pp. 390–91; *Daily Oklahoman*, August 1, 1931, August 2, 1931, and August 5, 1931; Rister, *Oil!*, pp. 263–65; Turner, "Regulation of the Oklahoma Oil Industry," pp. 181–93.

29. Turner, "Regulation of the Oklahoma Oil Industry," pp. 196–270; Blue Clark, "Oil and Gas Conservation," p. 391; Williamson et al., *American Petroleum Industry*, vol. 2, pp. 542–43; J. Stanley Clark, *The Oil Century: From the Drake Well to the Conservation Era*, pp. 182–85.

# Chapter 14

1. *Tulsa Tribune*, October 7, 1923; Leslie Brooks to Kenny A. Franks, March 9, 1978, OPC; Fred S. Clinton, "The Beginning of the International Petroleum Exposition and Congress," *Chronicles of Oklahoma*, vol. 26, no. 4 (Winter, 1948–49), pp. 479—80.

2. Clinton, "Petroleum Exposition," pp. 480–81; Brooks to Franks, March 9, 1978; Roberta Ironside, *An Adventure Called Skelly: A History of Skelly Oil Company Through Fifty Years 1919–1969*, pp. 36–37; *Tulsa Tribune*, October 7, 1923.

3. *Tulsa Tribune*, October 7, 1923; Clinton, "Petroleum Exposition," pp. 480–81.

4. Clinton, "Petroleum Exposition," pp. 481–82; *Tulsa Tribune*, October 7, 1923.

5. *Tulsa Tribune*, October 7, 1923; Brooks to Franks, March 9, 1978; Clinton, "Petroleum Exposition," pp. 481–82.

6. Clinton, "Petroleum Exposition," p. 482; *Tulsa Tribune*, October 8, 1923; Brooks to Franks, March 9, 1978.

7. Brooks to Franks, March 9, 1978; *Tulsa Tribune*, October 8–11, 1923; Clinton, "Petroleum Exposition," pp. 482–85.

8. Clinton, "Petroleum Exposition," pp. 485–87; *Tulsa*

*Tribune*, October 12–15, 1923.

9. *Tulsa Tribune*, October 10, 1923, and October 14, 1923; Clinton, "Petroleum Exposition," pp. 486–87.

10. Rister, *Oil!*, pp. 207–18; *Tulsa Daily World*, October 2–8, 1924.

11. *Tulsa Daily World*, October 9, 1924, and October 10, 1924; *Oil and Gas Journal*, October 9, 1924.

12. *Tulsa Daily World*, October 1, 1925, October 2, 1925, and October 4, 1925.

13. *Tulsa Daily World*, October 4, 1925, and October 5, 1925.

14. *Tulsa Daily World*, October 8, 1924, and October 5–10, 1925.

15. *Ibid.*, October 10, 1925; Brooks to Franks, March 9, 1978; *Tulsa Tribune*, September 24, 1927, September 25, 1927, September 26, 1927, September 27, 1927, September 28, 1927, September 29, 1927, September 30, 1927, October 1, 1927, and October 2, 1927.

16. *Tulsa Tribune*, October 2, 1927, October 20, 1928, October 21, 1928, October 22, 1928, October 23, 1928, October 24, 1928, October 25, 1928, October 26, 1928, October 27, 1928, and October 28, 1928.

17. *Ibid.*, October 5, 1929, October 6, 1929, October 7, 1929, October 8, 1929, October 9, 1929, October 10, 1929, October 11, 1929, October 12, 1929, and October 13, 1929; J. Stanley Clark, *The Oil Century*, pp. 176–88; Rister, *Oil!*, pp. 248–69.

18. Brooks to Franks, March 9, 1978; *Tulsa Tribune*, October 4, 1930, October 5, 1930, October 6, 1930, October 7, 1930, October 8, 1930, October 9, 1930, October 10, 1930, October 11, 1930, October 12, 1930, and May 11, 1934.

19. *Tulsa Tribune*, May 11, 1938, May 12, 1938, May 13, 1938, May 14, 1938, May 15, 1938, May 16, 1938, May 17, 1938, May 18, 1938, May 19, 1938, May 20, 1938; J. Stanley Clark, *The Oil Century*, pp. 182–88; Rister, *Oil!*, pp. 256–69, 216–326; Brooks to Franks, March 9, 1978.

20. Brooks to Franks, March 9, 1978; *Tulsa Tribune*, May 20, 1934; May 16, 1936, May 17, 1936, May 18, 1936, May 19, 1936, May 20, 1936, May 21, 1936, May 22, 1936, May 23, 1936, May 24, 1936, May 8, 1938, May 9, 1938, May 14, 1938, May 15, 1938, May 16, 1938, May 17, 1938, May 18, 1938, May 19, 1938, May 20, 1938, May 21, 1938, and May 22, 1938.

21. *Tulsa Tribune*, May 17, 1938, May 19, 1938, May 14, 1940, May 19, 1940, May 20, 1940, May 21, 1940, May 22, 1940, May 23, 1940, May 24, 1940, May 25, 1940, and May 26, 1940; Brooks to Franks, March 9, 1978.

22. Brooks to Franks, March 9, 1978; *Tulsa Tribune*, May 14, 1948, May 15, 1948, May 16, 1948, May 17, 1948, May 18, 1948, May 19, 1948, May 20, 1948, May 21, 1948, May 22, 1948, May 23, 1948, and May 24, 1948; Clinton, "Petroleum Exposition," pp. 486–87.

23. Clinton, "Petroleum Exposition," pp. 487–88; *Tulsa Tribune*, May 17, 1948, May 13, 1953, and May 14, 1953; Brooks to Franks, March 9, 1978.

24. Brooks to Franks, March 9, 1978; *Tulsa Tribune*, May 13, 1953, May 14, 1953, May 15, 1953, May 16, 1953, May 17, 1953, May 19, 1953, May 20, 1953, May 21, 1953, May 22, 1953, May 23, 1953, and May 13, 1959; *Tulsa Daily World*, May 24, 1953.

25. *Tulsa Daily World*, May 24, 1953; Brooks to Franks, March 9, 1978; *Tulsa Tribune*, May 13, 1959, May 14, 1959, May 15, 1959, May 16, 1959, May 18, 1959, May 19, 1959, May 20, 1959, May 21, 1959, May 22, 1959, and May 23, 1959.

26. *Tulsa Tribune*, May 11, 1966, May 12, 1966, May 13, 1966, May 14, 1966, May 16, 1966, May 17, 1966, May 18, 1966, May 19, 1966, May 20, 1966, and May 21, 1966; Brooks to Franks, March 9, 1978; *Tulsa Daily World*, May 22, 1966.

27. *Tulsa Daily World*, May 23, 1971, and May 24, 1971; Brooks to Franks, March 9, 1978; *Tulsa Tribune*, May 15, 1971, May 17, 1971, May 18, 1971, May 19, 1971, May 20, 1971, May 21, 1971, and May 22, 1971.

28. *Tulsa Tribune*, May 14, 1976, May 17, 1976, May 18, 1976, May 19, 1976, May 20, 1976, May 21, 1976, and May 22, 1976; *Tulsa Daily World*, May 22, 1976; Brooks to Franks, March 9, 1978; Interview, James O. Kemm, April 28, 1980, OPC.

29. Brooks to Franks, March 9, 1978; Clinton, "Petroleum Exposition," pp. 479–80.

## Chapter 15

1. Turner, "Regulation of the Oklahoma Oil Industry," pp. 181–93; Rister, *Oil!*, pp. 261–69; *Daily Oklahoman*, June 5, 1931.

2. Interview, Charles P. Dimit, September 9, 1960, Archives, PPC.

3. Turner, "Regulation of the Oklahoma Oil Industry," pp. 196–200; *Daily Oklahoman*, June 5, 1931.

4. *Daily Oklahoman*, August 5, 1931.

5. *Ibid.*; Turner, "Regulation of the Oklahoma Oil Industry," p. 200.

6. Turner, "Regulation of the Oklahoma Oil Industry," p. 200; *Daily Oklahoman*, August 5, 1931.

7. *Daily Oklahoman*, August 5, 1931; Turner, "Regulation of the Oklahoma Oil Industry," p. 200.

8. Turner, "Regulation of the Oklahoma Oil Industry," pp. 181–200; William H. Murray, *Memoirs of Governor Murray and True History of Oklahoma*, vol. 2, pp. 502–509; Rister, *Oil!*, pp. 255–64; Clark, "Oil and Gas Conservation," p. 389; *Daily Oklahoman*, August 5, 1931.

9. *Daily Oklahoman*, August 6, 1931, and August 7, 1931; Turner, "Regulation of the Oklahoma Oil Industry," pp. 200–202.

10. Murray, *Memoirs of Governor Murray*, vol. 2, pp. 507–509; *Daily Oklahoman*, August 6, 1931.

11. *Daily Oklahoman*, August 6, 1931, and August 11, 1931; J. Stanley Clark, *The Oil Century*, pp. 235–39; Turner, "Regulation of the Oklahoma Oil Industry," pp. 200–203.

12. Turner, "Regulation of the Oklahoma Oil Industry," pp. 201–203; *Daily Oklahoman*, October 9 and 10, 1931; Murray, *Memoirs of Governor Murray*, vol. 2, pp. 510–13; Ris-

ter, *Oil!*, p. 264; J. Stanley Clark, *The Oil Century*, pp. 237–39.

13. J. Stanley Clark, *The Oil Century*, pp. 238–39; *Daily Oklahoman*, June 22, 1932; Rister, *Oil!*, p. 265; Williamson et al., *American Petroleum Industry*, vol. 2, pp. 542–43; Turner, "Regulation of the Oklahoma Oil Industry," pp. 203–206.

14. Turner, "Regulation of the Oklahoma Oil Industry," pp. 206–13; *Daily Oklahoman*, June 22, 1932.

15. *Oil and Gas Journal*, September 1, 1933, and September 22, 1933; Turner, "Regulation of the Oklahoma Oil Industry," pp. 209–18.

16. Turner, "Regulation of the Oklahoma Oil Industry," pp. 221–42; Rister, *Oil!*, pp. 264–65; *Harlow's Weekly*, February 18, 1933; *Daily Oklahoman*, March 5, 1933.

17. *Daily Oklahoman*, March 5, 1933; State of Oklahoma, *Official Session Laws 1933*, pp. 278–304; *Oil and Gas Journal*, March 9, 1933; Turner, "Regulation of the Oklahoma Oil Industry," pp. 241–42.

18. Turner, "Regulation of the Oklahoma Oil Industry," p. 242; Williamson et al., *American Petroleum Industry*, vol. 2, pp. 542–43; *Oil and Gas Journal*, April 13, 1933; *Daily Oklahoman*, February 16, March 24, 1933, March 28, 1933, and April 11, 1933.

19. Bass, "Herbert Hiram Champlin," p. 47.

20. *Oil and Gas Journal*, April 13, 1933; State of Oklahoma, *Official Session Laws 1933*, pp. 278–80.

21. State of Oklahoma, *Official Session Laws 1933*, pp. 280–83; *Oil and Gas Journal*, April 13, 1933.

22. *Oil and Gas Journal*, April 13, 1933; State of Oklahoma, *Official Session Laws 1933*, pp. 282–84.

23. State of Oklahoma, *Official Session Laws 1933*, pp. 284–88; *Oil and Gas Journal*, April 13, 1933.

24. *Oil and Gas Journal*, April 13, 1933; State of Oklahoma, *Official Session Laws 1933*, pp. 288–94.

25. State of Oklahoma, *Official Session Laws 1933*; pp. 288–97; *Oil and Gas Journal*, April 13, 1933.

26. *Oil and Gas Journal*, April 13, 1933; State of Oklahoma, *Official Session Laws 1933*, pp. 297–301.

27. State of Oklahoma, *Official Session Laws 1933*, pp. 301–304; *Oil and Gas Journal*, April 13, 1933; Turner, "Regulation of the Oklahoma Oil Industry," p. 242.

28. Turner, "Regulation of the Oklahoma Oil Industry," pp. 243–47; *Oil and Gas Journal*, April 13, 1933; *Harlow's Weekly*, April 22, 1933.

29. C. W. Van Eaton, ed., *Harlow's Session Laws of 1935*, pp. 271–77; Frank Eagin and Everett Eagin, eds., *Official Session Laws 1941*, pp. 211–18; Turner, "Regulation of the Oklahoma Oil Industry," pp. 242–44, 247, 260–65.

30. Turner, "Regulation of the Oklahoma Oil Industry," pp. 185–90, 246–47, 260–70, 297–98; Rister, *Oil!*, p. 264; *Daily Oklahoman*, July 11, 1931; *Oil and Gas Journal*, April 13, 1933; J. Stanley Clark, *The Oil Century*, pp. 182–86; Williamson et al., *American Petroleum Industry*, vol. 2, pp. 542–43, 560.

## Chapter 16

1. Turner, "Regulation of the Oklahoma Oil Industry," pp. 61–62, 250–51; J. Stanley Clark, *The Oil Century*, pp. 194–95; Powell, "The Oil Industry and the Depression," p. 70.

2. Powell, "The Oil Industry and the Depression," pp. 70–71; J. Stanley Clark, *The Oil Century*, pp. 190–91; Turner, "Regulation of the Oklahoma Oil Industry," pp. 250–51; Williamson et al., *American Petroleum Industry*, vol. 2, pp. 540–41.

3. Williamson et al., *American Petroleum Industry*, pp. 545–46; J. Stanley Clark, *The Oil Century*, pp. 190–92; Powell, "The Oil Industry and the Depression," pp. 61–62.

4. Powell, "The Oil Industry and the Depression," p. 66; J. Stanley Clark, *The Oil Century*, pp. 191–92; Williamson et al., *American Petroleum Industry*, vol. 2, pp. 547–48.

5. Williamson et al., *American Petroleum Industry*, pp. 547–48; Powell, "The Oil Industry and the Depression," pp. 66–67; J. Stanley Clark, *The Oil Century*, pp. 191–92; Turner, "Regulation of the Oklahoma Oil Industry," pp. 188–89.

6. Turner, "Regulation of the Oklahoma Oil Industry," p. 189; J. Stanley Clark, *The Oil Century*, pp. 192–93; Williamson et al., *American Petroleum Industry*, vol. 2, pp. 547–48.

7. Williamson et al., *American Petroleum Industry*, p. 458; J. Stanley Clark, *The Oil Century*, pp. 193–94; Turner, "Regulation of the Oklahoma Oil Industry," p. 189.

8. Turner, "Regulation of the Oklahoma Oil Industry," pp. 250–51; J. Stanley Clark, *The Oil Century*, pp. 193–94; Powell, "The Oil Industry and the Depression," pp. 71–73; Williamson et al., *American Petroleum Industry*, vol. 2, pp. 548–49.

9. Williamson et al., *American Petroleum Industry*, vol. 2, pp. 548–49; J. Stanley Clark, *The Oil Century*, pp. 194–95; Turner, "Regulation of the Oklahoma Oil Industry," pp. 251–52.

10. Turner, "Regulation of the Oklahoma Oil Industry," p. 252; J. Stanley Clark, *The Oil Century*, pp. 194–95; Williamson et al., *American Petroleum Industry*, vol. 2, pp. 548–49; Powell, "The Oil Industry and the Depression," pp. 86–87.

11. Powell, "The Oil Industry and the Depression," pp. 87–102; Turner, "Regulation of the Oklahoma Oil Industry," p. 252.

12. Turner, "Regulation of the Oklahoma Oil Industry," pp. 252–53; Powell, "The Oil Industry and the Depression," pp. 102–104.

13. Turner, "Regulation of the Oklahoma Oil Industry," pp. 252–54; J. Stanley Clark, *The Oil Century*, p. 195; Williamson et al., *American Petroleum Industry*, vol. 2, p. 459.

14. Williamson et al., *American Petroleum Industry*, pp. 549–50; J. Stanley Clark, *The Oil Century*, pp. 195–96; Turner, "Regulation of the Oklahoma Oil Industry," p. 254.

15. Turner, "Regulation of the Oklahoma Oil Industry," p. 255; J. Stanley Clark, *The Oil Century*, p. 197; Williamson et al., *American Petroleum Industry*, vol. 2, p. 550; *Oil and Gas Journal*, December 6, 1934; Interstate Oil Compact Commission, *A Study of Conservation of Oil and Gas in the United States 1964*, p. 8.

16. Interstate Oil Compact Commission, *Conservation of Oil and Gas*, p. 8; Turner, "Regulation of the Oklahoma Oil Industry," p. 255; *Oil and Gas Journal*, December 6, 1934.

17. *Oil and Gas Journal*, January 3, 1935, and February 21,

1935; J. Stanley Clark, *The Oil Century*, pp. 197–98; Interstate Oil Compact Commission, *Conservation of Oil and Gas*, p. 8.

18. Interstate Oil Compact Commission, *Conservation of Oil and Gas*, p. 8; *Oil and Gas Journal*, February 21, 1935; Turner, "Regulation of the Oklahoma Oil Industry," p. 255; J. Stanley Clark, *The Oil Century*, pp. 255–56; Williamson et al., *American Petroleum Industry*, vol. 2, pp. 550–51.

19. Williamson et al., *American Petroleum Industry*, p. 551; Interstate Oil Compact Commission, *Conservation of Oil and Gas*, p. 8; J. Stanley Clark, *The Oil Century*, pp. 256–57; *Oil and Gas Journal*, February 21, 1935.

20. *Oil and Gas Journal*, February 21, 1935; Williamson et al., *American Petroleum Industry*, vol. 2, p. 551; J. Stanley Clark, *The Oil Century*, p. 257.

21. J. Stanley Clark, *The Oil Century*, pp. 197–98; Interstate Oil Compact Commission, *Conservation of Oil and Gas*, p. 8; Interview, Margaret Ray, May 17, 1978, OPC; *Oil and Gas Journal*, February 21, 1935; Litton, *History of Oklahoma*, vol. 2, p. 195.

22. Turner, "Regulation of the Oklahoma Oil Industry," p. 256; Van Eaton, *Harlow's Session Laws of 1935*, pp. 272–73.

23. Van Eaton, *Harlow's Session Laws of 1935*, pp. 273–77; Turner, "Regulation of the Oklahoma Oil Industry," p. 256.

24. Turner, "Regulation of the Oklahoma Oil Industry," pp. 256–57; Interstate Oil Compact Commission, *Conservation of Oil and Gas*, pp. 8–9, Interview, Ray, May 17, 1978.

25. C. W. Van Eaton, ed., *Session Laws of 1935*, pp. 232–36; Eagin and Eagin, *Official Session Laws 1941*, pp. 215–20; State of Oklahoma, *Official Session Laws 1945*, pp. 155–72; Turner, "Regulation of the Oklahoma Oil Industry," pp. 260–70, 285–87.

26. Turner, "Regulation of the Oklahoma Oil Industry," p. 257; State of Oklahoma, *Official Session Laws 1947*, pp. 331–34; Blakely M. Murphy, ed., *Conservation of Oil & Gas: A Legal History, 1948*, pp. 269–422; Robert E. Sullivan, ed., *Conservation of Oil & Gas: A Legal History 1958*, pp. 3–4.

## Chapter 17

1. Litton, *History of Oklahoma*, vol. 2, p. 191; Jack T. Conn and William B. Osborn to Kenny A. Franks, OPC; Williamson et al., *American Petroleum Industry*, vol. 2, pp. 556–58.

2. Williamson et al., *American Petroleum Industry*, pp. 539, 558–61; Litton, *History of Oklahoma*, vol. 2, pp. 197–98.

3. Interview, Forest Brokaw, June 30, 1978, OPC; Williamson et al., *American Petroleum Industry*, vol. 2, pp. 612–15.

4. Williamson et al., *American Petroleum Industry*, pp. 613–17; Interview, Brokaw, June 30, 1978.

5. Interview, Brokaw, June 30, 1978; Williamson et al., *American Petroleum Industry*, vol. 2, pp. 618–19.

6. Williamson et al., *American Petroleum Industry*, pp. 565–66, 747–48; Guy H. Woodward and Grace Steele Woodward, *The Secret of Sherwood Forest: Oil Production in England During World War II*, pp. xi–xii; Litton, *History of Oklahoma*, vol. 2, p. 192.

7. Litton, *History of Oklahoma*, vol. 2, p. 192; Williamson et al., *American Petroleum Industry*, vol. 2, pp. 749–62.

8. Williamson et al., *American Petroleum Industry*, pp. 762–98; Litton, *History of Oklahoma*, vol. 2, pp. 192, 198–200.

9. Interview, Melvin B. Heine, August 16, 1960, Archives, PPC.

10. Woodward and Woodward, *Secret of Sherwood Forest*, pp. 12–14, 24–25.

11. *Ibid.*, pp. 24–28.
12. *Ibid.*, pp. 28–32.
13. *Ibid.*, pp. 34–39.
14. *Ibid.*, pp. 49–62.
15. *Ibid.*, pp. 63–94.
16. *Ibid.*, pp. 94–107.
17. *Ibid.*, pp. 107–11.
18. *Ibid.*, pp. 111–35.
19. *Ibid.*, pp. 135–81.
20. *Ibid.*, pp. 181–209.
21. *Ibid.*, pp. 209–31.
22. *Ibid.*, pp. 231–47.

23. Rister, *Oil!*, p. 340; *Oil and Gas Journal*, June 5, 1941; United States Department of the Interior, *Mineral Yearbook 1941*, p. 1044.

24. United States Department of the Interior, *Mineral Yearbook 1941*, pp. 1044–45; *Oil and Gas Journal*, June 5, 1941, June 12, 1941, and June 19, 1941; Rister, *Oil!*, p. 340.

25. Rister, *Oil!*, pp. 340–41; United States Department of the Interior, *Mineral Yearbook 1941*, pp. 1044–45; Litton, *History of Oklahoma*, vol. 2, pp. 191–92, 198.

26. Litton, *History of Oklahoma*, vol. 2, p. 191; Rister, *Oil!*, pp. 340–41.

27. Rister, *Oil!*, p. 341; Litton, *History of Oklahoma*, vol. 2, p. 191.

28. Litton, *History of Oklahoma*, vol. 2, p. 191; *Oil and Gas Journal*, April 15, 1943; *Daily Oklahoman*, April 6, 1943; Rister, *Oil!*, p. 341.

29. Rister, *Oil!*, p. 341; Litton, *History of Oklahoma*, vol. 2, p. 191; *Daily Oklahoman*, April 6, 1943, and April 16, 1943; *Edmond Sun*, May 6, 1943; *Oil and Gas Journal*, April 22, 1943, and April 29, 1943.

30. Williamson et al., *American Petroleum Industry*, vol. 2, p. 4; Interview, E. E. Young and John B. Noble, July 6, 1978, OPC.

31. Rister, *Oil!*, pp. 341–42; Litton, *History of Oklahoma*, vol. 2, pp. 191–93, 198.

32. Litton, *History of Oklahoma*, vol. 2, p. 192; *Daily Oklahoman*, April 14, 1943.

33. Litton, *History of Oklahoma*, vol. 2, p. 192; Rister, *Oil!*, p. 353; Turner, "Regulation of the Oklahoma Oil Industry," pp. 276–77.

34. Turner, "Regulation of the Oklahoma Oil Industry," pp. 272–77; Rister, *Oil!*, pp. 378–80; Litton, *History of Oklahoma*, vol. 2, p. 198.

35. Litton, *History of Oklahoma*, vol. 2, p. 191; Turner, "Regulation of the Oklahoma Oil Industry," pp. 273–82.

36. Interview, Donald Thompson, June 29, 1978, OPC.

37. Williamson et al., *American Petroleum Industry*, vol. 2, pp. 795–821; Litton, *History of Oklahoma*, vol. 2, pp. 191–92; *Daily Oklahoman*, February 24, 1955.

## Chapter 18

1. Arthur W. McCray and Frank W. Cole, *Oil Well Drilling Technology*, p. 3, 324–25; J. Stanley Clark, *The Oil Century*, pp. 18–28; Brooks to Franks, March 9, 1978.

2. Morris et al., *Historical Atlas of Oklahoma*, p. 16; Litton, *History of Oklahoma*, vol. 2, p. 200.

3. Litton, *History of Oklahoma*, vol. 2, p. 200; Thomas Nuttall, *A Journal of Travels into the Arkansas Territory During the Year 1819*, in Reuben Gold Thwaites, ed., *Early Western Travels, 1748–1846*, vol. 13; Morris et al., *Historical Atlas of Oklahoma*, p. 16.

4. Morris et al., *Historical Atlas of Oklahoma*, p. 17; United States House of Representatives, *Exploration of the Red River of Louisiana, in the Year 1852: By Randolph B. Marcy*, 33rd cong., 1st sess.; Litton, *History of Oklahoma*, vol. 2, pp. 200–201; Territory of Oklahoma, *Department of Geological and Natural History Second Biennial Report*, pp. 37–38.

5. Territory of Oklahoma, *Department of Geological and Natural History Second Biennial Report*, p. 38; Litton, *History of Oklahoma*, vol. 2, p. 201.

6. Litton, *History of Oklahoma*, vol. 2, pp. 201–202; Territory of Oklahoma, *Session Laws of 1899*, pp. 173–75; Territory of Oklahoma, *Department of Geological and Natural History Second Biennial Report*, p. 1.

7. Territory of Oklahoma, *Department of Geological and Natural History Second Biennial Report*, pp. 1–41; Litton, *History of Oklahoma*, vol. 2, pp. 200–202.

8. Litton, *History of Oklahoma*, vol. 2, p. 202; State of Oklahoma, *Session Laws of 1907–1908* (n.p., n.d.), pp. 431–33; Rister, *Oil!*, pp. 192–93.

9. Rister, *Oil!*, pp. 192–93; Litton, *History of Oklahoma*, vol. 2, pp. 202–203.

10. Litton, *History of Oklahoma*, vol. 2, pp. 202–203; Rister, *Oil!*, pp. 184, 192–94; Steiger to Franks, March 14, 1979.

11. The College Blue Book, *Degrees Offered by College and Subject*, pp. 498–99; James O. Kemm to Kenny A. Franks, May 24, 1978, OPC; James A. Clark, *The Chronological History of the Petroleum and Natural Gas Industries*, pp. 143, 158, 180, 258.

12. Williamson et al., *American Petroleum Industry*, vol. 2, pp. 14, 100; Steiger to Franks, March 14, 1978, and March 8, 1979.

13. James A. Clark, *Chronological History*, p. 112; Rister, *Oil!*, pp. 139, 205–206; Brooks to Franks, March 9, 1978.

14. Steiger to Franks, March 14, 1978.

15. *Ibid.*; Williamson et al., *American Petroleum Industry*, vol. 2, pp. 316–17, 326.

16. Steiger to Franks, March 14, 1978.

17. Interview, Conn, July 11, 1978; Interview, Payne, August 5, 1978; Interview, Urschel, September 20, 1978.

18. Steiger to Franks, March 8, 1979.

19. Brooks to Franks, March 9, 1978; James A. Clark, *Chronological History*, p. 114.

20. James A. Clark, *Chronological History*, pp. 105–77; Turner, "Regulation of the Oklahoma Oil Industry," pp. 65–107.

21. Williamson et al., *American Petroleum Industry*, vol. 2, pp. 554–55; James A. Clark, *Chronological History*, pp. 189–217.

22. James A. Clark, *Chronological History*, p. 137; Brooks to Franks, March 9, 1978; Clinton, "Petroleum Exposition," pp. 479–88.

23. Litton, *History of Oklahoma*, vol. 2, p. 198; James A. Clark, *Chronological History*, p. 93.

24. James A. Clark, *Chronological History*, pp. 97–142; Rister, *Oil!*, pp. 194–95.

25. Rister, *Oil!*, pp. 327, 384–85; McCray and Cole, *Oil Well Drilling Technology*, pp. 38–40; James A. Clark, *Chronological History*, pp. 116, 145.

26. James A. Clark, *Chronological History*, pp. 123–204; Brooks to Franks, March 9, 1978; McCray and Cole, *Oil Well Drilling Technology*, pp. 443–44.

27. McCray and Cole, *Oil Well Drilling Technology*, pp. 65–66; Brooks to Franks, March 9, 1978; James A. Clark, *Chronological History*, pp. 149–243.

28. James A. Clark, *Chronological History*, pp. 128–80; Williamson et al., *American Petroleum Industry*, vol. 2, pp. 90–101, 342.

29. J. Stanley Clark, *The Oil Century*, pp. 219–20; Brooks to Franks, March 9, 1978; Interview, McGee, August 31, 1978; James A. Clark, *Chronological History*, pp. 117–222.

30. James A. Clark, *Chronological History*, pp. 124–219; Williamson et al., *American Petroleum Industry*, vol. 2, pp. 631–32.

31. Brooks to Franks, March 9, 1978; James A. Clark, *Chronological History*, pp. 176–246.

32. James A. Clark, *Chronological History*, pp. 180–85; Williamson et al., *American Petroleum Industry*, vol. 2, pp. 550–51.

33. "Historical Markers Co-Sponsored by the Oklahoma Petroleum Council and the Oklahoma Historical Society," OPC; Interview, Kemm, March 14, 1979.

34. Brooks to Franks, March 9, 1978; Litton, *History of Oklahoma*, vol. 2, p. 172.

## Chapter 19

1. D. W. Calvert, "New Leaflet Tells Facts About Oil, Gas in State," January, 1977, OPC; Rex Hudson, "A Brief History of Halliburton Services, 1916–1977," OPC; Interview, Malcolm E. Rosser III, June 7, 1978, OPC; Interview, Rex Hudson, June 7, 1978, OPC; J. Evetts Haley, *Erle P. Halliburton: Genius with Cement*, pp. 9–15.

2. Haley, *Erle P. Halliburton*, pp. 1–11; Interview, Rosser, June 7, 1978; Hudson, "History of Halliburton Services," pp. 1–2.

3. Hudson, "History of Halliburton Services," pp. 2–4; Interview, Rosser, June 7, 1978; Haley, *Erle P. Halliburton*, pp. 11–16.

4. Haley, *Erle P. Halliburton*, pp. 15–20; Hudson, "History of Halliburton Services," pp. 4–6.

5. Hudson, "History of Halliburton Services," pp. 5–7; Interview, Rosser, June 7, 1978; Haley, *Erle P. Halliburton*, pp. 20–27.

6. Haley, *Erle P. Halliburton*, pp. 21–24; Interview, Rosser, June 7, 1978.

7. Interview, Rosser, June 7, 1978; Interview, Hudson, June 7, 1978; Hudson, "History of Halliburton Services," pp. 4–7; Haley, *Erle P. Halliburton*, pp. 33–35.

8. Haley, *Erle P. Halliburton*, pp. 37–39; Interview, Rosser, June 7, 1978; Interview, Hudson, June 7, 1978.

9. Interview, Hudson, June 7, 1978; Haley, *Erle P. Halliburton*, pp. 39–43; Interview, Rosser, June 7, 1978; Hudson, "History of Halliburton Services," pp. 21–34, 48.

10. Hudson, "History of Halliburton Services," p. 25; Halliburton Services, "Offshore Service Brochure," OPC; Halliburton Services, *Halliburton Offshore Systems Worldwide*, OPC; Interview, Hudson, June 7, 1978; Interview, Rosser, June 7, 1978.

11. Interview, Rosser, June 7, 1978; Halliburton Company, *1977 Annual Report*, OPC, p. 4; Hudson, "History of Halliburton Services," pp. 35–43.

12. "History, Organization and Officers of Reading & Bates," OPC; Interview, Lew Fitzgerald, June 21, 1978, OPC; ". . . and the Wolf Never Came Back!" clipping, OPC.

13. ". . . and the Wolf Never Came Back!"; Interview, Fitzgerald, June 21, 1978; "History, Organization and Officers of Reading & Bates."

14. "History, Organization and Officers of Reading & Bates"; ". . . and the Wolf Never Came Back!"; Interview, Fitzgerald, June 21, 1978.

15. Interview, Fitzgerald, June 21, 1978; Reading & Bates, *George M. Reading*, OPC; "History, Organization and Officers of Reading & Bates."

16. ". . . and the Wolf Never Came Back!"; Reading & Bates, *Rig Locator, May 1, 1978*, OPC.

17. Reading & Bates, *Rig Locator, May 1, 1978*; Reading & Bates, *1977 Annual Report*, OPC, p. 11; Reading & Bates, *W. D. Kent, ibid.*; Reading & Bates, *J. W. Bates, ibid.*; Reading & Bates, *Rig Locator, May 1, 1978*.

18. Reading & Bates, *Rig Locator, May 1, 1978*; Interview, Fitzgerald, June 21, 1978; Reading & Bates, *1977 Annual Report*; preceding p. 1.

19. Oklahoma Petroleum Council "Outstanding Oklahoma Oil Man Award, 1969," OPC; Interview P. C. Lauinger, June 21, 1978, *ibid.*; Gene T. Kinney, "The Journal's First 75 Years," *Petroleum 2000*, p. 527.

20. Kinney, "The Journal's First 75 Years," p. 527; Interview, Lauinger, June 21, 1978.

21. Interview, Lauinger, June 21, 1978; Kinney, "The Journal's First 75 Years," p. 527.

22. Kinney, "The Journal's First 75 Years," pp. 527–30; Interview, Lauinger, June 21, 1978.

23. Interview, Lauinger, June 21, 1978; Kinney, "The Journal's First 75 Years," p. 530.

24. Kinney, "The Journal's First 75 Years," p. 530; Interview, Lauinger, June 21, 1978.

25. Interview, Lauinger, June 21, 1978; Kinney, "The Journal's First 75 Years," p. 530.

26. Kinney, "The Journal's First 75 Years," p. 530; Oklahoma Petroleum Council, "Outstanding Oklahoma Oil Man Award, 1969"; Petroleum Publishing Company, *Books, 1978*, OPC; Kinney, "The Journal's First 75 Years," p. 530.

27. The Williams Companies, *1977 Annual Report*, OPC.

28. *Ibid.*

29. Calvert, "New Leaflet Tells Facts About Oil, Gas in State."

## Chapter 20

1. Calvert, "New Leaflet Tells Facts About Oil, Gas in State"; Litton, *History of Oklahoma*, vol. 2, p. 194; D. W. Calvert, "Oil, Gas Pay Big Portion of Total Taxes in State," October, 1977, OPC; Sloan K. Childers, "Thousands in State Own Shares in Big Oil Firms," February, 1975, OPC.

2. Oklahoma Petroleum Council, "The Petroleum Industry in Oklahoma," December, 1974, OPC.

3. *Ibid.*

4. Interview, John March, June 8, 1978, OPC; John March, "The Samuel Roberts Noble Foundation, Inc., October 4, 1974," OPC.

5. March, "The Samuel Roberts Noble Foundation, Inc."; Interview, March, June 8, 1978.

6. Interview, March, June 8, 1978; March, "The Samuel Roberts Noble Foundation, Inc."

7. March, "The Samuel Roberts Noble Foundation, Inc."; Interview, March, June 8, 1978.

8. Interview, March, June 8, 1978; March, "The Samuel Roberts Noble Foundation, Inc.," October 4, 1974.

9. March, "The Samuel Roberts Noble Foundation, Inc."; Interview, March, June 8, 1978.

10. Interview, March, June 8, 1978; March, "The Samuel Roberts Noble Foundation, Inc."

11. March, "The Samuel Roberts Noble Foundation, Inc."; The Samuel Roberts Noble Foundation, Inc., *Annual Report for 1977*, OPC.

12. Interview, Norman L. Crockett, June 12, 1978, OPC; Interview, Ward Merrick, Jr., June 8, 1978, OPC.

13. Interview, Ward Merrick, Jr., June 8, 1978; Ardmore Junior Chamber of Commerce, *The History of Carter County*, unpaged.

14. Interview, Walter Neustadt, Jr., June 8, 1978, OPC; Interview, Ivar Ivask, June 12, 1978, OPC; "Presentation of the Neustadt International Prize for Literature," OPC; "Third Revised Charter of the Neustadt International Prize for Literature," *World Literature Today*, vol. 51, no. 4 (Autumn, 1977), pp. 566–67.

15. Kirkpatrick Foundation, Inc., "Articles of Agreement, May 17, 1955," Kirkpatrick Collection, Oklahoma Heritage Association; Kirkpatrick Oil Company Board of Directors, "Minutes," Kirkpatrick Collection, Oklahoma Heritage Association, Kirkpatrick Foundation, Inc., *1955–1975 Report*, unpaged.

16. Kirkpatrick Foundation, Inc., *1955–1975 Report*; Kirkpatrick Oil Company Board of Directors, "Minutes."

17. Kirkpatrick Foundation, Inc., *1955–1975 Report*; Interview, Mary Jo Sturm, September 29, 1977, OPC; Kenny A. Franks, "Always Underway: A Biography of John and Eleanor Kirkpatrick," OPC.

18. Franks, "Always Underway"; Kirkpatrick Foundation, Inc., *1955–1975 Report*; Interview, Sturm, September 29, 1977.

19. Oklahoma Heritage Association, "Oklahoma Heritage House Tour Guide," OPC.

20. *Ibid*.

21. *Ibid*.; Interview, Paul F. Lambert, June 13, 1978, OPC.

22. Oklahoma Petroleum Collection, "Outstanding Oklahoma Oil Man Award, 1970," OPC; "A Place To See," OPC; *Daily Oklahoman*, October 26, 1974.

23. *Daily Oklahoman*, October 26, 1974; "A Place To See."

24. Marianna O. Lewis, ed., *The Foundation Directory*, pp. 427–28; Ina Hall, "W. K. Warren: A Profile of a Man Who Has Already Built His Monuments," *Tulsa* (May, 1973).

25. Hall, "W. K. Warren."

26. *Ibid*.

27. Lewis, *The Foundation Directory*, p. 424; "John Elmer Mabee," Hall of Fame File, Oklahoma Heritage Association; Newspaper Clippings, John Elmer Mabee Folder, OHA.

28. "Alfred E. Aaronson," Hall of Fame File, Oklahoma Heritage Association; "Script for Appreciation Dinner for Alfred E. Aaronson," OHA; Alfred E. Aaronson, "Keeping Gilcrease Museum for Tulsa," *American Scene, Gilcrease Museum Dedication Issue*, vol. 5, no. 2 (1963), pp. 4–5; David Randolph Milsten, "Tulsa Pride," in the same periodical, p. 17.

29. "Alfred E. Aaronson"; "Script for Appreciation Dinner for Alfred E. Aaronson"; "Alfred E. Aaronson Human Relations Collection, Tulsa City-County Library System," OPC.

30. "Waite Phillips," Hall of Fame File, Oklahoma Heritage Association; *Philnews*, November 28, 1939.

31. *Philnews*, November 28, 1939; *Oklahoma's Orbit* (Oklahoma City); "Waite Phillips"; "The Woolaroc Story," Vertical Files, OHS; *Tulsa Museums* (Tulsa: Metropolitan Tulsa Chamber of Commerce, n.d.).

32. Atlantic Richfield Company, *Participation II*, OPC; Atlantic Richfield Company, *Annual Report*, 1977, OPC, pp. 31–32; Interview, William J. Hamilton, June 21, 1978, *ibid*.

33. Interview, Robert Weppler, June 21, 1978, OPC; Interview, John Steiger, June 21, 1978, OPC; Interview, Doris Cobb, June 21, 1978, OPC; Cities Service Company, *Henry L. Doherty Educational Foundation*, OPC.

34. Interview, Sloan Ken Childers, June 20, 1978, OPC; Interview, Marshall, June 19, 1978; Continental Oil Company, *Conoco*, pp. 216–217.

35. Calvert, "Oil, Gas Pay Big Portion of Total Taxes in State"; Lewis, *The Foundation Directory*, pp. 423–28.

## Chapter 21

1. Franks, "Always Underway"; Oklahoma Petroleum Council, "Outstanding Oklahoma Oil Man Award, 1961," OPC.

2. Oklahoma Petroleum Council, "Outstanding Oklahoma Oil Man Award, 1961."

3. *Ibid*.

4. *Daily Oklahoman*, October 7, 1962; Oklahoma Petroleum Council, "Outstanding Oklahoma Oil Man Award, 1962," OPC.

5. Oklahoma Petroleum Council, "Outstanding Oklahoma Oil Man Award, 1962."

6. Oklahoma Petroleum Council, "Outstanding Oklahoma Oil Man Award, 1963," OPC.

7. *Ibid*.

8. Oklahoma Petroleum Council, "Outstanding Oklahoma Oil Man Award, 1964," OPC.

9. Oklahoma Petroleum Council, "Outstanding Oklahoma Oil Man Award, 1965," OPC.

10. *Ibid*.

11. Interview, Payne, August 5, 1978; Oklahoma Petroleum Council, "Outstanding Oklahoma Oil Man Award, 1966," OPC.

12. Oklahoma Petroleum Council, "Outstanding Oklahoma Oil Man Award, 1966."

13. Oklahoma Petroleum Council, "Outstanding Oklahoma Oil Man Award, 1967," OPC.

14. *Ibid*.

15. Oklahoma Petroleum Council, "Outstanding Oklahoma Oil Man Award, 1968," OPC.

16. *Ibid*.

17. Oklahoma Petroleum Council, "Outstanding Oklahoma Oil Man Award, 1969."

18. *Ibid*.

19. Oklahoma Petroleum Council, "Outstanding Oklahoma Oil Man Award, 1970," OPC.

20. *Ibid*.

21. Oklahoma Petroleum Council, "Outstanding Oklahoma Oil Man Award, 1971," OPC.

22. *Ibid*.

23. Oklahoma Petroleum Council, "Outstanding Oklahoma Oil Man Award, 1972," OPC.

24. *Ibid*.

25. Oklahoma Petroleum Council, "Outstanding Oklahoma Oil Man Award, 1973," OPC.

26. *Ibid*.

27. Oklahoma Petroleum Council, "Outstanding Oklahoma Oil Man Award, 1974," OPC; Interview, John E. Kirkpatrick, July 23, 1977, July 30, 1977, August 6, 1977, August 13, 1977, and October 26, 1977, Kirkpatrick Collection.

28. Interview, Kirkpatrick, July 23, 1977, July 30, 1977, August 6, 1977, August 13, 1977, and October 26, 1977; Oklahoma Petroleum Council, "Outstanding Oklahoma Oil Man Award, 1974."

29. Oklahoma Petroleum Council, "Outstanding Oklahoma Oil Man Award, 1975," OPC.

30. *Ibid*.

31. Oklahoma Petroleum Council, "Outstanding Oklahoma Oil Man Award, 1976," OPC.

32. *Ibid*.

33. Oklahoma Petroleum Council, "Outstanding Oklahoma Oil Man Award, 1977," OPC.

34. *Ibid*.

35. Oklahoma Petroleum Council, "Outstanding Okla-

homa Oil Man Award, 1978," OPC.

36. *Ibid.*

## Chapter 22

1. American Petroleum Institute, *Petroleum Facts and Figures, 1963*, p. 36; American Petroleum Institute, *Petroleum Facts and Figures, 1967*, pp. 44–45; Turner, "Regulation of the Oklahoma Oil Industry," pp. 273–80.

2. Turner, "Regulation of the Oklahoma Oil Industry," pp. 279–82; Litton, *History of Oklahoma*, vol. 2, p. 198; American Petroleum Institute, *Petroleum Facts and Figures, 1967*, p. 45; Morris et al., *Historical Atlas of Oklahoma*, p. 70; American Petroleum Institute, *Petroleum Facts and Figures, 1963*, p. 36.

3. American Petroleum Institute, *Petroleum Facts and Figures, 1963*, p. 12; Turner, "Regulation of the Oklahoma Oil Industry," pp. 280–92; American Petroleum Institute, *Petroleum Facts and Figures, 1967*, pp. 11, 15, 31, 71.

4. State of Oklahoma, *Official Session Laws 1945*, pp. 162–70; Sullivan, *Conservation of Oil & Gas* (1958), pp. 186–91.

5. Sloan K. Childers, "Energy Problem Growing as Oil Imports Increase," April, 1976, OPC; Sloan K. Childers, "Business Woes Caused by Federal Regulators," August, 1976, OPC; D. W. Calvert, "Nation More Vulnerable to Another Oil Embargo," October, 1976, OPC.

6. Sloan K. Childers, "5 Sooner Oil Fields Rank Among Top 100 in Nation," April–May, 1975, OPC; Sloan K. Childers, "Drilling Boom Puts State Second in Oil, Gas Search," June, 1975, OPC; Sloan K. Childers, "Oil Company Profits Drop While Capital Needs Grown," August, 1975, OPC; Sloan K. Childers, "Leading Oil Firms Report Big Earnings Drop in '75," March, 1975, OPC.

7. D. W. Calvert, "State Editors Point Way to Solving Energy Woes," March, 1977, OPC; D. W. Calvert, "Proved Oil Reserved Drop, Show Need for Incentives," June, 1977, OPC.

8. Calvert, "Proved Oil Reserved Drop, Show Need for Incentives," June, 1977, OPC.

9. Barbara Shirley, "An Interview with Ed A. Smith of Service Drilling Co. . . . on June 14, 1977," OPC; Interview, Thompson, June 29, 1978; Interview, Osborn, July 17, 1979; Interview, Smith, June 21, 1978.

10. Interview, Smith, June 21, 1978; *Oklahoma City Saturday Oklahoman and Times*, June 24, 1978; Shirley, "An Interview with Ed A. Smith."

11. Shirley, "An Interview with Ed A. Smith"; Interview, Duncan, July 13, 1978; Interview, Young and Noble, July 6, 1978; Interview, Beard, July 10, 1978; Interview, R. M. Tillman, June 19, 1978, OPC; Interview, Sloan K. Childers, June 20, 1978, OPC.

12. Interview, Sloan K. Childers, June 20, 1978; Interview, Osborn, July 17, 1978; Interview, Smith, June 21, 1978; Interview, Stanley Learned, June 9, 1978, OPC.

13. Interview, Beard, July 10, 1978; Interview, Conn, July 10, 1978.

14. Interview, Smith, June 21, 1978; Interview, Neustadt, June 8, 1978; Interview, T. H. McCasland, Sr., June 7, 1978, OPC.

15. Interview, David G. Campbell, August 15, 1978, OPC; "Birth of a Gas Field," *Chilton's Oil and Gas Energy*, vol. 2, no. 2 (February, 1976), p. 34; Interview, Kemm, August 14, 1979.

16. Interview, Payne, August 5, 1978; Interview, McCasland, June 7, 1978; Interview, Duncan, July 13, 1978; Interview, Childers, June 20, 1978; Interview, McGee, August 31, 1978; Interview, Jack Maurer, June 7, 1978; Frank B. Taylor to James O. Kemm, March 2, 1979, OPC; James O. Kemm to Kenny A. Franks, March 5, 1979, OPC.

17. Interview, Rosser, June 7, 1978; Interview, Young and Noble, July 6, 1978; Interview, Merrick, June 8, 1978; Interview, Smith, June 21, 1978.

18. Interview, Smith, June 21, 1978; Interview, McCasland, June 7, 1978; Interview, Duncan, July 13, 1978; Interview, Payne, August 5, 1978; Interview, Maurer, June 7, 1978; Interview, Collins, July 10, 1978.

19. Interview, Collins, July 10, 1978; Interview, McCasland, June 7, 1978; Interview, Neustadt, June 8, 1978; Interview, Duncan, July 13, 1978.

20. Interview, Payne, August 5, 1978; Interview, Conn, July 11, 1978; Interview, Urschel, September 20, 1978.

21. Interview, Neustadt, June 8, 1978; Interview, McCasland, June 7, 1978; Interview, Collins, July 10, 1978; Interview, Duncan, July 13, 1978; Interview, Merrick, June 8, 1978.

22. Interview, Merrick, June 8, 1978; Interview, Collins, July 10, 1978, Interview, Neustadt, June 8, 1978; Interview, Duncan, July 13, 1978; Interview, McCasland, June 7, 1978.

23. Interview, McCasland, June 7, 1978; Interview, Osborn, July 17, 1978; Interview, Learned, June 9, 1978; Interview, Smith, June 21, 1978; Interview, Neustadt, June 8, 1978; Interview, Childers, June 20, 1978; Interview, McGee, August 31, 1978.

24. Interview, Kemm, March 14, 1979; Kemm to Franks, March 13, 1979, OPC.

25. Kemm to Franks, March 13, 1979, OPC; Interview, Kemm, March 14, 1979.

26. Interview, Kemm, March 14, 1979; Kemm to Franks, March 31, 1979.

27. Interview, McGee, August 31, 1978; Litton, *History of Oklahoma*, vol. 2, pp. 172 197–200; Interview, Duncan, July 13, 1978; Interview, Conn, July 10, 1978.

# Bibliography

### Oklahoma Petroleum Collection

Much of the material used in this book was taken from the Oklahoma Petroleum Collection of the Oklahoma Heritage Association, a collection that contains many manuscripts describing the oil industry within the state. Materials from the collection used in preparing this book include: personal recollections from many of the actual participants; various reports and information published by oil companies operating throughout the region; several unpublished histories of petroleum firms; speeches of oil company representatives and executives; many company newsletters and magazines; various newspaper releases prepared by individual companies and the Oklahoma Petroleum Council; biographical sketches of many state oilmen; citations and awards presented to various firms and individuals; copies of the "Outstanding Oklahoma Oil Man Awards" presented annually by the Oklahoma Petroleum Council; newspaper clippings concerning the development of the state's petroleum industry; letters and correspondence acquired from various individuals connected with the Oklahoma oil community; material from many of the state foundations and institutions that owe their existence to the petroleum industry; and various statistical data. This is one of the most comprehensive collections dealing with the Oklahoma petroleum industry, and no research on the subject is complete without referring to this collection.

### Manuscript Material

Bartlesville, Oklahoma. Phillips Petroleum Company. Archives.
Norman, Oklahoma. University of Oklahoma. Western History Collections. Heydrick Collection.
Oklahoma City. Oklahoma Heritage Association. Hall of Fame File.
Oklahoma City. Oklahoma Heritage Association. Kirkpatrick Collection.
Oklahoma City, Oklahoma. Oklahoma Historical Society. Indian Archives.
Oklahoma City, Oklahoma. Oklahoma Historical Society. Library Vertical Files.
Oklahoma City, Oklahoma. Oklahoma Historical Society. Museum.
Oklahoma City, Oklahoma. Oklahoma Historical Society. Oil in Oklahoma Series.
Oklahoma City, Oklahoma. Oklahoma State Archives. Roy M. McClintock Collection.

### Interviews

Archives. Phillips Petroleum Company.
  Alden, R. C. June 15, 1959.
  Barber, William M. July 1, 1959.
  Cordell, Oscar L. November 17, 1960.
  Dimit, Charles P. September 9, 1960.
  Heine, Melvin B. August 16, 1960.
  Koopman, H. E. and Heine, Melvin B. August 12, 1960.
  Peters, V. D. November 17, 1960.
  Sommerville, O. S. May 7, 1963.
  Stewart, Ralph L. August 1, 1960.
  Traywick, Earnest E. July 20, 1960.
  Trower, Harry A. September 9, 1960.
  Williams, J. Sam. May 7, 1963.
"Indian-Pioneer History." Grant Foreman, ed., 113 vols. Indian Archives. Oklahoma Historical Society.
  Fugate, Andrew Jackson. June 18, 1937.
  Rider, Louisa. April 26, 1938.
Kirkpatrick Collection. Oklahoma Heritage Association.
  Kirkpatrick, John E. July 23, 1977, July 30, 1977, August 6, 1977, August 13, 1977, and October 26, 1977.
  Sturm, Mary Jo. September 29, 1977.
Oklahoma Petroleum Collection. Oklahoma Heritage Association.
  Beard, William M. July 10, 1978.
  Beecher, Charles E. June 20, 1978.
  Broakaw, Forest. June 30, 1978.
  Campbell, David G. August 15, 1978.
  Childers, Sloan K. June 20, 1978.
  Cobb, Doris, June 21, 1978.

Collins, Robert J. July 10, 1978.
Conn, Jack T. July 11, 1978.
Crockett, Norman L. June 12, 1978.
Duncan, Jack. July 13, 1978.
Endacott, Paul. June 20, 1978.
Fitzgerald, Lew. June 21, 1978.
Hamilton, William J. June 21, 1978.
Hudson, Rex. June 7, 1978.
Ivask, Ivar. June 12, 1978.
Johnson, Louise Straight. June 20, 1978.
Kemm, James O. March 14, 1979, and April 28, 1980.
Lambert, Paul F. June 13, 1978.
Lauinger, P. C. June 21, 1978.
Learned, Stanley. July 9, 1978.
McCasland, T. H. June 7, 1978.
McGee, Dean A. August 31, 1978.
March, John. June 8, 1978.
Marshall, Keely. June 19, 1978.
Maurer, Jack. June 7, 1978.
Merrick, Ward, Jr. June 8, 1978.
Neustadt, Walter, Jr. June 8, 1978.
Osborn, William B., Jr. July 17, 1978.
Payne, William T. August 5, 1978.
Ray, Margaret, May 17, 1978.
Rosser, Malcolm, III. June 7, 1978.
Smith, Edward A. June 14, 1977 and July 21, 1978.
Steiger, John. June 21, 1978.
Thompson, Donald. June 29, 1978.
Tillman, R. N. June 19, 1978.
Urschel, Charles. September 20, 1978.
Weppler, Robert. June 21, 1978.
Young, E. E. and Noble, John B. July 6, 1978.

## Theses and Dissertations

DeBerry, Drue Lemuel. "The Ethos of the Oklahoma Oil Boom Frontier, 1905–1929." Master's thesis, University of Oklahoma, 1970.

Forbes, Charles G. "The Origin and Early Development of the Oil Industry in Oklahoma." Ph.D. dissertation, University of Oklahoma, 1939.

Frazier, J. Vere, Jr. "History of the Glennpool Oil Field." Master's thesis, University of Oklahoma, 1951.

Gilbert, James L. "Three Sands: Oklahoma Oil Field and Community of the 1920's." Master's thesis, University of Oklahoma, 1967.

Powell, Richard Hays. "The Oil Industry and the Depression from the Development of Greater Seminole Through the Passage of the Oil Code." Master's thesis, University of Oklahoma, 1968.

Robinson, Gilbert L. "History of the Healdton Oil Field." Master's thesis, University of Oklahoma, 1937.

Turner, Alvin O. "The Regulation of the Oklahoma Oil Industry." Ph.D. dissertation, Oklahoma State University, 1977.

## Newspapers

*Bartlesville Enterprise*. Bartlesville, Oklahoma.
*County Democrat*. Tecumseh, Oklahoma.
*Cushing Democrat*. Cushing, Oklahoma.
*Cushing Independent*. Cushing, Oklahoma.
*Cycler*. Enid, Oklahoma.
*Daily Ardmoreite*. Ardmore, Oklahoma.
*Daily Chieftain*. Vinita, Oklahoma.
*Daily Oklahoman*. Oklahoma City, Oklahoma.
*Edmond Sun*. Edmond, Oklahoma.
*Harlow's Weekly*. Oklahoma City, Oklahoma.
*Kansas City Times*. Kansas City, Missouri.
*Mounds Enterprise*. Mounds, Oklahoma.
*Muskogee Weekly Phoenix*. Muskogee, Oklahoma.
*Oil and Gas Journal*. Tulsa, Oklahoma.
*Oil Investor's Journal*. Tulsa, Oklahoma.
*Oklahoma State Capital*. Guthrie, Oklahoma.
*Oklahoma's Orbit*. Oklahoma City, Oklahoma.
*Osage Journal*. Pawhuska, Oklahoma.
*Philnews*. Bartlesville, Oklahoma.
*Saturday Oklahoman and Times*. Oklahoma City, Oklahoma.
*Seminole County News*. Seminole, Oklahoma.
*Shawnee Weekly Herald*. Shawnee, Oklahoma.
*The Triangle*. Cleveland, Oklahoma.
*Tonkawa News*. Tonkawa, Oklahoma.
*Tulsa Daily World*. Tulsa, Oklahoma.
*Tulsa Democrat*. Tulsa, Oklahoma.
*Tulsa Tribune*. Tulsa, Oklahoma.
*Wewoka Capital-Democrat*. Wewoka, Oklahoma.

## Published Materials

Aaronson, Alfred E. "Keeping Gilcrease Museum for Tulsa." *American Scene, Gilcrease Museum Dedication Issue*, vol. 5, no. 2 (1963), pp. 4–5.

Allhands, James L. "History of the Construction of the Frisco Railway Lines in Oklahoma." *Chronicles of Oklahoma*, vol. 3, no. 3 (September, 1925), pp. 229–239.

American Association of Petroleum Geologists. *Trek of the Oil Finders: A History of Exploration for Petroleum*. Tulsa: American Association of Petroleum Geologists, 1975.

American Petroleum Institute. *Petroleum Facts and Figures*. New York: J. J. Little and Ives Company, 1928.

———. *Petroleum Facts and Figures*. Baltimore: Lord Baltimore Press, 1931.

———. *Petroleum Facts and Figures, 1963*. New York: American Petroleum Institute, 1963.

———. *Petroleum Facts and Figures, 1967*. New York: American Petroleum Institute, 1967.

Ardmore Junior Chamber of Commerce. *The History of Carter County*. Fort Worth: University Supply and Equipment Company, 1957.

Ball, Max W.; Ball, Douglas; and Turner, Daniel S. *This Fascinating Oil Business*. New York: Bobbs-Merrill, 1965.

Barnes, Cassius M. *Report of the Governor of Oklahoma to the Secretary of the Interior for the Fiscal Year Ended June 30, 1898*. Washington: Government Printing Office, 1898.

———. *Annual Report of the Governor of Oklahoma to the Secretary of the Interior for the Fiscal Year Ended June 30, 1988*. Washington: Government Printing Office, 1899.

Bass, Henry B. "Herbert Hiram Champlin." *Chronicles of Oklahoma*, vol. 33, no. 1 (Spring, 1955), pp. 43–48.

Beckwith, H. T. "Osage County." *Oklahoma Geological Survey Bulletin No. 40*, vol. 3 (July, 1930), pp. 211–67.

"Birth of a Gas Field." *Chilton's Oil and Gas Energy*, vol. 2, no. 2 (February, 1976), pp. 33–34.

Bowles, Charles E. "Oklahoma Petroleum: an Industrial Survey." *Oklahoma Geological Survey Bulletin No. 40*, vol. 1 (July, 1930), pp. 79–99.

Boyle, J. Phillip. "Hughes County." *Oklahoma Geological Survey Bulletin No. 40*, vol. 3 (July, 1930), pp. 611–27.

Burwell, Kate Pearson. "The Richest People in the World." *Sturm's Statehood Magazine*, vol. 2, no. 4 (June, 1906), pp. 88–96.

"Carter County." *Oklahoma Geological Survey Bulletin No. 19*, pt. 2 (April, 1917), pp. 67–102.

Clark, G. C. and Cooper, C. L. "Kay, Grant, Garfield and Noble Counties." *Oklahoma Geological Survey Bulletin No. 40*, vol. 2 (July, 1930), pp. 67–105.

Clark, J. Stanley. *The Oil Century: From the Drake Well to the Conservation Era*. Norman: University of Oklahoma Press, 1958.

Clark, Blue. "The Beginning of Oil and Gas Conservation in Oklahoma, 1907–1931." *Chronicles of Oklahoma*, vol. 55, no. 4 (Winter, 1977–1978), pp. 375–91.

Clark, James A. *The Chronological History of the Petroleum and Natural Gas Industries*. Houston: Clark Book Company, 1963.

Clinton, Fred S. "The Beginning of the International Petroleum Exposition and Congress." *Chronicles of Oklahoma*, vol. 26, no. 4 (Winter, 1948–1949), pp. 479–88.

———. "First Oil and Gas Well in Tulsa County." *Chronicles of Oklahoma*, vol. 30, no. 3 (Autumn, 1952), pp. 312–32.

*College Blue Book. Degrees Offered by College and Subject*. New York: Macmillan Co., 1977.

Continental Oil Company. *Conoco: The First One Hundred Years*. New York: Dell Publishing Company, 1975.

Corden, Seth K. and Richards, W. B., comps. *The Oklahoma Red Book*. 2 vols. Oklahoma City: Democrat Printing Company, 1912.

"Creek County." *Oklahoma Geological Survey Bulletin No. 19*, pt. 2 (April, 1917), pp. 150–92.

Douglas, Clarence B. *The History of Tulsa, Oklahoma: A City with a Personality*. 3 vols. Chicago: S. J. Clarke Publishing Company, 1921.

Eagin, Frank and Eagin, Everett, eds. *Official Session Laws 1941*. Guthrie: Co-Operative Publishing Company, 1941.

Emerson, Earle E. *Playing My Part*. Oklahoma City: Quintella Printing Company, 1978.

Ferguson, Thompson B. *Annual Report of the Governor of Oklahoma to the Secretary of the Interior for the Fiscal Year Ended June 30, 1902*. Washington: Government Printing Office, 1902.

———. *Report of the Governor of Oklahoma to the Secretary of the Interior, 1904*. Washington: Government Printing Office, 1904.

———. *Report of the Governor of Oklahoma to the Secretary of the Interior for the Year Ended June 30, 1905*. Washington: Government Printing Office, 1905.

Finney, Frank F., Sr. "The Indian Territory Illuminating Oil Company." *Chronicles of Oklahoma*, vol. 37, no. 2 (Summer, 1959), pp. 149–61.

Fischer, LeRoy H., ed. *Territorial Governors of Oklahoma*. Oklahoma City: Oklahoma Historical Society, 1975.

Forbes, Gerald, "Southwestern Oil Boom Towns." *Chronicles of Oklahoma*, vol. 17, no. 4 (December, 1939), pp. 393–400.

———. "History of the Osage Blanket Lease." *Chronicles of Oklahoma*, vol. 19, no. 1 (March, 1941), pp. 70–81.

Frantz, Frank. *Report of the Governor of Oklahoma to the Secretary of the Interior for the Year Ended June 30, 1906*. Washington: Government Printing Office, 1906.

Glasscock, Carl B. *Then Came Oil: The Story of the Last Frontier*. New York: Bobbs-Merrill, 1938.

Goetzmann, William H. *Army Exploration in the American West, 1803–1863*. New Haven: Yale University Press, 1959.

Gregory, Robert. *Oil in Oklahoma*. Muskogee: Leake Industries, Inc., 1976.

Haley, J. Evetts. *Erle P. Halliburton: Genius with Cement*. Duncan: Halliburton Services, 1959.

Hall, Ina, "W. K. Warren: A Profile of a Man Who Has Already Built His Monuments." *Tulsa* (May, 1973), pp. 74–78.

Hill, Edward Allison. *Story of the Mid-Continent Oil and Gas Field*. Tulsa: Burkhart Printing and Stationary Company, 1914.

Hofsommer, Donovan L., ed. *Railroads in Oklahoma*. Oklahoma City: Oklahoma Historical Society, 1977.

Hutchinson, L. L. "Preliminary Report on the Rock Asphalt, Asphaltite, Petroleum and Natural Gas in Oklahoma." *Oklahoma Geological Survey Bulletin No. 2* (March, 1911), pp. 1–256.

Interstate Oil Compact Commission. *A Study of Conservation of Oil and Gas in the United States*. Oklahoma City: Interstate Oil Compact Commission, 1965.

Ironside, Roberta. *An Adventure Called Skelly: A History of Skelly Oil Company Through Fifty Years 1919–1969*. New York: Appleton-Century-Crofts, 1970.

Kappler, Charles J. *Indian Affairs: Laws and Treaties*. 5 vols. Washington: Government Printing Office, 1904–1941.

Kelder, Bill, ed. *A History of Apco Oil Corporation and Its Predecessor Company Anderson-Prichard Oil Corporation*. Oklahoma City: Apco Oil Corporation, 1977.

Killian, D. P. "Henry Vernon Foster, 1875–1939." *Chronicles of Oklahoma*, vol. 20, no. 4 (December, 1942), pp. 441–43.

Kinney, Gene T. "The Journal's First 75 Years." *Petroleum 2000*. Tulsa: The Petroleum Publishing Company, 1977, pp. 527–30.

Kirkpatrick Foundation, Inc. *1955–1975 Report*. Oklahoma City, n.d.
Levorsen, A. I. "Geology of Seminole County." *Oklahoma Geological Survey Bulletin No. 40*, vol. 3 (July, 1930), pp. 289–353.
Lewis, Marianna O., ed. *The Foundation Directory*. New York: The Foundation Center, 1977.
Litton, Gaston, *History of Oklahoma at the Golden Anniversary of Statehood*. 4 vols. New York: Lewis Historical Publishing Company, Inc., 1957.
Logsdon, William G. *The University of Tulsa*. Norman: Oklahoma Heritage Association and University of Oklahoma Press, 1977.
McCray, Arthur W. and Cole, Frank W. *Oil Well Drilling Technology*. Norman: University of Oklahoma Press, 1959.
Mathews, John J. *The Osages: Children of the Middle Waters*. Norman: University of Oklahoma Press, 1961.
Mills-Bullard, Bess, comp. "Digest of Oklahoma Oil and Gas Fields." *Oklahoma Geological Survey Bulletin No. 40*, vol. 1 (July, 1928), pp. 101–276.
Milsten, David Randolph. "Tulsa Pride." *American Scene Magazine, Gilcrease Museum Dedication Issue*, vol. 5, no. 2 (1963), p. 17.
Miner, H. Craig. *The Corporation and the Indian*. Columbia: University of Missouri Press, 1976.
Morris, John W.; Goins, Charles R.; and McReynolds, Edwin C. *Historical Atlas of Oklahoma*. Norman: University of Oklahoma Press, 1977.
Murphy, Blakely M., ed. *Conservation of Oil & Gas: A Legal History, 1948*. Chicago: American Bar Association, Section of Mineral Law, 1949.
Murray, William H. *Memoirs of Governor Murray and True History of Oklahoma*. 3 vols. Boston: Meador Publishing Company, 1945.
Nuttall, Thomas. *A Journal of Travels into the Arkansa Territory During the Year 1819*. Early Western Travels, 1748–1846, edited by Reuben Gold Thwaites, vol. 13. 32 vols. Chicago: Arthur H. Clark Company, 1904–1907.
Oklahoma Corporation Commission. *Seventh Annual Report*. Guthrie: Co-Operative Publishing Company, 1914.
———. *Eighth and Ninth Annual Reports*. Oklahoma City: Harlow Publishing Company, 1916.
———. *Tenth Annual Report*. Oklahoma City: Warden Company, 1917.
"Oklahoma County." *Oklahoma Geological Survey Bulletin No. 19*, pt. 2 (April, 1917), pp. 361–66.
Rister, Carl C. *Oil! Titan of the Southwest*. Norman, University of Oklahoma Press, 1949.
Shirk, George H. *Oklahoma Place Names*. Norman, University of Oklahoma Press, 1974.
Sinclair Oil and Gas Company. *A Great Name in Oil: Sinclair Through Fifty Years*. New York: McGraw-Hill, 1966.
Snider, L. C. *Oil and Gas in the Mid-Continent Fields*. Oklahoma City: Harlow Publishing Company, 1920.
Snyder, Henry G., comp. *The Compiled Laws of Oklahoma 1909*. Kansas City: Pipes-Reed Book Company, 1909.
State of Oklahoma. *Directory of Oklahoma, 1975*. Oklahoma City: Impress Graphics, Inc., 1975.
———. *Emergency Laws Passed by First Legislature State of Oklahoma 1907–1908*. Ardmore: The Ardmoreite Press, n.d.
———. *Session Laws of 1907–1908*. N.p., n.d.
———. *Session Laws of 1909*. Oklahoma City: Oklahoma Engraving and Printing Company, 1909.
———. *Session Laws of 1913*. Guthrie: Co-Operative Publishing Company, 1913.
———. *Session Laws of 1915*. Oklahoma City: Warden Printing and Publishing Company, 1915.
———. *Official Session Laws 1933*. Guthrie: Co-Operative Publishing Company, 1933.
———. *Official Session Laws 1945*. Guthrie: Co-Operative Publishing Company, 1945.
———. *Official Session Laws 1947*. Guthrie: Co-Operative Publishing Company, 1947.
Stewart, Roy P. and Woods, Pendleton. *Born Grown: An Oklahoma City History*. Oklahoma City: Metro Press, 1974.
Sullivan, Robert E., ed. *Conservation of Oil & Gas: A Legal History 1958*. Chicago: American Bar Association, Section of Mineral and Natural Resources Law, 1960.
Teague, Margaret W. *History of Washington County and Surrounding Area*. 2 vols. Bartlesville: Bartlesville Historical Commission, 1968.
Territory of Oklahoma. *Department of Geological and Natural History Second Biennial Report*. Guthrie: The State Capital Company, 1902.
———. *Session Laws of 1895*. N.p., 1895.
———. *Session Laws of 1899*. Guthrie: State Capital Printing Company, 1899.
———. *Session Laws of 1903*. Guthrie: The State Capital Company, 1903.
———. *Session Laws of 1905*. Guthrie: The State Capital Company, 1905.
"Third Revised Charter of the Neustadt International Prize for Literature." *World Literature Today*, vol. 51, no. 4 (Autumn, 1977), pp. 566–67.
Tinker, George E. "Who Gets the Oil?" *The Osage Magazine*, vol. 1 (May, 1910), pp. 53–60.
Tomlinson, D. W. "Carter County." *Oklahoma Geological Survey Bulletin No. 40*, vol. 2 (July, 1930), pp. 284–310.
Travis, A. "Oklahoma County." *Oklahoma Geological Survey Bulletin No. 40*, vol. 2 (July, 1930), pp. 433–61.
*Tulsa Museums*. Tulsa: Metropolitan Tulsa Chamber of Commerce, n.d.
Tyson, Carl N.; Thomas, James H.; and Faulk, Odie B. *The McMan: The Lives of Robert M. McFarlin and James A. Chapman*. Norman: Oklahoma Heritage Association and University of Oklahoma Press, 1977.
United States Department of the Interior. *Mineral Yearbook 1941*. Washington: Government Printing Office, 1943.
———. *Regulations to Govern the Leasing of Lands in the Osage Reservation, Oklahoma, for Oil and Gas Mining Purposes*. Washington: Government Printing Office, 1912.
———. *Report, 1901, Indian Affairs*. Washington: Government Printing Office, 1902.
———. *Reports of the Department of the Interior for the Fiscal Year*

*Ended June 30, 1915, Administrative Reports.* 2 vols. Washington: Government Printing Office, 1916.

———. *Report of the Commissioner of Indian Affairs for the Fiscal Year Ended June 30, 1922.* Washington: Government Printing Office, 1922.

United States Geological Survey. *Mineral Resources of the United States, Part II, Nonmetals, 1906–1915.* Washington: Government Printing Office, 1923.

United States House of Representatives. *Exploration of the Red River of Louisiana, in the Year 1852: By Randolph B. Marcy.* 33rd Congress, 1st Session. Washington: A. O. P. Nicholson, 1854.

Van Eaton, E. W., ed. *Harlow's Session Laws of 1935.* Oklahoma City: Harlow Publishing Corporation, 1935.

———. *Session Laws of 1935.* Oklahoma City: Harlow Publishing Corporation, 1935.

Weirich, T. E. "Pottawatomie County." *Oklahoma Geological Survey Bulletin No. 40*, vol. 3 (July, 1930), pp. 587–99.

Williamson, Harold F.; Andreano, Ralph L.; Daum, Arnold R.; and Klose, Gilbert C. *The American Petroleum Industry.* 2 vols. Evanston, Illinois: Northwestern University Press, 1959.

Woodward, Guy H. and Woodward, Grace Steele. *The Secret of Sherwood Forest: Oil Production in England During World War II.* Norman: University of Oklahoma Press, 1973.

Wright, Muriel H. "First Oklahoma Oil Was Produced in 1859." *Chronicles of Oklahoma*, vol. 4, no. 4 (December, 1926), pp. 322–28.

# Index

Aaronson, Alfred E.: 226
A&P Railroad: *see* Atlantic and Pacific Railroad
A. E. Denny No. 1: 105
AAPG: *see* American Association of Petroleum Geologists
Abernathy, Jack H.: 238
Acetylene welding, development of: 203
Ada, Okla.: 3, 183, 246
Adams, H. H.: 27
Adams, K. S. ("Boots"): 231
Adams, Lewis: 28
Adams, Thomas J.: 28
Adams, Wash: 28
Adams, William A.: 10
Agrico Chemical Company: 216
Alabama, state of: 181, 216
Alaska, state of: 107, 181, 206
Alcord Oil Company: 111
Alden, R. C.: 38
Allen Dome field: 124
Allen field: 115
Allied Steel Products Corporation: 236
Allred, James V.: 176
Almeda field: 64
Almeda Oil Company: 23, 58
Alum Bluff, Choctaw Nation: 9
Amarillo, Texas: 128
Amerada Oil Company: 93, 119
Amerada Petroleum Company: 135
American Association of Petroleum Geologists (AAPG): 200, 206
American Iron and Machine Company: 136
American Petroleum Institute (API): 175–76
American Pipe Line Company, 63–64
*American Saturday Night:* 151
Ames, C. B.: 127
Amoco Research Center: 248
Amos-B No. 2: 124
Anadarko Basin: 243–44, 247
Anadarko Shelf: 243
Anderson, Frank: 129
Anderson, G. E.: 128
Anderson, J. Steve: 114

Apco Refining Company: 114
API: *see* American Petroleum Institute
Apple, Sam A.: 80, 229
Arab oil embargo: 240, 247
Arbuckle Fault: 183
Arbuckle Limestone: 129
Arbuckle Mountains: 195
Ardmore Independent Oil Producers Association: 82
Ardmore, Okla.: 78, 81, 219, 221–22, 228, 232
Ardmore Producers Association: 141
Ardmore Refinery: 141
Ardmore Refining Company: 82
Arizona, state of: 107, 181, 209, 216
Arkansas, state of: 107, 154, 176, 180, 216–17
Arkoma Basin: 243
Armstrong, L. D.: 151
Armstrong, Marvin: 127
Armstrong, Walt: 35
Arutunoff, Armais: 205
Ash Can No. 1: 132
Ash Can No. 2: 132
Asher field: 115
Asphalt Oil, Mining, and Manufacturing Company: 36
Asphaltus Spring, Okla.: 35
Associated Producers Company: 45
Atchison, Topeka, and Santa Fe Railroad Company (AT&SF): 23, 79–80
Atlantic and Pacific Railroad: 23
Atlantic field: 104
Atlantic Oil Company: 208
Atlantic Oil Producing Company: 104
Atlantic Petroleum Company: 109
Atlantic Richfield Company: 228, 248
Atlantic Richfield Foundation: 228
Atoka Agreement: 21
Atoka, Okla.: 9–10
Augusta field, Kans.: 199
Automobile, influence on the oil industry: 23–25
Avant field: 58
Aylesworth, Allison: 28

Bailey, Harvey: 96
Bain, H. Foster: 195
Baker and Strawn Oil Company: 105
Baker, Clyde: 209
Bald Hill field: 47
Baldo, Lucio: 153
Bale, Hubert E.: 236
B&B Drilling Company: 209
Baptist Medical Center Burn Center: 228
Barber, William M.: 92
Barkley, Sam: 92, 97
Barnes, Cassius M.: 25
Barnsdall Corporation: 125
Barnsdall field: 63
Barnsdall Oil Company: 58–61, 188
Barnsdall, T. N.: 59, 230
Barrett, Charles F.: 147
Barrow, Claude V.: 127, 130
Bartles, Jacob: 16
Bartlesville Energy Technology Center: 206
Bartlesville field: 203
Bartlesville, Okla.: 16, 18, 22, 25, 36, 57–58, 64, 92, 105, 138, 200, 203, 206, 227–28, 230–31, 233, 248
Bartlesville Research Center: 145
Bartlesville Sand: 15, 21, 76–77, 101
Bates, Albert L.: 96
Bates, J. W., Jr.: 209–10
Bates, J. W., Sr.: 209, 211
Beard, J. G.: 89
Beard Oil Company: 89
Beard, William M.: 90
Beaumont, Texas: 35, 212, 236
Beck, Henry: 95
Beebe field: 115
Beecher, Charles E.: 145, 147
Beekly, A. L.: 151
Beggs, George: 109
Beggs, Okla.: 49
Bell City Lime: 202
Bell, John R.: 194
Berryhill, D. L.: 28
Berryhill, Earl: 43
Bethel field: 115, 120
Betsy Foster No. 1: 115–18

275

Better Method Oil Well Cementing Company: 207; *see also* Halliburton Services
B.F. Walker, Inc.: 219
Big Chief Drilling Company: 95, 231, 238
Bigheart, James: 14
Billings field: 107
Bird Creek–Flat Rock field: 60
Bishop's Alley, in Seminole, Okla.: 95
Blackman, Carl: 151
Blackwell field: 107
Blackwell Oil and Gas Company: 111
Blackwell, Okla.: 35, 108
Blake, Isaac Elder: 107
Bland, C. W.: 27–29
Bland, Sue A.: 26–27, 29, 30
Bline, C. J.: 147
Bloom, C. L.: 15, 18, 20–21
Bodine City, Okla.: 130
Boiler Makers Union: 75
Bolene Refining Company: 88
Bolene, Victor: 109
Bonel Rye: 220
*Books Abroad* International Prize for Literature: 221
Boomtowns: 89–100
Boone, Nathan: 194
Borden, O. V.: 151
Borger, Texas: 39, 205
Boston field: 58
Bourland, B. D.: 189
Bourne, A. F.: 151
Bourque, A. V.: 151–52
Bovaird, D. D.: 235–36
Bovaird, Inc.: 235
Bovaird, William J.: 235
Bowery, The, in Kiefer, Okla.: 95
Bowlegs field: 115
Boyd Springs, Okla.: 3
Boyle, Patrick C.: 213
Boyles, Axtell J.: 175
Braman field: 114, 246
Brazil: 211, 216
Breckenridge, Clifton R.: 30
Brinton field: 49
Briody, James: 156
Bristow, Okla.: 68, 93
Brock No. 2: 42
Brown, Ann Noble: 219
Brown, E. E.: 139
Brown, J. F.: 156
Brown Refining Company: 77
Bruner No. 1: 119
Bryan County, Okla.: 3
Bryant, F. W.: 151
Buchanan, D. E.: 151
Buel, Joseph W.: 10
Bunker, C. D.: 195
Bunn, John R.: 128
Burbank field: 67, 102–104, 106–107, 204
Burbank, Okla.: 102

Burbank Sand: 104
Burgess Sand: 161
Burkburnett field, Texas: 207
Burns, Buckner: 5
Burns, Joe: 47
Burton, George E.: 202
Buthram, Frank: 196
Bushyhead, Dennis W.: 8
Butler, Pa.: 12, 28
Byrd, Edward: 10, 12

Cabot Shops: 205
Caddo County, Okla.: 196
Caddo Petroleum Company: 114
Calbert, D. W.: 241
California, state of: 26, 30, 66, 100, 107, 125, 174, 176, 180, 206–207, 209, 216
Cameron, E. D.: 196
Campbell, David G.: 243
Campbell, W. E.: 43
Campion, John: 151
Canada: 107, 114, 209, 219
Caney River: 15–16, 18
Cantilever derrick, development of: 205
Carey, William V.: 16
Carmichael Horizon: 12
Carpenter, Everett: 59, 128, 196, 198–99
Carr City field: 115
Carter County, Okla.: 78, 105, 189, 246
Carter Oil Company: 64, 94, 101–103, 105, 203
Cass, Jennie: 18
Castlebury No. 1: 193
Caton, Low: 47
Cement field: 114
Cement, Okla.: 196
Champlin, Herbert H.: 109, 168
Champlin, Joe: 229
Champlin Petroleum Company: 248
Champlin Refining Company: 109, 162–62, 168, 190
Chandler, Okla.: 35
Chanute Refining Company: 79
Chapman, James A.: 229
Chattanooga Horizon: 112
Chelsea, Okla.: 5, 10
Chelsea Refining Company: 77
Cherokee Indians: 3–18, 23, 49, 51, 60, 64, 235
Cherokee Lease: 10, 12
Chesley, Frank: 41–42
Chicago, Ill.: 69, 126, 155, 175, 189, 232
Chickasaw Indians: 3–16, 21, 49, 52, 79–80, 229
Chickasaw Oil Company: 5
Choctaw Indians: 3–16, 21, 23, 49, 52
Choctaw Oil and Refining Company: 8, 10
Cimarron Pipe Line Company: 109, 191
Cities Service Foundation: 228
Cities Service Oil Company: 59–60, 104, 145, 191, 198–200, 228–30
Cities Service Pipe Line Company: 229
Civil War: 5, 195
Cleveland County, Okla.: 128, 134
Cleveland Oil, Gas and Manufacturing Company: 40
Cleveland field: 35, 59
Cleveland, Okla.: 35, 40, 53
Cleveland Sand: 77
Clifton, Okla.: 79
Clinton, Fred S.: 27–28, 151
Coalton field: 47
Cobb, Guy P.: 31, 33
Cochran, Ad D.: 193
Cochrane, Jesse: 16
Cochran Springs, Cherokee Nation: 5
Coe, R. W.: 85
Colbert, Winchester: 5
Colcord, Charles F.: 30, 139
Coline Oil Company: 84, 128, 132
Colline, Dr. Robert J.: 246
Colonel Robinson's Hotel, in Red Fork, Okla.: 27
Colonial Refining Company: 77
Colorado, state of: 30, 99, 173, 176, 180
Columbia Drilling Machine: 79
Comar Oil Company: 103, 111, 246
Combs, Evelyn Hefner: 223, 229
Commission to the Five Civilized Tribes: *see* Dawes Commission
Connally, J. A.: 156
Connally, Thomas T.: 176
Conn, Jack T.: 246
Conoco: *see* Continental Oil Company
Conservation legislation: House Bill No. 2, 171, 180; House Bill No. 78, 54; House Bill No. 168, 141; House Bill No. 172, 181; House Bill No. 187, 171, 181; House Bill No. 188, 171; House Bill No. 238, 55–56; House Bill No. 274, 171; House Bill No. 339, 239; House Bill No. 390, 181; House Bill No. 395, 141, 144; House Bill No. 481, 168–71, 181; House Bill No. 483, 171; House Bill No. 723, 139; Oil Inspector Act, 53; Senate Bill No. 10, 171; Senate Bill No. 11, 54; Senate Bill No. 124, 172, 181; Senate Bill No. 130, 139; Senate Bill No. 208, 171, 181; Senate Bill No. 346, 181, 171; Standard Gas Measurement Law, 239–40
Consumers' Refining Company: 77
Continental Oil and Transportation Company: 107
Continental Oil Company: 107–108, 125, 193, 228, 233, 242, 248
Coody's Bluff, Cherokee Nation: 10
Cook, Nellie: 154
Coolidge, Calvin: 156, 173
Cordell, Oscar L.: 148, 151
Cornelius, E. H.: 151
Cornelius, I. E.: 152

# INDEX

Corsicana Oil Company: 81, 84–85
Cosden and Company: 152; *see also* J. S. Cosden Company
Cosden Oil and Gas Company: 109
Cosden Oil Company: 109, 118–119, 183
Cosden Refining Company: 77
Cotteral, John H.: 165
Crawford, J. E.: 152
Creek County, Okla.: 95
Creek Indians: 26, 28, 30–35, 43, 47, 51, 64, 67, 68
Creek Oil Company: 42
"Creekology": 200
Crescent, Okla.: 88
Cristie, Bob: 188
Critchlow, J. M.: 80–81, 97
Crockett, Norman: 220
Cromwell field: 92, 115, 118, 120
Cromwell Oil and Gas Company: 128–29
Cromwell, Okla.: 95–96
Crosbie Company: 84
Crossman, Luther: 29
Crossman, P. L.: 27, 28, 30
Cruce Hobar Sand: 235
Crystal Oil Company: 80–81, 83–84
Cudahy, J. M.: 44
Cudahy, John: 21
Cudahy, Michael: 21
Cudahy Oil Company: 12, 15–16, 21–22
Cumberland, Okla.: 79
Curtis Act: 21–22, 50–51
Curtis, Charles: 50
Cushing field: 15, 56, 67–78, 82, 89, 101, 105, 139, 141, 193, 196, 200, 203
Cushing, Okla.: 35, 44, 92, 97, 99, 115, 139
Cushing Refining Company: 77
Cyril Refining Company: 114

Daeson, Oak: 10
Dague, A. B. C.: 151
Dallas, Texas: 77, 80, 176, 179, 186
Dalley, Claude: 128
Darby Petroleum Company: 124
Darden, Robert M.: 5
Davidson, John: 34
Davis, B. F.: 116
Davis, M. V.: 129
Davis, Samuel: 28
Dawes Act: 51
Dawes Commission: 28, 30–31, 33–34, 36, 67, 229
Dean A. McGee Eye Institute: 223–24
De Barr, Edwin: 195
Decker, Charles E.: 196
Deer Creek field: 109
De Golyer, Everette L.: 200
Delaney, W. A.: 183
Delano, Columbus: 5
Delaware Indians: 16
Demit, Charles P.: 162

Denver, Colo.: 44, 107, 219
Derden, J. H.: 128
Deusta, Richardo A.: 153
Devonian Oil Company: 104
Diamond coring rig: 202
Dickerson, Judge J. T.: 84
Dierks Lumber Company: 89
Diesel, Rudolph: 23
Dill field: 115
Disney, Wesley E.: 176
Dixie Oil Company: 118
Dodd, Ray: 41
Dodge, Henry: 194
Doherty, Henry L.: 59, 104, 145, 147, 158, 200
Domes field: 64
Donohue, C. H.: 34
Dora field: 115
Dott, Robert H.: 196
Douthit, Herman: 189
Drake, Edwin L.: 3, 156, 159, 194–95
Drake, N. F.: 195
Drumright, Aaron: 70–71
Drumright field: *see* Cushing field
Drumright, Okla.: 73–74, 95, 97
Duluth-Oklahoma Company: 108
Duncan, Okla.: 89, 208–209, 235, 248
Dundee Petroleum: 83
Dunn, J. M.: 28
Durant, Alexander R.: 8–9
Dwyer, Martin: 160
D-X Oil Company: 183

Eakring-Duke's Woods field, England: 189
Eakring field, England: 188
Earlsboro field: 115, 120–21, 124
Earlsboro, Okla.: 203
Earlsboro Sand: 90, 121
Eason Oil Company: 88, 246
East Cromwell field: 115
East Earlsboro field: 115
East Little River field: 115
East St. Louis, Ill.: 39, 44, 126
East Seminole field: 115
East Texas field, Texas: 125
Edgecomb Metals Company: 217
Edmond field: 183
Edmond, Okla.: 81, 84
Ehrhardt, Paul H.: 153, 156
Elbon Rye: 220
El Dorado field, Kans.: 60, 199
Eldred, J. C.: 42
Elgin field, Kans.: 63
Elliott, D. M.: 156
Ellsworth, Henry: 194
Emerson City, Okla.: 130
Emerson, Earl E.: 95–96
Empire Companies: 145
Empire Gas and Fuel Company: 104, 196
Empire Pipe Line Company: 132
Enabling Act, 1906: 53
Endicott Horizon: 112

Endicott, Paul: 102
England, Oklahomans drilling in: 186–89
English, A. Z.: 28
Enid, Okla.: 105, 109, 168
Entex, Inc.: 238
Espy, W. E.: 151
Eufaula, Okla.: 25
Evans, A. Grant: 196
Evans, J. H. ("Uncle Joe"): 156
Everett, C. T.: 151
Extension North field: 235

Fain-Porter Drilling Company: 186–87, 189
Farren, Charles F.: 151
Faucett, H. W.: 7–10
Federal Oil Commission Board: 200
Fellows, W. J.: 40
Ferguson, Frank: 40
Ferguson, Thompson B.: 35
Ferris, Scott: 53
Fife, Timmie: 29
Filley, E. R.: 151
Firestone Oil Company: 129
Fisher, Frank, 101–102
Fish field: 115
Fite, F. B.: 28
Fitts field: 183, 206, 230
Fitts, John: 183
Five Civilized Tribes: 3–16, 21–22, 36, 52
Fixico No. 1: 121–22
Flanagan, F. G.: 10
Flanagan, W. J.: 73
Fleetborn Oil Company: 117
Fleet, Jim: 116
Florida, state of: 66, 133, 181, 216–17
Floyd, Charles "Pretty Boy": 95
Foraker field: 104
Foraker, Okla.: 101
Formby field, England: 189
Fort Sill, Okla.: 35
Fort Smith, Ark.: 21, 194
Fort Worth, Texas: 77, 80
Foster, Edwin B.: 57
Foster, Henry: 14–15, 57, 61–63, 137
Foster Petroleum Company: 129–30
Fowler, E. L.: 156
Fowler, Jacob: 194
Franchot, D. W.: 202
Francis field: 115
Franklin Building, Oklahoma City: 132
Franklin Wirt: 80, 82, 132, 144, 174, 229
Frank Phillips Foundation: 228
Franks Manufacturing Company: 159, 205
Franks Manufacturing Division, Cabot Shops: 205
Frazier, Robert H.: 243
French, M. C.: 49, 139
Frensland Oil Company: 119
Frisco Railroad: *see* St. Louis and San Francisco Railroad

Fulkerson, R. A.: 70

Gailey No. 2: 42
Gailey No. 3: 42
Galbreath, Robert: 30, 41–44, 47, 139
Galey, J. H.: 16
Galt, Edward: 80
Gano, J.K.: 69
Garber field: 108–109
Garber, Okla.: 199
Gardner, J. H.: 151–52
Gardner Oil Company: 117
Gardner, James H.: 200
Garfield County, Okla.: 108
Garland, R. F.: 121–22
Garvin County, Okla.: 189, 240
Gas City field: 235
Gas Processors Association: 206
Gavin, Katherine: 154
Geneva-Pearl Oil Company: 84
Geolograph, development of: 204–205
Geophysical Research Corporation: 203
*George M. Reading* (ship): 210–11
Georgetown University: 233
Georgia, state of: 181
Getty Oil Company: 248
Getty Refining and Marketing Company: 248
Geyer, E. Park ("Spot"): 109
Gibbons, Floyd: 136
Gilcrease, Thomas: 226–27
Ginter, R. L.: 151
Glad, Paul: 220
Glasscock, Frank: 151
Glenn, Elma: 43
Glenn, Hugh: 194
Glenn, Ida: 41–42
Glenn Pool field: 35, 41–47, 49, 52–53, 69, 75, 101, 115, 193, 212, 244
Goddard, C. B.: 85
Goddard Health Center: 220
Goddard, H. H.: 151
Going Snake District, Cherokee Nation: 10
Golden Trend field: 240
Gonafels, Manuel: 153
Goodrich, H. B.: 79
Goodwell Oil Company: 109
Gore, Thomas B.: 176
Gotebo, Okla.: 35
Gould, Charles N.: 59, 195–96, 198–99
Grady County, Okla.: 235
Graham, N. R.: 151
Grand Saline: 3
Granite Mountain: 35
Grant County, Okla.: 109
Gray field: 115
Grayson County, Texas: 9
Grayson field: 115
Great Depression: 158, 183, 192, 202
Greater Seminole field: 95, 114–26, 138, 162, 244
Great Lakes Pipe Line Company: 125

Gregg, E. P.: 9
Gregg, Josiah: 194
Greiner, A. J.: 35
Grey, W. H.: 152
Guffey & Galey: 64
Guffey-Gillespie Oil Company: 181
Guffey, James M.: 16
Gulf Coast: 46, 208, 210
Gulf of Mexico: 107, 114, 203, 209, 211
Gulf Oil Company: 46, 77, 93, 208, 216
Gulf Pipe Line Company: 44–45, 64, 77
Gulf Production Company: 45
Gunsburg-Foreman Company: 84
Guthrie, E. Bee: 151
Guthrie, Okla.: 25, 35, 165
Gutowsky, Ace: 189–91
Gwynne, L. D.: 156
Gwynne, R. D.: 151
Gypsy Oil Company: 85, 93–101, 103, 121, 196, 204

H. C. Price Company: 203
HF alkylation process, development of: 205
Hall, Celia: 129
Halliburton Energy Institute: 209
Halliburton, Erle P.: 203, 207–209
Halliburton, Jet-Mixer: 207
Halliburton Oil Well Cementing Company: 203, 207–208
Halliburton Oil Well Cementing Company, Ltd.: 208–209
Halliburton Services: 207–209, 217, 248; *see also* Better Method Oil Well Cementing Company
Halliburton, Vida: 207
Hamilton Switch field: 49
Hamilton, W. R.: 151
Hannah, James W.: 84
Hardridge, Eli E.: 29
Hardy, Summers: 151
Harper, F. E.: 133
Harper Oil Company: 133–34
Harrington, Barney: 150
Hartman, T. J.: 151–52
Haseman, W. P.: 202
Haskell, Charles N.: 54, 61, 196
Hawkeye Oil Company: 237
Hayner, J. M.: 151–52
Head, K. C.: 204
Headlton field: 78–88, 101, 105, 115, 142–44, 193, 229, 240, 244
Healdton Oil and Gas Company: 109
Healdton oil-field fire: 83–86
Healdton, Okla.: 86–87, 141
Healdton Sand Zone: 87
Hefner Company: 237
Hefner, Evelyn: *see* Evelyn Hefner Combs
Hefner Production Company: 237
Hefner, Robert A., Jr.: 229, 236–37
Hefner, Robert A., Sr.: 222–23, 229
Hefner, William: 223, 229
Heggem, Alf G.: 151–52

Heine, Melvin B.: 186
Heller, Martin: 10
Helmerich and Payne Oil Company: 231
Helmerich, Walter: 231, 246
Hennessey-Garber formation: 128
Henry L. Doherty Education Foundation: 228
Henry, Pat: 10
Hewitt field: 105
Hewitt-Wilson field: 207
Heydrick, James A.: 33
Heydrick, Jesse A.: 28–29, 33–34
Hickman, Bertha: 101
Hickory Creek field: 61
Hill, Robert T.: 195
Hinderliter, Frank: 151
Hitchcock, Edward: 195
Hitchcock, Ethan A.: 30, 34
Hevick, L. C.: 127
Hold, Ed: 87
Holdenville field: 115
Holiness Mission, in Three Sands field: 97
Hollow, Rosalind: 154
Holmes, A. C.: 151
Holson, Samuel: 9
Hominy field: 63
Hoover, Herbert: 173–74
Hoover Sand: 104
Hotson Horizon: 112
Houdry's catalytic cracking system, 183–85
Houdry, Eugene: 183–85
House No. 1: 124
Hoy farm in Garber field: 108
Hoy Horizon: 112–13
Hoy Sand: 199
Hubbard field: 114
Hudgeons, Jerry: 243
Huffman, O.A.: 129–30
Hughes County, Okla.: 115
Hughes, Richard: 151
Humble Oil Company: 208
Humes, R. P.: 151
Humphrey's Petroleum Company: 109
Hunter, I. W.: 156
Hunton Limestone: 118, 121, 123
Hurley, A. W.: 151
Hutchison, L. L.: 196
Hynes, George A.: 10

Ickes, Harold L.: 174, 176, 186
Ida Glenn No. 1: 35, 41–42
Idaho, state of: 107, 181, 216
Illinois, state of: 176, 180, 216
Independent Development League: 82
Independent Oil and Gas Company: 118, 120–21, 124
Independent Petroleum Association of America (IPAA): 174
Independent Producers Association: 157
Independent Producers League (IPL): 139, 141
Indian Territory Illuminating Oil Company

## INDEX

(ITIO): 36, 39, 57–59, 61–64, 118, 121, 124, 128, 130, 132, 134–35, 137, 204
Indiana, state of: 181, 216
Ingram No. 1: 121
International Petroleum Exposition and Congress (IPE, IPEC): 151–62, 202, 225, 228
International Refining Company: 77
IPAA: *see* Independent Petroleum Association of America
IPL: *see* Independent Producers League
ITIO: *see* Indian Territory Illuminating Oil Company
Ivask, Ivar: 221
I. X. Ranch: 28

Jackson, L. B.: 151–52
James No. 1: 121
James, Thomas: 194
Jane Oil Refining Company: 77
Jarrett, W. R.: 96
J. E. & L. J. Mabee Foundation, Inc.: 225–26
Jenks, Okla.: 45
J. H. Smith School Land Well No. 1: 110–11
Jim Wallace No. 1: 118
Jodlowski, John: 128
Johnson, Roy M.: 80
Johnstone, William: 16, 18
John S. Wicks Company: 28–29
Jones, B. B.: 68-69
Jones, M.: 69
Jones, P. M.: 187
Jones Sand: 77
Jones, W. Alton: 191
J. S. Cosden Company: 77
*J. W. Bates* (ship): 210–11

Kansas City, Mo.: 36, 44, 155
Kansas, state of: 5, 20, 26, 50, 54, 106, 109, 155, 173, 176, 180, 198, 208
Kansas Torpedo Company: 40
Karcher, J. C.: 202
Katherine Oil Company: 75
Katy Railroad: *see* Missouri, Kansas & Texas Railroad
Kay County Gas Company: 109
Kay County, Okla.: 103, 106, 108–109, 246
Keeler, George: 16
Keeler, William Wayne: 233–34
Kelly, George ("Machine Gun"): 96
Kemph, Frank: 208
Kennamer, Franklin E.: 165
Kennedy, Cordelia Ann: 154
Kennedy, J. A.: 156
Kennedy, John Q. A.: 29
Kennedy, L. E.: 151
Kentucky, state of: 71, 181, 216
Keokuk field: 115

Kerlyn Oil Company: 233
Kerr-McGee Corporation: 132, 203, 223, 233, 248
Kerr-McGee Foundation, Inc.: 228
Kiefer field: 99
Kiefer, Okla.: 44, 95, 202
King, B. D.: 183
King, James: 9
King, J. Berry: 176
Kingfisher County, Okla.: 239–40
Kingston, Okla.: 79
Kingwood Oil Company: 49
Kiowa, Comanche, and Apache Reservation: 35
Kiowa, Okla.: 44
Kirk, Charles T.: 196
Kirkpatrick and Bale: 236
Kirkpatrick, Eleanor: 221–22, 236
Kirkpatrick Fine Arts Building: 222
Kirkpatrick Foundation: 221–22
Kirkpatrick, Joan: 222
Kirkpatrick, John E.: 221–22, 229, 236
Kirkpatrick Oil Company: 221–22
Kirkpatrick Planetarium: 222
Kitner No. 5: 135
Klinge, Judge S. H.: 154
Klumph, W. E.: 156
Knox field: 235
Konawa field: 115
Koopman, H. H.: 64
Kroll, Cornelius: 151
Kruse, Dr. Paul F.: 220

La Fortune, Joseph A., Sr.: 236
Lamont, Robert P.: 157, 174
Land, J. H.: 29
Landon, Alf M.: 176
Landon field: 61
Landon, Ollie: 73
Larkin, Charles A.: 196
Latrobe, Charles: 194
Lauinger, Frank T.: 213–14
Lauinger, P. C., Jr.: 216
Lauinger, P.C., Sr.: 215–16, 233
Lawrence, J. W.: 129
Lawson, C. N.: 156
Layton Horizon: 112
Layton Sand: 18, 77, 104
Learned, Stanley: 91–92, 104, 243
Leavenworth, Henry: 194
Leede, Edward H.: 243
Leede Exploration: 243
Lehigh, Okla.: 9
Leland Stanford University: 195, 229
Lemason, C. M.: 151
Leslie Brooks and Associates: 206
Levorsen, A. I.: 230
Lewis, W. L.: 151
Lexington, Okla.: 35
Lincoln County, Okla.: 134
Lindsey, Lilah D.: 28
Lions Clubs of Oklahoma Eye Bank: 223

Lipe, Major: 16
Little River field: 115, 124
Little Star (Secretary of Cherokee Nation): 16
Livingston, Julius: 239
Livingston Oil Company: 119
Lloyd Noble Arena: 220
Logan County, Okla.: 239
Long, I. G.: 151
Long, Stephen Harriman: 194
Louisiana, state of: 114, 176, 181, 187, 208, 210, 216–17, 219
Lower Hoover Horizon: 112
Lowery, William "Uncle Bill": 40
Lucien field: 183
Lucky field: 47
Luther Oil and Gas Company: 128
Luther, Okla.: 128
Lyle, Ora: 73
Lynn, William: 10

Mabee, John Elmer: 225–26
Mabee, Lottie J.: 225
M. B. & K. Oil Company: 61
McBirney, J. H.: 151
McBride, A. P.: 15, 18, 20–21, 57
McCasland Foundation: 228
McCasland, T. H., Sr.: 105, 235, 246
McClure, H. O.: 151–52
McClure, Mary: 81
McClure, W. G.: 156
McCormick, D. O.: 156
McCurtain, Edmund: 8
McCurtain, E. M.: 9
McDowell, Robert W.: 237–38
McElroy, H. E.: 151
McFadden, W. H.: 156
McFarlin, R. L.: 151
McFarlin, Robert: 229
McGee, A. T.: 80
McGee, Dean A.: 132, 203, 223–24, 244; early life, 233; education, 233; oil business, 233; civic affairs, awards, 233
McGivern, Hugh B.: 156
McGraw, J. J.: 151–52
McGuinn, Frank: 42
McIntyre, Edward F.: 151
McKee, Sam: 111
McKelvey, J. S.: 151
McLean, John G.: 228
McLoud, Okla.: 35
McMan Oil Company: 75, 119
McRill, Albert L.: 147
Mack Oil Company: 105, 235
Madalene field: 104
Mad House Saloon, in Kiefer, Okla.: 95
Magnolia Oil Company: 75, 81, 84–85, 87, 118, 141, 124
Magnolia Petroleum Company: 208
Magnolia Pipe Line Company: 82, 141
Malone, E. L.: 93
Maloney, Thomas: 69

Manion field: 104
Manning, Everett: 151
March, John: 220
Marcy, Randolph B.: 194–95
Marion Institute: 236
Markham, John H.: 156
Markham Sand: 235
Marland, Ernest W.: 101–102, 106–11, 113, 174, 176, 180, 229
Marland Oil and Refining Company: 64
Marland Oil Company: 101–102, 106, 108
Marland Refining Company: 106, 109
Marlow, Okla.: 88
Martial law in the oil fields: 147–50, 162–72
Martin, Frank: 128
Maryland, state of: 181
Matson, G. L.: 151, 204
Maud field: 115
Mayer, T. F.: 151
Mayo Hotel, Tulsa, Okla.: 156, 229
Maysville, Okla.: 205
Mediterranean Sea: 211–12
Melton, W. A.: 151–52
Merchants Oil and Gas Company: 127
Merrick Chair in Western American History, University of Oklahoma: 220
Merrick Computer Building, University of Oklahoma: 233
Merrick Foundation: 220–21, 223
Merrick, Ward, Jr.: 220, 244
Merrick, Ward, Sr.: 82–83, 220, 232–33
Methodist Church, in Three Sands, Okla.: 97
Mexican Eagle Oil Company, Ltd.: 200
Meyers, Charles: 151
Michigan, state of: 176, 180
Mid-Continent Gas Company: 226
Mid-Continent Oil and Gas Field: 20, 43, 125, 144, 148, 193, 200, 203, 207, 209, 212, 243
Mid-Continent Petroleum Company: 226
Mid-Continent Petroleum Corporation: 125, 237
Mid-Continent Port Arthur Pipe Line and Refining Company: 45
Mid-Continent Refiners Corporation: 237
Mid-Continent Region: *see* Mid-Continent Oil and Gas Field
Midco Petroleum Company: 109
Middle Hoover Horizon: 112
Mid-Kansas Oil and Gas Company: 115, 119, 148
Midland Oil Company: 102
Midland Valley Railroad: 45
Military Petroleum Advisory Board: 234
Millard, William J.: 105
Miller, Bluford: 28
Miller Brothers: 106–107
Miller, George: 106
Miller, Joe: 106
Miller, Shorty: 41

Miller, Zack: 106
Milliken Refining Company: 77
Million Dollar Elm, Pawhuska, Okla.: 66, 102
Minger, J. E.: 151
Minnehome Oil and Gas Company: 64
Minnetoka Oil and Gas Company: 40
Miskell, P. M.: 151
Mission field: 115
Mississippi Horizon: 112
Mississippi Sound: 104
Mississippi, state of: 181, 189, 210
Missouri, Kansas & Texas Railroad: 5, 10, 23, 70
Missouri, state of: 5, 154, 216
Moffitt, D. W.: 151
Molhousen, H. B.: 195
Montana, state of: 107, 181, 216
Montee, F. E.: 116
Moore, C. T.: 128
Moore, J. C.: 40
Moore, Lee C.: 205
Moore, Morris: 129–30
Moore, Okla.: 193
Morgan and Flynn Oil Company: 121
Morgan, George D.: 128
Morgan, Tom A.: 243
Morris field: 47
Morton, Rogers C. B.: 160
Mosier, M. H.: 42
Mounds, Okla.: 42
Mount Pleasant Dome field: 75
Munn, M. J.: 196
Murdock, Okla.: 108
Murray, Cicero I.: 167
Murray, C. M.: 151
Murray, William A. "Alfalfa Bill": 147–48, 150, 162–68, 173–74
Muskogee, Okla.: 12, 21, 25, 28, 43, 54
Musselman Sand: 77
Mutual Oil and Gas Company: 127
Myers Dome field: 63

National Association of the Independent Oil Producers: 152
National Exploration Company: 101
National Guard in the oil fields: 147–50, 162–72
National Industrial Recovery Act (NIRA): 174–75
National Oil Company: 10
National Stripper Well Association: 266
Natura field: 49
Naval Reserve Officers Training Corps: 222
Nebraska, state of: 89, 109, 181
Nelagoney field: 64
Nellie Johnstone No. 1: 15–26
Neodesha, Kans.: 16, 21, 33, 36, 64, 77
Neosho Crossing, Cherokee Nation: 3
Neustadt International Prize for Literature: 221
Neustadt, Walter: 221

Neustadt, Walter, Jr.: 221, 243
Nevada, state of: 26, 107, 181, 216
Newalla, Okla.: 127
Newby, Jerry B.: 128
New Deal: 173–74
New Jersey, state of: 109
Newkirk Horizon: 112
Newkirk, Okla.: 35, 98
New Mexico, state of: 107, 173, 176, 180, 208
New Spring Place, Cherokee Nation: 3
New State Refining Company: 77
New York City, N.Y.: 187, 215, 220
New York, state of: 10, 181, 243
Nichols, F. B.: 204
Nicoma Park, Okla.: 135
Niles, Alva J.: 152
Noble, Ann: *see* Ann Noble Brown
Noble Corporation: 187
Noble County, Okla.: 109–10, 183
Noble Drilling Company: 186–87, 189, 219
Noble, Ed: 219
Noble, Lloyd: 186–89, 218–19, 229
Noble, Sam: 219
North Bethel field: 115
North Canadian River: 138, 194
North Carolina, state of: 181
North Dakota, state of: 181, 216
North End Club, in Seminole, Okla.: 95
North St. Louis field: 115
North Sea: 188, 209, 211
North Searight field: 115
Norman, Okla.: 135
Nowata County, Okla.: 203
Numa Oil Company: 75
Nuttall, Thomas: 194

O'Brien, Martin C.: 156
OCB: *see* Oil Conservation Board
OCC: *see* Oklahoma Corporation Commission
Office of Petroleum Administration for Defense: 234
Office of Petroleum Coordination: 186
Office of Petroleum Coordinator for National Defense: 186
Office of Price Administration: 187
Offshore drilling, development of: 203
Ogden, Utah: 107
O'Hern, D. W.: 196
Ohio, state of: 41, 106, 181, 216
*Oil & Gas Journal*: 212–16, 233
Oil Code: 175
Oil Conservation Board (OCB): 173–74
Oil-pool analyzer, development of: 203
Oil State Petroleum Company: 88
Oil States Advisory Committee (OSAC): 173–74
Okesa field: 58
Okfuskee County, Okla.: 115, 119
Oklahoma A & M: 219, 231

## INDEX

Oklahoma Center for Arts and Sciences: 222
Oklahoma Christian College: 221
Oklahoma City Community Foundation: 222
Oklahoma City field: 92, 95, 127–38, 141, 147, 162, 166–68, 175
Oklahoma City, Okla.: 35, 45, 82, 89, 92, 127–28, 180–81, 203, 222, 227, 236–37, 248; city council, 132, 147; symphony, 222; zoo, 222
Oklahoma City University: 222
Oklahoma City Wilcox Pool Engineering Association: 238
Oklahoma Constitution: 54–55, 97, 162
Oklahoma Corporation Commission (OCC): 75–76, 82, 139, 162–72, 175, 182, 201–202, 239; Order No. 813, 75, 141, 201; Order No. 814, 82; Order No. 829, 76; Order No. 920: 142, 201; Order No. 937: 143–44; Order No. 1299: 144–45
Oklahoma County, Okla.: 128, 134
Oklahoma Geological Commission: 195
Oklahoma Geological Survey: 108, 195–96
Oklahoma Hall of Fame Galleria: 223
Oklahoma Heritage Archives: 223
Oklahoma Heritage Association: 222–23, 237
Oklahoma Heritage Center: 237
Oklahoma Heritage House: 222
Oklahoma Medical Research Foundation: 222
Oklahoma Natural Gas Company: 202
Oklahoma Natural Gas, Light, and Heat Company: 35
Oklahoma Panhandle: 194
Oklahoma Petroleum Council (OPC): 229–39, 241
Oklahoma petroleum industry: regulation of, 47–56, 139–50, 162–72; innovations of, 194–206; related industries, 207–17; legacy of, 218–28; future of, 239–48
Oklahoma Science and Arts Foundation: 222
Oklahoma, state of: legislature, 139, 141, 168, 171; supreme court, 170, 222
Oklahoma State Oil Company: 45
Oklahoma State University: 220–21, 228
Oklahoma Territorial Department of Geology and Natural History: 195–96
Oklahoma Territorial Geologist: 195–96
Oklahoma Territorial University: 195
Okmulgee, Okla.: 47, 49, 126, 193
Olson, A. O.: 219
Olympic field: 115
O'Neal, C. R.: 136
101 Ranch: 106
101 Ranch Oil Company: 106–107
OPC: see Oklahoma Petroleum Council
Oregon, state of: 181
Organic Act: 23

Organization of Petroleum Exporting Countries: 160
OSAC: see Oil States Advisory Committee
Osage Agency: 66
Osage Allotment Act: 64–66
Osage City field: 58
Osage County, Okla.: 101–102, 104–106, 227, 247
Osage field: 202
Osage Hills: 59
Osage Indians: 14–16, 18, 20, 23, 25, 36, 56–67, 102–103
Osage Oil Company: 57
Osage Reservation: 75, 101, 104
Osborne, H. E.: 156
Osborn, William B., Jr.: 95, 121, 241
Osborn, William B., Sr.: 116
Oswego Horizon: 112
Oswego Lime: 18
Oswego Sand: 104
Otto, Nikolaus: 23
"Outstanding Oklahoma Oil Man Award": 229–39
Overlees, Frank M.: 16
Owen, Albert P.: 28
Owen, Robert L.: 10

PAB: see Petroleum Administration Board
Pacific Coast: 107
Pacific railway survey: 195
Paine-Kirkpatrick Wing, National Cowboy Hall of Fame and Western Heritage Center: 222
Palmer (early-day driller in Chickasaw Nation): 79
Paola, Kans.: 7
Partridge, George C.: 156
Partridge, H. E.: 156
Paschal, E. A.: 128
Patterson, M. K., Jr.: 220
Patton Company: 36
Pawhuska field: 64
Pawhuska, Okla.: 25, 66, 102–104
Pawnee County, Okla.: 25, 35, 40, 189
Payne County, Okla.: 25
Payne, William T.: 95, 231–33, 246
Pearsonia field: 64
Peavine-Wilcox field: 134
Pennsylvania, state of: 5, 29, 68–70, 74, 86, 106–107, 111, 159, 181, 213
Perkins Cementing, Inc.: 207
Perrine, Irving: 196
Perry, E. R.: 152
Perry, G. M.: 18
Perryman, L. C.: 28
Perry, Okla.: 211
Pershing field: 64
Petrochemicals: 199
Petroleum Administration Board (PAB): 175
Petroleum Equipment Institute: 206
Petroleum Experiment Station: 206

Petroleum Industry Council for National Defense: 186
Petroleum Industry War Council (PIWC): 186
Petroleum Publishing Company: 212–17
Pettee, William: 30
Petters, V. D.: 113
Pettit field: 64
Pew, J. Edgar: 156
Philbrook Art Center: 228
Philgas: 38
Phillips, Frank: 37–39, 103–104, 148, 227–28, 231
Phillips, John: 12
Phillips, L. E.: 37–38
Phillips, Orie L.: 165
Phillips Petroleum Company: 37–39, 64, 92, 102–104, 113–14, 125–26, 132, 138, 148, 156, 162, 180, 191, 203, 205, 228, 233, 243, 248
Phillips Petroleum Foundation, Inc.: 228
Phillips, Waite: 227–29
Phoenix Oil Company: 14–15, 20–21, 23, 36, 57
Pickens County, Chickasaw Nation: 79
Pierce Oil Corporation: 77
Pike, Zebulon Montgomery: 194
Pine, Clyde: 243
Pine, W. B.: 121
PIWC: see Petroleum Industry War Council
Plains Development Company: 80
Pneumatic rotary drilling, development of: 205
Polo field: 205
Ponca City field: 107
Ponca City, Okla.: 35, 97, 99, 105–106, 108, 111, 176, 248
Ponca Indians: 106–107
Pond Creek field: 61
Pontotoc County, Okla.: 5, 115, 122, 183
Portable well servicing unit, development of: 205
Port Arthur, Texas: 44–45, 77, 81
Porter, Frank: 186
Porter, Hollis P.: 151
Porter, Pleasant: 32–33
Postle field: 240
Post Oak Oil Company: 238
Pottawatomie County, Okla.: 35, 115, 121–22, 124, 134, 189
Pourtales, Albert-Alexandre de: 194
Pozmix: 203
Prairie Oil and Gas Company: 36, 42, 44–45, 49, 64, 71, 77, 109, 115, 121
Prairie Pipe Line Company: 118, 203
Prairie View Church, Tonkawa field: 111
Pressure drilling, development of: 205–206
Preston, Alex: 49
Preston field: 47
Prichard, L. H.: 114
Producers and Refiners Corporation: 237
Propane, underground storage of: 202

281

Prue field: 67, 104
Pure Oil Company: 125, 208
Putnam City, Okla.: 127

Quapaw field: 61

Rainbow Dance Hall, Seminole: 95
Ramona, Okla.: 36, 203
Ramsey, Asa E.: 151
Ramsey, W. R.: 132
Randal, Lillian: 154
Ray, Charles D.: 243
Ray's Well Servicing: 205
Reading & Bates Offshore Drilling Company: 209–12, 217
Reading, George M.: 209
Reavis, Holland S.: 212
Rector, Sara: 71
Reda Pump Company: 205
Red Fork field: 12, 16, 23, 26–34, 115, 193
Red Fork Land and Investment Company: 35
Red Fork, Okla.: 41–42, 44
Red River Oil Company: 81, 84
Reece Drilling Company: 129
Reed, Lulu: 73
Reflection seismograph, development of: 202–203
Regulation of the oil industry: 47–56, 139–50, 162–72
Reisling, May: 154
Rex Oil Company: 84
Reynolds, John M.: 14
Richards, E. A.: 151
Riggs, R. J.: 128
Riley, Ralph C.: 151
Ritchie, W. R.: 151
Roach, W. P.: 156
Rock Island Railroad: 128
Rockefeller, John D.: 156
Roe, I. G.: 151
Roop, C. W.: 128
Roosevelt, Franklin D.: 173–75, 186
Roosevelt, Theodore: 53
Rosanna field: 115
Rosser, Eugene P.: 187–89
Rosser, I. G.: 151
Ross, Finley: 10
Ross, John: 3
Ross, Lewis: 3
Ross, Robert B.: 16
Roxana Petroleum Company: 111, 132
Roxana Petroleum Corporation: 61, 109
Roxanna Oil Company: 209
Rucker, Alvin: 129

Sacred Heart field: 115
Sageeyah Switch, Cherokee Nation: 10
Saint Anthony's Hospital Dental Clinic: 222
Saint Francis Hospital: 225–32

St. John's Hospital: 225
St. Louis & San Francisco Railroad (S. L. & S. F.): 10, 23, 27–28
St. Louis field: 115, 124
St. Louis, Okla.: 155
Samedan Oil Company: 219
Samuel Roberts Noble Foundation: 219–20
Sanders, A. J.: 156
Sandfork Gas and Petroleum: 36
Sand Springs Home Oil Company: 61
Santa Fe Railroad: 127, 133
Santa Fe Trail: 194
Sapulpa, James: 28
Sapulpa, Okla.: 12, 23, 27, 29, 99
Sapulpa Refining Company: 103
Sapulpa, William A.: 28
Sartori, J. A.: 151
Sasakwa, Okla.: 95, 116
Sayre, Okla.: 202
Schell, John: 40
Schwab, Charles M.: 156
"Scottie the Baptist" (Baptist preacher): 97
Searight field: 115, 120, 124
Searight, F. J.: 123–24
Secondary recovery, development of: 203
Seminole City field: 115, 120–24
Seminole County, Okla.: 114–26
Seminole field: see Greater Seminole field
Seminole Indians: 3–16, 114, 116
Seminole Oil District: 115
Seminole, Okla.: 95, 97, 202, 119, 121, 123
Seminole Plan: 148
Seminole Sand: 118, 121–24
Seneca Oil Company: 238
Service Drilling Company: 93
Shaffer, C. B.: 69–70
"Shaffer rant": 138
Shamrock Dome field: 75
Shartel, John: 127
Shawnee field: 115
Shawnee, Okla.: 35, 228
Shell Oil Company: 121, 209
Shepherd Oklahoma Heritage Library: 223
Sherwin, R. S.: 195
Sherwood Forest, England: 187–88
Shidler, Okla.: 38
Shot-gun houses: 93
Sho-Vel-Tum field: 240
Shumard, George C.: 194–95
Sibley, George Champlin: 194
Simley, E. L.: 28
Simms, Jake: 98
Simpson Bromide Sand: 189
Simpson Formation: 120
Simpson, W. P.: 130
Sinclair, Earl W.: 44
Sinclair, Harry: 229
Sinclair, Harry Ford: 44, 162, 165
Sinclair Oil and Gas Company: 44, 108, 124, 130–32, 135, 163–64, 199, 202, 238

Sinclair Oil and Refining Corporation: 44
Sinclair Refining Company: 109
Skelly Oil Company: 103, 125, 151, 207
Skelly Oil Company Foundation: 228
Skelly, William G.: 151–52, 159, 207, 229
Skiatook, Okla.: 60
Skinner Sand: 77
"Sky Pilot" (Methodist preacher: 97
S. L. & S. F. Railroad: see St. Louis & San Francisco Railroad
Slick City, Okla.: 246
Slick No. 1A: 112
Slick, Tom: 68–70, 96, 125, 130, 200
Smith, A. J.: 3
Smith, Carl D.: 196
Smith E. A.: 93
Smith, Edward A.: 148, 241
Smith, E. H.: 10
Smith, Harry: 151
Smith, R. H.: 115–17
Smith Sand: 118
Sneed, Earl: 151, 161
Snyder, A. L.: 97
Snyder, L. C.: 196
Society of Exploration Geophysists: 206
Society of the Sacred Mission: 187
Socony-Vacuum Oil Company: 184
Sohio Oil Company: 190–91
Sommerville, O. S.: 37
Sooner Trend field: 239–40, 247
South Canadian River: 120, 183
South Carolina, state of: 181
South Dakota, state of: 181, 216
South Elgin field: 61, 63
Southern Oklahoma Memorial Hospital Association: 220
Southwell, C. A. P.: 186
Southwestern Association of Petroleum Geologists: 200
Southwestern Oil Company: 109
Southwestern Petroleum Company: 113
Spencer, Okla.: 127
Sperry, Okla.: 60
Spindletop, Texas: 25, 35
Squirrel Sand: 77
Stamper No. 1: 132
Standard Oil: 46, 77, 93, 107, 109, 191
Standard Oil Company of New Jersey: 228
Standley, Herbert: 132
Stanford University: 237
State Lease 764 No. 1: 203
Stephens County, Okla.: 235, 240
Stephenson, Alex: 156
Sterling, Ross: 173
Stewart, D. C.: 82
Stewart, Ralph L.: 92
Straight, Herbert R.: 59–60, 229–30
Straight, R. J.: 229–30
Streeter, E. J.: 127
Stripper wells: 202
Sue A. Bland No. 1: 28

Suitcase Sand: 104
Sun Oil Company: 81, 183–85, 208
Sunray D-X Oil Company: 237–38
Sunray Oil Company: 183, 237
Sykes Sand: 118

Taber, George H.: 228
Taff, Joseph A.: 195
Tallant, Okla.: 200
Taneha field: 46
Taylor, Charles H.: 196, 200
Telescoping aluminum mast, development of: 205
Tennessee, state of: 181
Terrebonne Parish, La.: 203
Texaco, Inc.: 248; *see also* Texas Company
Texarkana, Texas: 174
Texas Company, The: 45–46, 77, 105, 189, 208; *see also* Texaco, Inc.
Texas County, Okla.: 240
Texas Panhandle: 125
Texas Railroad Commission: 162
Texas, state of: 9, 26, 46, 81, 88, 96, 109, 114, 117, 154–55, 166, 173–74, 176, 180, 187, 195, 208–09, 216, 219, 247
Thomas, Elmer: 175–76
Thomas field: 114
Thomas Gilcrease Foundation: 228
Thomas Gilcrease Institute of American History and Art: 227
Thomas, J. Elmer: 200
Thornton, C. E.: 209–10
Three-in-One Oil and Gas Company: 105
Three Sands field: *see* Tonkawa field
Tidal Oil Company: 104
Tiger Creek Avenue, Drumright, Okla.: 95
Tillman, R. M.: 243
Titusville, Pa.: 3, 80, 194
Todd, Mrs. E. G.: 40
Tom B. Slick Company: 112
Tom B. Slick Oil Company: 118
Tonkawa Horizon: 112
Tonkawa, Okla.: 97, 115
Tonkawa (Three Sands) field: 97–99, 106–107, 109–14, 199, 204
Trapp, M. E.: 155–56
Traugh field: 115
Traywick, Earnest E.: 37
Trout, L. E.: 128
Trowbridge, C. M.: 156
Trower, Harry A.: 125
Tucker, E. T.: 151
Tuloma Oil Company: 226
Tulsa Fuel and Manufacturing Company: 47
Tulsa, Okla.: 12, 23, 26–27, 35, 41, 44, 64, 69, 99, 105, 148, 151–61, 183, 191, 200, 201, 205–206, 209, 211–12, 216, 225–29, 232, 237, 248
Turkey Pen Hollow field: 47
Turner, Roy J.: 134

Tuskahoma: Okla.: 9
Twin State Company: 84
Twin State Oil Company: 81

Udden, J. A.: 151
Unassigned Lands: 23, 127
Uncle Bill No. 1: 40
Uncle Sam Refinery: 64
Union Agency: 54
United States: Geological Survey, 87, 195; Constitution, 97, 163–64; Supreme Court, 107, 166, 176; Bureau of Mines, 175, 203; Military Academy, 236; Naval Academy, 236; Department of Energy, 241–43
United States Oil and Gas Company: 10, 12
University of Kansas: 104, 231, 233
University of Oklahoma: 186, 196, 198, 200, 202, 220–22, 235, 237–38
University of Tulsa: 196, 198, 228; School of Petroleum Engineering, 196, 198
Universal Oil Products: 205
Upper Hoover Horizon: 112–13
Upshaw, A. M.: 3
Urschel, Charles F.: 96, 246
U-7 Zone: 132, 147

Vacuum Oil Company: 183
Valerius, M. M.: 151
Vanderver, W. A.: 151–52
Van Orstrand, C. E.: 204
Van Vleet, A. H.: 195
Van Winkle, Lee: 30
Vensel, Dorothy: 154
Vensel, Verne: 154
Vernon field: 114
Viersen, A. A.: 193
Viersen and Cochran Drilling Company: 193
Viersen, Sam K., Jr.: 193
Viersen, Sam K., Sr.: 193
Viger, Fred G.: 156
Vinita, Okla.: 23, 28, 77
Vivian Bilby Foundation, Inc.: 228

Wagner, C. L.: 128
Wagner, J. H.: 156
Wagner, Noah E.: 203
Wagner No. 1: 190
Walker, A. L.: 143
Walker, B. F.: 219
Walker, Don: 187–89
Walker, F. A.: 5
Walnut Grove, Oklahoma City: 132, 138
Walters, E. E.: 66, 102
Warren, J. S.: 151
Warren, William K.: 160, 225
Washington, D.C.: 279, 202, 215, 233
Washington, state of: 181
Watonga Trend field: 243
*W. D. Kent* (ship): 211

Welch, W. M.: 151
Well cementing, development of: 203
Well Surveys: 205
Wentz, Lew: 229
Wertzberger, D. P.: 205
West, Charles: 141
West Edmond field: 115, 134, 189–91
Western States Land and Development Company: 109
West Short Junction field: 193
West Texas oil fields: 125
West Tulsa, refineries in: 77
West Virginia, state of: 14, 43, 106, 181
Wetley field: 115
Wetumpka, Okla.: 118
Wewoka field: 115, 118
Wewoka Oil and Gas Company: 118
Wewoka, Okla.: 92, 115–16, 202
Wewoka Sand: 115
Wewoka Trading Company: 115
Wheeler, Frank: 68–69
Wheeler, Okla.: 79
Wheller Sand: 77
Whipple, Amiel Weeks: 195
White, Paul J.: 195
Whiteside, Allan: 151
Wicey field: 202
Wichita, Kans.: 59, 113–14
Wichita Mountains: 194–95
Wick, John S.: 28
Wiet, E. H.: 151
Wilber, Fred S.: 159
Wilbur, Ray L.: 173
Wilcox Horizon: 112
Wilcox Sand: 104, 112, 124, 130, 134–35
Wild Mary Sudik: 127, 134–38
Wildhorse Creek District: 102
Wildhorse field: 61
Wilkinson, James B.: 194
William K. Warren Foundation: 225
William K. Warren Medical Research Center: 225, 232
William M. Graham Oil and Gas Company: 64
Williams Brothers Company, The: 233
Williams Company, The: 216–17
Williams Exploration Company: 216
Williams, J. Sam: 138
Wilson, U. T. B.: 30
Wirick No. 1: 183
Wirt Franklin Petroleum Corporation: 132, 229
Wiser field: 58
Wood, H. L.: 102
Woodman, William: 10
Wood, Robert H.: 196
Woods Sand Horizon: 235
Woods, S. H.: 128
Woodward, Okla.: 228
Woolaroc Museum: 227–28
Workman Oil and Gas Company: 50
*World Literature Today*: 221

283

World War I: 64, 75, 78, 82, 87, 101, 105, 125, 139, 185, 232, 237, 246
World War II: 34, 159, 185–89, 192, 193, 209, 221, 236
Wright, Allen: 7–9
Wright, Clarence H.: 230–31
Wright, E. N.: 9–10
Wright, J. George: 31–33
Wrightsman, Charles J.: 69–70

Wynona field: 64
Wyoming, state of: 107, 176, 181, 216

X686 field: 64

Yale field: 75
Yargee, John I.: 28
Yellow Leaf Crossing, Cherokee Nation: 10
Yost, F. R.: 160

Youngblood No. 1: 123
Youngblood No. 3: 124
Young Men's Christian Association: 225
Youngstown field: 49
Y686 field: 64

Zink, John: 151